# AN INTRODUCTION TO COMMUNITY DEVELOPMENT

# AN INTRODUCTION TO COMMUNITY DEVELOPMENT

Edited by Rhonda Phillips and Robert H. Pittman

Routledge
Taylor & Francis Group

LONDON AND NEW YORK

First published 2009
by Routledge
2 Park Square, Milton Park, Abingdon, Oxon OX14 4RN

Simultaneously published in the USA and Canada
by Routledge
711 Third Avenue, New York, NY 10017

*Routledge is an imprint of the Taylor & Francis Group, an informa business*

© 2009 Selection and editorial matter, Rhonda Phillips and Robert H. Pittman; individual chapters, the contributors

Typeset in Garamond by Wearset Ltd, Boldon, Tyne and Wear

*British Library Cataloguing in Publication Data*
A catalogue record for this book is available from the British Library

*Library of Congress Cataloging-in-Publication Data*
An introduction to community development/edited by Rhonda Phillips and Robert H. Pittman.
p. cm.
Includes bibliographical references and index.
1. Community development. 2. Economic development. I. Phillips, Rhonda. II. Pittman, Robert H.
HN49.C6I554 2009
307.1′4—dc22  200802513

ISBN13: 978-0-415-77384-3 (hbk)
ISBN13: 978-0-415-77385-0 (pbk)
ISBN13: 978-0-203-88693-9 (ebk)

# Contents

# Figures

# Tables

# Boxes

# Contributors

**Beverly A. Browning** has been a grant-writing consultant and contract bid specialist for units of municipal government and other nonprofit organizations for 25 years. She has assisted clients throughout the United States in receiving awards of more than $200 million. She is the author of over two dozen grants-related publications, including *Grant Writing For Dummies*™ and *Perfect Phrases for Grant Proposals*. She serves on the Advisory Board for the University of Central Arkansas Community Development Institute. She facilitates two-day grant-writing boot camps throughout the United States.

**Deepak Chhabra** is a Senior Assistant Professor at Arizona State University. Prior to her appointment at Arizona State University, she served as an assistant professor at the University of Northern Iowa and California State University, Sacramento. She acquired her Ph.D. from North Carolina State University. She has presented research papers at global, national, and regional conferences. She has also published in leading refereed journals. These include the *Annals of Tourism Research, Journal of Travel Research, Tourism Analysis, Event Management, Leisure/Loiser, Journal of Heritage Tourism*, and the *Journal of Vacation Marketing*. She has also served as a project director and principal investigating officer on several statewide projects associated with socio-economic impacts of tourism.

**Jessica LeVeen Farr** is the Regional Community Development Manager for Tennessee for the Federal Reserve Bank of Atlanta, Nashville Branch. She works with banks, nonprofit organizations, and government agencies to develop programs that promote asset building, affordable housing, job creation and other related community development initiatives. Previously, she was an assistant vice president at Bank of America in Nashville in the Community Development Corporation (CDC), where she oversaw the single family housing development program. She received her Masters in City Planning from UNC-Chapel Hill with an emphasis on community economic development in 1999. She graduated from University of California, San Diego in 1993 with a BA in urban studies.

**Gary P. Green** is Professor in the Department of Rural Sociology at the University of Wisconsin-Madison and co-director of the Center for Community and Economic Development at the University of Wisconsin-Extension. He has been at the University of Wisconsin for 12 years and taught for eight years at the University of Georgia. His research, teaching, and outreach activities focus on community and regional development. His recent books include *Asset Building and Community Development* (Sage 2002) and *Amenities and Rural Development: Theory, Methods and Public Policy* (Edward Elgar 2005). He has consulted on community development issues in international settings, such as Ukraine, New Zealand, and South Korea.

**John Gruidl** is a Professor in the Illinois Institute for Rural Affairs at Western Illinois University where he teaches, conducts research, and creates new outreach programs in community and economic development. He earned a Ph.D. in Agricultural and Applied Economics from the

University of Wisconsin-Madison in 1989, with a major in the field of Community Economics. Gruidl has created and directed several successful programs in community development. From 1994 to 2005 he directed the award-winning Peace Corps Fellows Program, a community-based internship program for returned Peace Corps volunteers. He also helped to create the MAPPING the Future of Your Community, a strategic visioning program for Illinois communities. He currently serves as Director of the Midwest Community Development Institute, a training program for community leaders.

**Anna Haines** is an Associate Professor in the College of Natural Resources at the University of Wisconsin-Stevens Point and Director of the Center for Land Use Education at the University of Wisconsin-Extension. Prior to joining UWSP, she served as a Peace Corps volunteer and has worked for such organizations as the World Bank. Her research, teaching, and outreach activities focus on community land-use planning and management. Her recent international activities with the Global Environmental Management Education Center at UWSP have focused on small garden systems. She co-authored *Asset Building and Community Development* (Sage 2002) with Gary Green.

**Janet R. Hamer** is the Senior Community Development Manager with the Federal Reserve Bank of Atlanta, Jacksonville Branch. Her primary geographic areas of responsibility are north, central, and southwest Florida. She has over 20 years' experience in housing, community and economic development and urban planning. Prior to joining the Federal Reserve, she served as chief of housing services for the Planning and Development Department of the City of Jacksonville for three years. For the previous 18 years, she served as deputy director of the Community Development Department of the City of Daytona Beach. Before moving to Florida, she was employed as a regional planner in Illinois. She has a BA from Judson College, Elgin, Illinois and a MA in Public Affairs from Northern Illinois University, Dekalb, Illinois.

**William Hearn** is Director of Global Site Selection Consulting with CH2M Hill Lockwood Greene and President of Site Dynamics LLC, a consulting firm focused on developing online products for investors and economic development organizations. He has 20 years' consulting experience in this field and has conducted site selection studies for many different organizations. He has also advised states and regions on economic development policy. He received his Bachelor's degree in History and his Masters in City and Regional Planning from The University of North Carolina, Chapel Hill. He began his career at IBM Germany and worked at The United Nations Industrial Development Organization, and an investment consulting firm in Berlin, Germany. He also worked for the North Carolina Department of Commerce and spent ten years with another site selection firm prior to joining CH2M Hill.

**Ronald J. Hustedde** is a Professor in the Department of Community and Leadership Development at the University of Kentucky. He teaches graduate courses in community development and has an Extension (public outreach) appointment. He is a past president of the Community Development Society and has served on the board of directors of the International Association for Community Development (based in Scotland). His community development work has focused on public issues deliberation, public conflict analysis and resolution, leadership development and rural entrepreneurship. He has a Ph.D. in sociology from the University of Wisconsin-Madison with three other graduate degrees in community development, agricultural economics, and rural sociology.

**David R. Kolzow** is President of Team Kolzow, Inc., an economic development consulting firm in Franklin, Tennessee. Previously, he served as executive director of the Tennessee Leadership

Center in Nashville, Tennessee. Dr. Kolzow has over 30 years' consulting experience in site selection, real estate development planning, and community economic development with firms such as Lockwood Greene, Fluor Daniel, and PHH Fantus. He also served as chairman of the Department of Economic Development at the University of Southern Mississippi. He has authored a number of articles, as well as the books *Strategic Planning for Economic Development* (1992) and *Leadership: The Key Issue in Economic Development* (2002). He received his B.S. degree from Concordia University and his Ph.D. in Geography from Southern Illinois University in Carbondale.

**Joseli Macedo** has 20 years' experience in urbanism and architecture. She is an Urban and Regional Planning Professor and an urban planning consultant. She has worked in the United States and abroad, specializing in housing and community development, international development planning, and urban design. A member of the American Institute of Certified Planners, she worked for several years as a professional urban planner before obtaining her doctoral degree and joining academia. Currently, she is a faculty member at the University of Florida's Department of Urban and Regional Planning, where she teaches graduate-level studios and seminars, and serves as Undergraduate Coordinator. In addition to her academic work, she has completed several projects as a consultant, such as the Consolidated Plan for the City of Miami in the US, and the World Bank's sponsored Affordable Housing Needs Assessment Methodology Adaptation in Brazil for the Cities Alliance project.

**Deborah M. Markley** is Managing Director and Director of Research for the Rural Policy Research Institute's Center for Rural Entrepreneurship, a national research and policy center in Lincoln Nebraska. Her research has included case studies of entrepreneurial support organizations, evaluation of state industrial extension programs, and consideration of the impacts of changing banking markets on small business finance. She has extensive experience conducting field-based survey research projects and has conducted focus groups and interviews with rural bankers, entrepreneurs, business service providers, venture capitalists, small manufacturers, and others. Her research has been presented in academic journals, as well as to national public policy organizations and Congressional committees.

**Paul W. Mattessich** is Executive Director of Wilder Research, one of the largest applied social research organizations in the United States, dedicated to improving the lives of individuals, families, and communities. He has done research, lecturing, and consulting with nonprofit organizations, foundations, and government in North America, Europe, and Africa since 1973. His book on effective partnerships, *Collaboration: What Makes It Work* (2001, 2nd edn) is used worldwide, along with the online Wilder Collaboration Factors Inventory. His 1997 book, *Community Building: What Makes It Work* is widely recognized and used by leaders and practitioners in community and neighborhood development. He received his Ph.D. in Sociology from the University of Minnesota.

**Derek Okubo** is Senior Vice President for the National Civic League and oversees all programs out of the National Headquarters in Denver, Colorado. He has delivered extensive technical assistance on a variety of issues for local governments of all sizes throughout the United States. He has led or been a part of over 60 community-based planning processes around the country. He is also the author of numerous published articles and handbooks. Previously, he worked with Big Brothers of Metro Denver, Inc. where he designed and implemented a youth volunteer program that received national attention. He then received an appointment to the staff of Colorado's governors as a liaison to communities. He is a graduate of the University of Northern Colorado.

**Rhonda Phillips**, AICP, CEcD, is a Professor in the School of Community Resources and

Development at Arizona State University and a visiting Professor at SUNY Plattsburgh. She is a specialist in community and economic development, holding dual professional certifications in economic development (CEcD) and urban and regional planning (AICP) with over 20 years' experience with private, public, and nonprofit organizations at the international, national, state, and local levels. Her work with community indicators systems focuses on improving quality of life in communities and regions, including working with heritage and cultural regeneration projects in Northern Ireland as a Fulbright Scholar. She is Editor of *Community Development: Journal of the Community Development Society*, and author of several books, including *Concept Marketing for Communities* (2002) and *Handbook for Community Development* (2006).

**Robert H. Pittman** is Executive Director of the Strategic Growth Institute and the Community Development Institute at the University of Central Arkansas, where he also serves as Associate Professor in the College of Business. Prior to joining the faculty at UCA, he served as director of Business Location and Economic Development Consulting for Lockwood Greene, a worldwide engineering firm. He has over 20 years' experience in business location and economic development consulting for clients in the U.S. and abroad. A former deputy director of the International Development Research Council (now CoreNet), he is a widely published author and frequent speaker in the field of business location and economic development.

**Kenneth M. Reardon** is Professor and Director of the Graduate Program in City and Regional Planning at the University of Memphis where he is engaged in research, teaching, and outreach activities in the fields of neighborhood planning, community development, and community/university development partnerships. Prior to joining the University of Memphis faculty, he served as an associate professor and former chair of the Department of City and Regional Planning

at Cornell University where he played a key role in establishing the Cornell Urban Scholars Program, Cornell Urban Mentors Initiative, and the New Orleans Planning Initiative. He also served as an assistant and associate professor of City and Regional Planning at the University of Illinois at Urbana-Champaign where he initiated the East St. Louis Action Research Project. He currently serves on the editorial boards of the *Journal of Planning Literature* and the *Michigan Journal of Community Service Learning*.

**Richard T. Roberts** is a graduate of the University of Alabama. He has practiced economic development in three Southeastern states and worked for a variety of communities in Alabama, Mississippi, and Northwest Florida. During the past 30 years, he has managed both rural and metro chambers of commerce, economic development authorities and a convention and tourism organization. He has written marketing plans and existing industry programs and has established and organized multi-county as well a local economic development programs. He currently resides in Dothan, Alabama, and is employed by the Covington County Economic Development Commission.

**Tom Tanner** has worked extensively in the field of economic impact analysis and model building across the country, working for the Center for Agriculture and Rural Development at Iowa State University; the Center for Economic Development at the University of Wisconsin-Superior; Regional Economic Models, Inc. in Massachusetts; and at the Carl Vinson Institute of Government at the University of Georgia, where he has completed the design of the Regional Dynamics Economic Model, an economic modeling tool used to conduct local impact analysis across the country. Currently, he is Director of the Regional Dynamics and Economic Modeling Laboratory at Clemson University.

**John W. (Jack) Vincent II** is currently President of Performance Development Plus of Metairie, LA., a consulting company that he co-founded in 1993.

The firm specializes in community development, economic development and organizational development consulting. He has over 35 years' experience in various professional and technical positions and he is currently an owner and board member of a homeland security company. He holds a Bachelor's degree in Government from Southeastern Louisiana University, and a Master's degree from the University of New Orleans in Curriculum and Instruction (Adult Learning and Development). He is a graduate of the University of Central Arkansas's Community Development Institute in Conway, Arkansas, and is a certified Professional Community and Economic Developer.

**Monieca West** is an experienced economic and community development professional. In 2001, she retired from SBC-Arkansas where she had served as director of economic development since 1992. She is a past chair of the Community Development Society and the Community Development Council and has received numerous awards for services to community organizations and public education. In 2004, she returned to full-time work for the Arkansas Department of Higher Education where she is Program Manager for the Carl Perkins Federal Program which funds career and technical education projects at the postsecondary level. She also does freelance work in the areas of facilitation, leadership development, technical writing, association management, and webmaster services.

**Stephen M. Wheeler** joined the Landscape Architecture faculty at the University of California at Davis in 2007. He is also a member of UCD's Community Development and Geography Graduate Groups. He previously taught in the Community and Regional Planning Program at the University of New Mexico, where he initiated a Physical Planning and Design concentration, and the University of California at Berkeley, where he received his Ph.D. and Master of City Planning degrees. He received his Bachelor's degree from Dartmouth College. His areas of academic and professional expertise include urban design, physical planning, regional planning, climate change, and sustainable development. His current research looks at state and local planning for climate change, and evolving built landscape patterns in metropolitan regions.

# Acknowledgments

Like community development itself, creating a book on the subject is a team effort. We are thankful first of all to the chapter authors who have graciously consented to share their professional knowledge and experience with the readers. While the editors' and authors' names appear in the book, so many other community and economic development scholars and practitioners have played important roles in bringing this book to fruition. One of the most rewarding aspects of studying and practicing the discipline is learning from others along the way. This volume reflects the collective input over decades of countless community developers who may not be cited by name in the book but who have nonetheless had a profound impact on it.

The editors are also grateful to Arizona State University and the University of Central Arkansas for their support of community and economic development. Like most books, this one took more time than anticipated to complete, and we are appreciative of the patience and assistance offered by our employers and colleagues.

Of course you cannot have a book without a publisher, and we wish to express our thanks to Routledge and the Taylor & Francis Publishing Group. Their expert assistance throughout the process of creating this book is acknowledged and appreciated. And finally, we owe our families much gratitude, for without their support and encouragement this book would not exist.

Rhonda Phillips and Robert Pittman

# Editors' introduction

Community development is a complex and interdisciplinary field of study – one that is boundary spanning in its scope and multidimensional in its applications. Why is this? It's because community development not only concerns the physical realm of community, but also the social, cultural, economic, political and environmental aspects as well. Evolving from an original needs-based emphasis to one that is more inclusive and asset-based, community development is a now a distinct and recognized field of study. Today, scholars and practitioners of community development are better equipped to respond to the challenges facing communities and regions. Because its applications are wide-ranging yet always aimed at improving quality of life, it is important to understand the underlying foundations and theory of community development as well as the variety of strategies and tools used to achieve desired outcomes.

This text seeks to address the challenging and exciting facets of community development by presenting a variety of essential and important topics to help students understand its complexities. The chapter authors represent perspectives from both academe and practice, reflecting the applied nature of the discipline. Importantly, this book emphasizes the strong link between community development and economic development which is all too often overlooked in the literature. We believe a discussion of one is incomplete without a discussion of the other. Hopefully, this book will serve to more closely align the study and practice of these two inextricably related disciplines.

This text is presented in the spirit of community development as planned efforts to improve quality of life. With this goal in mind, 24 chapters covering a range of issues have been selected and organized into four major categories: (1) foundations; (2) preparation and planning; (3) programming techniques and strategies, and (4) issues impacting community development.

Part I: Foundations, provides an introduction and overview of the discipline as well as its underlying premises. In Chapter 1 we present the basic concepts and definitions of community development and how it relates to economic development, a central theme of this book. Chapter 2 distills a variety of ideas from different fields into a theoretical underpinning for community development. Hustedde offers seven contextual perspectives that provide this theoretical core: organizations, power relationships, shared meanings, relationship building, choice making, conflicts, and integration of paradoxes. Chapter 3 focuses on the concept of capacity building, both inside and outside the community. Haines explains the value of adopting an asset-based approach, and how it is dramatically different from the needs-based approaches of the past. Mattessich explains in Chapter 4 how social capital (or capacity) lies at the heart of community development. Analogous to other forms of capital, social capital constitutes a resource that may be used by communities to guide outcomes. The fifth and final chapter in this section outlines the foundation of processes and applications introducing students to community development as a practice. Echoing Chapter 1, Vincent explains that community development is closely linked to economic development in practice.

Part II: Preparation and planning, covers the variety of ways in which communities organize, assess, and plan for community development. In Chapter 6, Okubo takes the reader through the process of establishing goals and a vision for the future – essential activities for success in community development. Without this foundation, it is difficult to accomplish the desired outcomes. Chapter 7 addresses the all-important question, "How should we be organized?" West outlines different types of community-based organizations and their structures, and shows examples in practice. Chapter 8, by Kolzow, discusses the need for communities to effectively integrate skill development into their activities. The premise is that great leadership leads to the most desirable community development outcomes. Vincent's second contribution, Chapter 9, provides a broad perspective on the total community assessment process. It discuses comprehensive assessments and the areas that should be considered, including a community's physical, social and human infrastructure and capital. Chapter 10 by Green provides information on techniques such as asset inventories, identifying potential partners and collaborators, various survey instruments and data collection methods. The final chapter in this section, Chapter 11, by Hearn and Tanner, discusses how to asses the underlying strengths and weaknesses of the local economy. It provides an overview of economic impact analysis and how it may be used to allocate scarce community financial resources.

Part III: Programming techniques and strategies, gives several specific application areas for community development and how these areas may be approached. West's second contribution, Chapter 12, addresses the vital question of how to develop a quality workforce in the community. It provides examples of initiatives that communities have used to address this need. Pittman's Chapter 13 provides an overview of how to attract new businesses into a community and expand and retain businesses already there in order to strengthen the local economy. Creating recognition for the community, identifying the appropriate target audience, and the most effective marketing message are discussed. In Chapter 14, Pittman and Roberts explain the importance of focusing on businesses already present in communi-

ties. An existing business program can help communities in many direct and indirect ways and is often more effective in job creation than other approaches. Gruidl and Markley present entrepreneurship as a community development strategy in Chapter 15 as a vital component driving economic growth and job creation. The fundamentals for implementing a strategy of supporting entrepreneurs and creating a nurturing environment are outlined. Chhabra and Phillips' Chapter 16 explores ways in which communities can tap into the lucrative and growing tourism industry. A variety of models and approaches are reviewed. Chapter 17 by Macedo provides a basic understanding of how the housing typology, density and affordability affect housing and community development. Reardon's Chapter 18 discusses the model of participatory neighborhood planning. This model seeks to improve quality of life with comprehensive revitalization strategies grounded in an asset-based approach. Our final contribution to this part, Chapter 19, by Phillips and Pittman begins with the premise that progress evaluation is not only challenging but vital, and organizations must be able to assess and demonstrate the value and outcome of their activities. Specific types of evaluation are introduced in the context of practical application.

Part IV: Issues impacting community development, focuses on a few of the many and diverse issues relevant to community development theory and practice. Chapter 20 by Hamer and Farr gives an overview and explanation of the different types of community development financing from public and private sources. It includes definitions of key terms as well as ideas on structuring funding partnerships. Browning's Chapter 21 gives information on how to research and write grant proposals. Grants are a major component of funding for many community development organizations and this chapter provides specific ideas for improving the chances of garnering successful funding. Kolzow and Pittman's Chapter 22 begins with an overview of the increasing interconnectedness of the global economy. It continues with a discussion of the impacts of globalization on community development and strategies on how to respond. Chapter 23 by Wheeler provides a basic background on the concept of sustainability and how it applies to both the theory and practice of community development. It also gives examples of strategies that may be implemented to help increase sustainable approaches. The final chapter offers some concluding observations on issues covered in the book, and discusses the important role of community development in helping shape the future of our society.

As stated at the outset, community development is indeed a complex and interdisciplinary field, as evidenced by the breadth and scope of the chapters presented. We encourage students of community development to embrace the "ethos" of the community development discipline as one that focuses on creating better places to live and work and increasing quality of life for all.

Rhonda Phillips
Arizona State University

Robert Pittman
University of Central Arkansas

# PART I
# Foundations

# 1 A framework for community and economic development

## Rhonda Phillips and Robert H. Pittman

Community development has evolved over the past few decades into a recognized discipline of interest to both practitioners and academicians. However, community development is defined in many different ways. Most practitioners think of community development as an outcome – physical, social, and economic improvement in a community – while most academicians think of community development as a process – the ability of communities to act collectively and enhancing the ability to do so. This chapter defines community development as both a process and an outcome and explains the relationship between the two.

A related discipline, economic development, is also defined in different ways. This chapter offers a holistic definition of economic development that not only includes growing businesses and creating jobs but increases in income and standards of living as well. Economic development is also shown to be both an outcome and a process. The community and economic development chain shows the links, causal relationships, and feedback loops between community and economic development, and illustrates how success in one facilitates success in the other.

## Introduction

Community development has many varying definitions. Unlike mathematics or physics where terms are scientifically derived and rigorously defined, community development has evolved with many different connotations. Community development has probably been practiced for as long as there have been communities. It is hard to imagine the American colonies being successfully established in the seventeenth century without some degree of community development, even if the term had not yet come into existence.

Many scholars trace the origin of modern community development as a discipline to post-World War II reconstruction efforts to improve less developed countries (Wise 1998). Others cite the American "war on poverty" of the 1960s with its emphasis on solving neighborhood housing and social problems as a significant influence on contemporary community development (Green and Haines 2002). As the following box shows, the origins of community development are actually very old. A major contribution of community development was the recognition that a city or neighborhood is not just a collection of buildings but a "community" of people facing common problems with untapped capacities for self-improvement. Today, community is defined in myriad ways: in geographic terms, such as a neighborhood or town ("place based" or communities of place definitions), or in social terms, such as a group of people sharing common chat rooms on the Internet, a national professional association or a labor union (communities of interest definitions).

## BOX 1.1 EVOLUTION OF COMMUNITY DEVELOPMENT

Community development as a profession has deep roots, tracing its origins to social movements (it is, after all, about "collective" action) of earlier times such as the Sanitary Reform Movement of the 1840s and later housing reforms. Beyond North America, community development may be called "civil society," or "community regeneration," and activities are conducted by both government and non-governmental organizations (NGOs). There may or may not be regulation of organizations, depending on different countries' policy framework (for a review of community development in Europe see Hautekeur 2005). The Progressive Movement of the 1890s through the first few decades of the twentieth century was all about community development, although the term itself did not arise until mid-century.

During the 1950s and 1960s, social change and collective action again garnered much attention due to the need to rectify dismal conditions within poverty-stricken rural areas and areas of urban decline. The civil rights and antipoverty movements led to the recognition of community development as a practice and emerging profession, taking form as a means to elicit change in social, economic, political and environmental aspects of communities. During the 1960s, literally thousands of community development corporations (CDCs) were formed, including many focusing on housing needs as prompted by federal legislation providing funding for nonprofit-based community organizations. This reclaiming of citizen-based governing was also prompted in response to urban renewal approaches by government beginning with the US Housing Act of 1949. The richness of the CDC experience is chronicled in the Community Development Corporation Oral History Project by the Pratt Center for Community Development (www.prattcenter.net/cdcoralhistory.php). This includes one of the first CDCs in the US, the Bedford Stuyvesant Restoration Corporation in the city of New York.

Evolution of the discipline continued; in 1970 two journals were established, *Community Development* in the UK and *Community Development: Journal of the Community Development Society* in North America, as well as the establishment of academic programs with an emphasis on community development (typically, a public administration, public policy or urban planning degree with a concentration available in community development). Today, there are about 4000 CDCs in the US, with most focusing on housing development. However, many also include a full range of community development activities, with about 25 percent providing a comprehensive array of housing development, homeownership programs, commercial and business development, community facilities, open space/environmental, workforce and youth programs, and planning and organizing activities (Walker 2002), Other organizations practice community development too, including public sector ones as well as private for-profit companies and other nonprofits (see Box 1.1, "Who Practices Community and Economic Development" at the end of this chapter for more information). As the variety of topics in this book attests, community development has evolved from its roots in social activism and housing to encompass a broad spectrum of processes and activities dealing with multiple dimensions of community including physical, environmental, social and economic.

Community development has evolved into a recognized discipline drawing from a wide variety of academic fields including sociology, economics, political science, planning, geography, and many others. A quick Internet search reveals how much the field has evolved – the authors' search returned 19,200,000 hits for "community development." Today there are many academic and professional journals focusing on community development. The interest of researchers and practitioners from many different disciplines has contributed greatly to the growth and development of the field. However, community development's growth and interdisciplinary nature have led to the current situation where it is defined and approached in many different ways, and, all too often, "never the twain shall meet."

This chapter takes a broad approach to community development. While it is impossible in one chapter (or book) to completely cover such a large field, many different aspects of community development are included. In particular, the authors believe the strong interrelationship between community and economic development is often overlooked in research and practice. This interrelationship is one focus of this chapter and book.

The terms community development and economic development are widely used by academicians, professionals, and citizens from all walks of life and have almost as many definitions as users. Economic development is perhaps more familiar to laypersons. If random individuals on the street were asked what economic development is, some might define it in physical terms such as new homes, office buildings, retail shops, and "growth" in general. Others might define it as new businesses and jobs coming into the community. A few thoughtful individuals might even define it in socio-economic terms such as an increase in per capita income, enhanced quality of life, or reduction in poverty.

Ask the same individuals what community development is, and they would probably think for a while before answering. Some might say it is physical growth – new homes and commercial buildings – just like economic development. Others might say it is community improvement such as new infrastructure, roads, schools, and so on. Most respondents would probably define community and economic development in terms of an *outcome* – physical growth, new infrastructure, or new jobs. Probably no one would define them in terms of a *process* and many would not understand how they are interrelated. This is unfortunate because some of these passers-by are probably involved in community and economic development efforts, serving as volunteers or board members for chambers of commerce, economic development agencies, or charitable organizations.

The purpose of this introductory chapter is to provide meaningful descriptions of community development and economic development as both processes and outcomes, explore what they entail, and understand them as distinct yet closely related disciplines. First, the focus will be on community development, followed by a discussion of economic development, and finally, an examination of the relationship between the two.

## Community development

The beginning step in defining community development is to define "community." As mentioned previously, community can refer to a location (communities of place) or a collection of individuals with a common interest or tie whether in close proximity or widely separated (communities of interest). A review of the literature conducted by Mattessich and Monsey (2004) found many definitions of community such as:

> People who live within a geographically defined area and who have social and psychological ties with each other and with the place where they live.
>
> (Mattessich and Monsey 2004: 56)

> A grouping of people who live close to one another and are united by common interests and mutual aid.
>
> (National Research Council 1975 cited in Mattessich and Monsey 2004: 56)

A combination of social units and systems which perform the major social functions ... (and) the organization of social activities.
(Warren 1963 cited in Mattessich and Monsey 2004: 57)

These definitions refer first to people and the ties that bind them and second to geographic locations. They remind us that without people and the connections among them, a community is just a collection of buildings and streets. In this context, community development takes on the mantle of developing stronger "communities" of people and the social and psychological ties they share. Indeed this is how community development is defined in much of the literature. Discussions that reflect this aspect focus on community development as an educational process to enable citizens to address problems by group decision making (Long 1975 cited in Mattessich and Monsey 2004: 58). Or, they may describe community development as involvement in a process to achieve improvement in some aspect of community life where normally such action leads to the strengthening of the community's pattern of human and institutional relationships (Ploch 1976 cited in Mattessich and Monsey 2004: 59).

All of these concepts of community development focus on the *process* of teaching people how to work together to solve common problems. Other authors define community development more in terms of an action, result, or *outcome*: local decision making and program development resulting in a better place to live and work (Huie 1976 cited in Mattessich and Monsey 2004: 58); or a group of people initiating social action to change their economic, social, cultural and/or environmental situation (Christenson and Robinson 1989 cited in Mattessich and Monsey 2004: 57).

These conceptions show that community development should be considered as both a process and an outcome. Therefore, a working definition of community development in simple but broad terms is:

A *process*: developing and enhancing the ability to act collectively, and an *outcome*: (1) taking collective action and (2) the result of that action for improvement in a community in any or all realms: physical, environmental, cultural, social, political, economic, etc.

Having arrived at a comprehensive definition of community development, the focus can now shift to what facilitates or leads to community development. The community development literature generally refers to this as *social capital* or *social capacity*, which describes the abilities of residents to organize and mobilize their resources for the accomplishment of consensual defined goals (Christenson and Robinson 1989 cited in Mattessich and Monsey 2004: 61), or the resources embedded in social relationships among persons and organizations that facilitate cooperation and collaboration in communities (Committee for Economic Development 1995 cited in Mattessich and Monsey 2004: 62).

Simply put, social capital or capacity is the extent to which members of a community can work together effectively to develop and sustain strong relationships; solve problems and make group decisions; and collaborate effectively to plan, set goals, and get things done. There is a broad literature on social capital with some scholars making the distinction between *bonding capital* and *bridging capital* (Agnitsch et al. 2006). Bonding capital refers to ties within homogeneous groups (e.g., races, ethnicities, social action committees, or people of similar socio-economic status) while bridging capital refers to ties among different groups.

There are four other forms of "community capital" often mentioned in the community development literature (Green and Haines 2002: viii):

1 Human capital: labor supply, skills, capabilities and experience, etc.
2 Physical capital: buildings, streets, infrastructure, etc.
3 Financial capital: community financial institutions, micro loan funds, community development banks, etc.
4 Environmental capital: natural resources, weather, recreational opportunities, etc.

All five types of community capital are important. However, it is difficult to imagine a community making much progress without some degree of social capital or capacity. The more social capital a community has, the more likely it can adapt to and work around deficiencies in the other types of community capital. When doing community assessments (see Chapter 9), it is useful to think in terms of these five types of community capital.

So far working definitions of community, community development, and social capital have been provided. To complete the community development equation, it is necessary to identify how to create or increase social capital or capacity. This process is generally referred to as *social capital building* or *capacity building*: an ongoing comprehensive effort to strengthen the norms, supports, and problem-solving resources of the community (Committee for Economic Development 1995 cited in Mattessich and Monsey 2004: 60).

Notice that this sounds like the definitions of the process of community development given above. We have come full circle. The *process* of community development *is* social capital/capacity building which leads to social capital which in turn leads to the *outcome* of community development.

Figure 1.1 depicts the community development chain. The solid lines show the primary flow of causality. However, there is a feedback loop shown by the dotted lines. Progress in the outcome of community development (taking positive action resulting in physical and social improvements in the community) contributes to capacity building (the process of community development) and social capital. For example, better infrastructure (e.g., public transportation, Internet access) facilitates public interaction, communications, and group meetings. Individuals who are materially, socially, and psychologically better off are likely to have more time to spend on community issues because they have to devote less time to meeting basic human and family needs. Success begets success in community development. When local citizens see positive results (outcome), they generally become more enthused and plow more energy into the process because they see the payoff. Research has shown that there are certain characteristics of communities that influence their ability to do capacity building and create social capital (Mattessich and Monsey 2004). Chapter 4 (this volume) discusses some of these community characteristics.

## Economic development

As with community development, modern economic development grew in part from efforts to improve less developed countries and the American war on poverty (Malizia and Feser 1999). Immediately after World War II and certainly before, parts of the American South were not unlike third world countries with rampant poverty and unemployment due

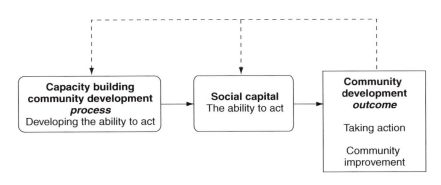

*Figure 1.1* **Community development chain**

to the decline of agricultural jobs. Many southern states developed programs to recruit industries from the northern US. with cheap labor and government incentives (e.g., tax breaks) as bait. In the 1960s, the emphasis on economic development was at the Federal level with Great Society programs aimed at eliminating "pockets of poverty" such as the southern Appalachian Mountains region. In the 1970s and 1980s, the emphasis shifted to states and localities. If they did not already have them, many communities created economic development organizations with public as well as private funding. In many communities, economic development was part of city or county government, while in other communities it was under the auspices of private nonprofit organizations such as local chambers of commerce.

Initially, most economic development agencies focused on industrial recruitment – enticing new companies to locate in their communities. State and local departments of economic development aggressively courted industry and increased the amount and variety of incentives. Soon it became apparent that thousands of economic development organizations were chasing a limited number of corporate relocations/expansions and playing a highly competitive game of incentives and marketing promotions. As communities realized that there were other ways to create jobs, they began to focus on internal opportunities such as facilitating small business development and ensuring that businesses already located in the community stayed and expanded there. Many communities also realized that by improving education, government services, the local labor supply, and the business climate in general, they could make themselves more attractive to industry (see discussion in Chapter 22 on second- and third-wave economic development strategies).

Like community development, economic development has evolved into a broad and multidisciplinary field. A national association of economic development professionals has offered the following definition (AEDC 1984: 18[1]):

(Economic development is) the process of creating wealth through the mobilization of human, financial, capital, physical and natural resources to generate marketable goods and services. The economic developer's role is to influence the process for the benefit of the community through expanding job opportunities and the tax base.

Most economic developers concentrate on creating new jobs. This is generally the key to wealth creation and higher living standards. Job creation generally involves the "three-legged stool" of recruiting new businesses, retaining and expanding businesses already in the community, and facilitating new business start-ups. As discussed previously, economic developers originally concentrated mainly on recruiting new businesses. Stiff competition for a limited number of new or expanded facilities in a given year led some communities to realize that another way to create jobs is to work with companies already in the area to maximize the likelihood that, if they need to expand existing operations or start new ones, they would do so in the community and not elsewhere. Even if an expansion is not involved, some businesses may relocate their operations to other areas for "pull" or "push" reasons (Pittman 2007). They may relocate to be closer to their customers, closer to natural resources, or for any number of strategic business reasons ("pull"). Businesses may also relocate because of problems with their current location such as an inadequate labor force, high taxes, or simply lack of community support ("push"). Although communities cannot influence most pull factors, they can act to mitigate many push factors. If the problem is labor, they can establish labor training programs. If the problem is high taxes, they can grant tax incentives in return for creating new jobs. Business retention and expansion has become a recognized subfield of economic development, and there are many guides on the subject (e.g., Entergy 2005). Chapter 14 addresses business retention and expansion in more detail.

There is also much that communities can do to facilitate the start-up of new local businesses. Some communities create business incubators where fledgling companies share support services, benefit from reduced rent, or even get free consulting assistance.

Financial assistance, such as revolving loan funds at reduced interest rates, is also a common tool used to encourage small business start-ups. Making a community "entrepreneurial friendly" is the subject of Chapter 15.

While the majority of new jobs in most regions are created by business retention and expansion and new business start-ups (Roberts 2006), many communities continue to define economic development in terms of recruiting new facilities. Some communities are stuck in the old paradigm of "smokestack chasing" – relying solely on recruiting a new factory when manufacturing employment is declining nationally and many companies are moving off-shore. These communities may have historically relied on traditional manufacturing

## BOX 1.2 GROWTH VS. DEVELOPMENT

Growth by itself could be either an improvement or a detriment to a community (Blair 1995: 14). For instance, a facility that paid very low wages might open in an area and the population and overall size of the economy might increase, but per capita incomes might fall, and the quality of life might suffer. Such growth could bring more congestion, pollution, and other negative externalities without a commensurate increase in public resources or commitment to address them. Of course, many communities would be happy to get any new facility regardless of the wage rate. A minimum wage job is better than no job, and there is a portion of the labor force in most communities that is a good match for minimum wage jobs (e.g., teenagers and adult low-skilled workers).

Since growth does not always equate with a better standard of living, a higher order concept of economic development is needed that better reflects the actual well-being of residents. Comprehensive economic development efforts, therefore, should be directed toward improving the standard of living through higher per capita income, better quality and quantity of employment opportunities, and enhanced quality of life. Increases in per capita income (adjusted for inflation) are often used as indicators of welfare improvements (Blair 1995: 14). There are many other indicators of welfare and quality of life for community residents such as poverty rates, health statistics, and income distribution (see Chapter 19 on community indicators), but per capita income or income per household is a common measurement of economic well-being. Whether a higher per capita income equates to a good quality of life depends on the individual. Some individuals would rather have an income of $20,000 per year in a scenic rural area than $100,000 per year in a large city. Moreover, per capita income is not necessarily a measure of purchasing power. The cost of living varies from place to place, and a dollar goes further where prices are low. Like community development, these descriptions portray economic development as both a process and an outcome – the process of mobilizing resources to create the outcome of more jobs, higher incomes, and an increased standard of living, however it is measured.

From this discussion it should be apparent that development has a very different connotation than growth. Development implies structural change and improvements within community systems encompassing both economic change and the functioning of institutions and organizations (Boothroyd and Davis 1993; Green and Haines 2002). Development is deliberate action taken to elicit desired structural changes. Growth, on the other hand, focuses on the quantitative aspects of more jobs, facilities construction, and so on – within the context that more is better. One should carefully distinguish, then, between indicators that measure growth versus development. By these definitions, a community can have growth without development and vice versa. The important point to note, however, is that development not only facilitates growth but also influences the kind and amount of growth a community experiences. Development guides and direct growth outcomes.

companies for the bulk of their employment (e.g., textile mill or garment factory), and when these operations shut down, the only thing they knew to do was to pursue more of the same. Elected officials, civic board members, and citizens in these communities need to be educated and enlightened to the fact that the paradigm has shifted. They should be recruiting other types of businesses such as service industries (that are not going off-shore themselves) while practicing business retention and expansion and creating new business start-up programs.

While economic development today is often defined by the "three-legged stool," there is much more to the profession. As shown in a recent survey (SEDC 2006), economic developers get involved in workforce development, permitting assistance, and many other issues, some of which are better defined as community development (Figure 1.2).

Many communities are beginning to realize that it is often better, especially when recruiting new companies, to practice economic development on a regional basis and combine resources with nearby communities. Rather than create an economic development agency providing the same services for every community, it is usually more efficient to combine resources and market a region collectively through one larger organization. Regardless of where a new facility locates or expands, all communities in the region will benefit as employees live in different areas and commute to work.

## The relationship between community and economic development

While conceptions and definitions of community and economic development vary, in practice they are inextricably linked on many levels and are highly synergistic. To understand these synergies, consider another definition of community development (Green and Haines 2002: vii):

> Community development is ... a planned effort to produce assets that increase the capacity of residents to improve their quality of life. These assets may include several forms of community capital: physical, human, social, financial and environmental.

Recall the previous definition of economic development (AEDC 1984):

> The process of creating wealth through the mobilization of human, financial, capital, physical and natural resources to generate

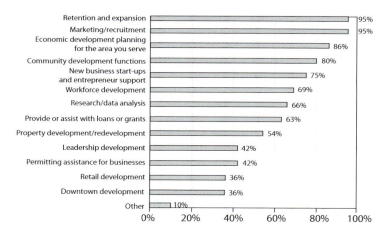

*Figure 1.2* **Economic development activities performed by SEDC members**

marketable goods and services. The economic developer's role is to influence the process for the benefit of the community through expanding job opportunities and the tax base.

These two definitions are clearly parallel. The purpose of community development is to *produce* assets that may be used to improve the community, and the purpose of economic development is to *mobilize* these assets to benefit the community. Both definitions refer to the same community capital assets: human, financial, and physical (environmental or natural resources). As mentioned above, a more modern holistic definition of economic development would include not only wealth and job creation but increasing the quality of life and standard of living for all citizens. This expanded definition is certainly compatible with community development. The definition of economic development does not include social capital per se, but while economic developers might not have used this term when this definition was created, it will be seen that social capital is important for economic development as well as for community development.

As most economic developers will attest, a key to success in economic development – new business recruitment, retention and expansion of existing businesses, and new business start-up – is to have a "development-ready" community. Most businesses operate in competitive markets, and one of the major factors influencing business profitability is their location. When making location decisions, businesses weigh a host of factors that affect their costs and profits such as:

- available sites and buildings;
- transportation services and costs (ground, water, air);
- labor cost, quality and availability;
- utility costs (electricity, natural gas);
- suitability of infrastructure (roads, water/sewer);
- telecommunications (Internet bandwidth);
- public services (police and fire protection).

Quality of life factors (e.g., education, health care, climate, recreation) are also important in many loca-tion decisions (Pittman 2006). If a community scores poorly in these factors, many companies would not consider it development ready. A weakness in even one important location factor can eliminate a community from a company's search list.

Whether a community is considered development ready depends on the type of business looking for a location. For example, important location criteria for a microchip manufacturing facility include a good supply of skilled production labor, availability of scientists and engineers, a good water supply, and a vibration-free site. A call center seeking a location would focus more on labor suitable for telephone work (including students and part-time workers), a good non-interruptible telecommunications network, and, perhaps, a time zone convenient to its customers. While location needs differ, a community lacking in the location factors listed above would be at a disadvantage in attracting or retaining businesses. Shortcomings in these or other location factors would increase a firm's costs and make it less competitive. In addition, companies are risk averse when making location decisions. If a company is comparing two similar communities where one has a prepared site that is construction ready with all utilities in place and the other offers a cornfield and a promise to have it developed in six months, it is apparent which the company will choose. Most companies would not want to incur the risk that development of a site would be delayed by construction problems, cost overruns, or simply by local politics. Lost production equates to lost revenues and profits.

There are many location factors that are subjective and not easily quantified. For example, it is not feasible to objectively measure factors such as work ethic, ease of obtaining permits, or a community's general attitude toward business (these and other factors are sometimes collectively referred to as the "business climate"). Yet in the final decision process after as many factors as possible have been quantified, these intangibles often determine the outcome. Most company executives prefer to live in desirable communities with good arts and recreation, low crime, and a neighborly, cordial atmosphere.

It should be apparent now how important community development is to economic development. If a community is not development ready in the physical sense of available sites, good infrastructure, and public services, it will be more difficult to attract new businesses and retain and expand existing ones. New businesses starting up in the community would be at a competitive disadvantage. Being development ready in this physical sense is an *outcome* of community development (taking action and implementing community improvements).

The *process* of community development also contributes to success in economic development. First, as has been discussed, the process of community development (developing the ability to act in a positive manner for community improvement) leads to the outcome of community development and a development-ready community. In addition, some of the intangible but important location factors may be influenced through the process of community development. Companies do not like to locate in divided communities where factions are openly fighting with one another, city councils are deadlocked and ineffective, and citizens disagree on the types of businesses they want to attract (or even if they want to attract any businesses). As a company grows, it will need the support of the community for infrastructure improvements, good public education, labor training, and many other factors. Communities that are not adept (or worse, are totally dysfunctional) at the process of community development are less likely to win the location competition. Furthermore, company executives would probably prefer not to live in such a place.

As discussed, economic development, like community development, is also a process. Establishing and maintaining a good economic development program is not easy. Significant resources must be devoted to hiring staff, providing suitable office facilities, and marketing the community. Most communities that are successful in economic development have strong support (financial and otherwise) from both the public and private sectors. Success in economic development does not come overnight. Some community residents mistakenly believe that if

a local economic developer gets on a plane and calls on corporations in distant cities, these companies will see how wonderful their fair city is and choose to locate there. There are two problems with this belief: (1) there are thousands of other "fair cities" competing for new facilities, and (2) relocation and expansion decisions happen only sporadically for most companies.

Communities that are successful in economic development devote the appropriate resources to the effort, design good programs, and stay with them for the long-haul. Over time a good economic development program pays dividends. If communities approach economic development in a start-and-stop fashion, frequently changing programs as, say, new mayors take office, the likelihood of success is significantly lower. A well-planned and widely supported economic development program based on consensus building through the process of community development has a much higher likelihood of success. Osceola, Arkansas, is an inspiring example of good community development leading to success in economic development in just this manner (see Box 1.3).

Shaffer, Deller, and Marcouiller (2006: 61) describe the relationship and synergy between community development and economic development as follows:

> We maintain that community economic development occurs when people in a community analyze the economic conditions of that community, determine its economic needs and unfulfilled opportunities, decide what can be done to improve economic conditions in that community, and then move to achieve agreed-upon economic goals and objectives.

They also point out that the link between community development and economic development is sometimes not understood or appreciated:

> Economic development theory and policy have tended to focus narrowly on the traditional factors of production and how they are best

**BOX 1.3 OSCEOLA, ARKANSAS: COMMUNITY DEVELOPMENT TURNS A DECLINING COMMUNITY AROUND**

Osceola is a town of about 9000 population in the Mississippi Delta in Northeast Arkansas. Like many rural communities, Osceola and Mississippi County grew up around the agriculture industry. As farm employment declined, the city attracted relatively low-skilled textile manufacturing jobs which ultimately disappeared as the industry moved off-shore. In 2001 a major employer, Fruit of the Loom, shut down its Osceola plant leaving the community in a crisis. Not only was unemployment rampant, the local schools were classified as academically distressed and were facing a state takeover. Osceola's mayor stated: "we almost hit the point of no return" (Shirouzu 2006).

The remaining businesses were having difficulty finding labor with basic math skills for production work. To address this problem, executives from these businesses began to work with the city administration to find a solution. They decided to ask the state for permission to establish a charter school that could produce well-educated students with good work skills. After many community meetings and differences of opinion, the community united behind the effort and opened a charter school.

Shortly after the school was established, Denso, a Japanese-owned auto parts company, was looking for a manufacturing site in the southern states. City representatives showed Denso the charter school and repeatedly told the company how the community had come together to solve its labor problem. Denso decided to locate the plant in Osceola, creating 400 jobs with the potential to grow to almost 4000. While Denso cited the improved labor force as a reason for selecting Osceola, the company was also impressed with how the community came together and solved its problem. An executive from Denso was quoted as saying, "It was their aggressiveness that really impressed us."

Denso was also concerned about the community's commitment to continuous improvement in the schools. Regarding this issue, the Denso executive was quoted as saying, "Is there a future here? Are they doing things that are going to drive them forward? Do they have that commitment to do it? We saw that continuously in Osceola." Denso located in Osceola not only because of its labor force, but because it was convinced the community would continue to practice good community development principles and move forward. Osceola's success story was featured on the front page of the national edition of *The Wall Street Journal* (Shirouzu 2006).

allocated in a spatial world. We argue that community economic development must be broader than simply worrying about land, labor and capital. This broader dimension includes public capital, technology and innovation, society and culture, institutions, and the decision-making capacity of the community.

(Shaffer et al. 2006: 64)

The authors make it clear that community development and economic development are inextricably linked, and if scholars and practitioners of economic development do not address community develop-

ment, they are missing an important part of the overall equation.

Now the diagram begun in Figure 1.1 can be completed to show the holistic relationship between the process and outcome of community development and economic development (see Figure 1.3).

The community development chain is as depicted in Figure 1.1: capacity building (the process of community development) leads to social capital which in turn leads to the outcome community development. In addition, communities with social capacity (the ability to act) are inherently more capable of creating good economic development

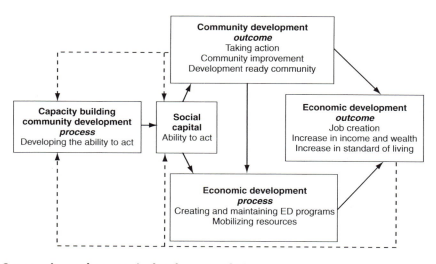

**Figure 1.3 Community and economic development chain**

programs should they choose to do so. When these communities take action (community development outcome), they create and maintain effective economic development programs that mobilize the community's resources. They also improve their physical and social nature and become more development ready, which leads to success in business attraction, retention and expansion, and start-up.

Citizens should understand the community and economic development chain in order to move their communities forward efficiently and effectively. While community developers might not believe they are practicing economic development and vice versa, in reality, they are all practicing community economic development.

---

## BOX 1.4 DIFFERENT DEFINITIONS OF COMMUNITY ECONOMIC DEVELOPMENT

Community development and economic development are frequently used interchangeably, and the term "community economic development" is often seen as well. Shaffer and colleagues (2006) used it to describe the integration of the community and economic development processes. Some authors, however, also use it to refer to "local economic development" encompassing growth (economic), structural change (development), and relationships (community). It is often seen in Canada and the UK (see e.g., Haugthton (1999) or Boothroyd and Davis (1993)).

## Who practices community and economic development?

The short (and correct) answer to this question is that all citizens who are interested in moving their communities forward should consider themselves practitioners of community and economic development. As discussed in this book (e.g., Chapters 5 and 6), to be successful and sustainable, community and economic development plans should be based on input from all socio-economic groups – everyone should feel that they have a voice in their community's future. Furthermore, successful implementation of community and economic development programs requires the engagement and collective action of all citizens.

In practice, community and economic developers can broadly be classified into two basic groups: paid professionals and volunteers. As we have discussed, community development is a broad field encompassing all aspects of society – housing, health care, education, transportation, and so on. Therefore, in theory any paid professional in the public or private sector who is working to improve their part of the community in any of these fields is a community developer. Note that the private sector includes private for-profit entities and private nonprofit organizations. A more practical definition of a community development professional would be anyone working in a government, nonprofit or other organization whose job definition involves improving certain aspects of a community for benefit of the community itself. For example, a person working for city government might have the title "community development specialist" and be responsible for improving one or more aspects of the community such as housing or health care availability.

This indistinct boundary between full-time professionals whose job description includes community development and other professionals working in health care, education, and so on that have community development impacts underscores the close relationship between community development and these related disciplines. When professionals in all these fields are more aware of the community development impacts of their jobs, the overall community development effort is strengthened. For example, a city planner should certainly design a land-use plan with the safety and convenience of community residents in mind, but should also be aware of the potential impacts of the plan on economic development, housing affordability, or any number of overall community development issues.

The economic development profession, on the other hand, is usually more distinct and readily defined. It is important to note that both disciplines have recognized professional certifications. As discussed previously, economic development involves activities such as creating new jobs and increasing household or per capita incomes – eliciting structural change within the area's economy. To accomplish this, economic developers may recruit new companies, work to retain and expand existing companies, facilitate new business start-ups, and engage in related activities. Economic developers are employed by a variety of public and private organizations, usually with clearly defined economic development responsibilities. In some communities, economic development activities are concentrated in the public sector (e.g., a city or county department of economic development) while in others they may be concentrated in the private sector (e.g., chamber of commerce). In most communities, economic development responsibilities are shared by the public and private sectors (see IEDC (2006) for more information on different models of economic development organization and service delivery).

Finally, there are the community and economic development volunteers. These include residents participating in community and economic development activities such as public meetings, planning sessions or community initiatives as well as board members of community and economic development organizations. Success in community and economic development requires dedicated, well-trained professionals and volunteers alike working together effectively for the community's benefit.

## BOX 1.5 COMMUNITY AND ECONOMIC DEVELOPMENT TRAINING AND CERTIFICATION

As community and economic development have evolved into recognized academic and professional disciplines, training and certification programs in the field have grown as well. There are thousands of local community and economic development training programs offered by state and local governments and organizations in countries around the world. Many universities offer undergraduate and graduate degrees in community and/or economic development. In North America, two major professional training programs are the Economic Development Institute offered through the University of Oklahoma (http://edi.ou.edu/) and the Community Development Institute offered through several universities and agencies around the US under the auspices of the Community Development Council (www.cdcouncil.com). The Certified Economic Developer (CEcD) program is administered by the International Economic Development Council (www.iedconline.org). The Community Development Council offers certification in community and economic development to full-time professionals (Professional Community and Economic Developer – PCED) and to community volunteers (Certified Community Development Partner – CEDP).

## Conclusion

Some communities do not believe they can influence their own destiny and improve their situation. They have a fatalistic attitude and feel they are victims of circumstances beyond their control – a closed factory, a downsized military base, or a natural disaster. In reality, they can build a better future. For every community that fails to act, there is another that is pro-actively applying the tools of community and economic development to better itself. Successful communities realize that community develop-ment is a group effort involving all citizens, not just the mayor, chamber of commerce president, or economic development professional.

Sometimes it takes an event, such as a local plant closing, to shock a community into action (as in Osceola, Arkansas). On the other hand, many communities realize that change is inevitable and choose to be prepared by practicing good community and economic development. In today's dynamic global economy, maintaining the status quo is rarely an option – either a community moves itself forward or by default moves backward.

## CASE STUDY: TUPELO, MISSISSIPPI

### From agriculture to manufacturing to auto assembly

Tupelo-Lee County, Mississippi, is often cited as a prime example of community transformation. It has been the subject of studies, articles, and even books on community and economic development (see e.g., Grisham 1999). Tupelo transformed itself from a community heavily reliant on agriculture and garment industry jobs into a dynamic, growing community with thousands of skilled manufacturing jobs. Immediately after World War II there were 2000 people employed in low-skill manufacturing jobs. By 1992, in the Tupelo-Lee County region, there were 92,000 manufacturing jobs and almost 350,000 total jobs (Martin 1996). This transformation was accomplished by applying good community and economic development principles: assessing the situation, working together to develop a plan, and making the community development ready and attractive to industry.

However, Tupelo-Lee County and the surrounding region eventually faced another challenge as many of their manufacturing industries began to downsize or move off-shore. Applying the same community and economic development principles in 2007, the state and region won the intense competition for a Toyota vehicle assembly plant which will create 2000 new manufacturing jobs and upward of 5000 total new jobs. Three counties in the region joined forces and set their sites on recruiting an auto assembly plant. They made the area more attractive by passing bonds, improving schools, and developing an inter-governmental agreement to share the costs and revenues from the new facility (Meridian Star 2007). The approach to community and economic development that Tupelo-Lee County has taken throughout the years, as described by a local official, illustrates many of the principles described in this chapter:

1    One economic development agency for Tupelo-Lee County (the Community Development Foundation); no competing ED organizations
2    Working together and presenting a united front under the Community Development Foundation
3    Strong private sector leadership

- Three leadership programs for young adults
- Welcome all newcomers

4    Belief that community development precedes economic development: must be "development ready"
5    Recognize importance of regionalism in economic development: CREATE Foundation for NE Mississippi
6    Positive media coverage on economic development

- Newspaper publisher/owner George McClane set up CREATE Foundation

7    Positive labor climate and race relations
8    Strategic planning with regular updates
9    Strong public/private partnerships

- Economic development
- Education: private funding to supplement public education

10   Patience – a realization that community and economic development takes time and they must "stay the course"

(Source: Lewis Whitfield, Tupelo community leader, presentation to 2005 Community Development Institute Central.)

## Keywords

Community, community development, economic development, growth, social capital, capacity building, business recruiting, business retention and expansion, new business start-ups, community and economic development chain.

## Review questions

1 What are the five types of community capital? Which are strong or weak in your community?

2 How is community development both a process and an outcome?

3 What is the difference between growth and development?

4 What is the difference between bonding (social) capital and bridging capital?

5 What are the "three legs of the stool" in traditional economic development? What other activities do economic developers do?

6 Why is it important for a community to be "development ready?"

7 How is community development related to economic development?

## Note

1 The American Economic Development Council merged with the Council for Urban Economic Development in 2001 to become the International Economic Development Council.

## Bibliography

Agnitsch, K., Flora, J. and Ryan, V. (2006) "Bonding and Bridging Capital: The Interactive Effects on Community Action," *Journal of the Community Development Society*, 37(1): 6–52.

American Economic Development Council (AEDC) (1984) *Economic Development Today: A Report to the Profession*, Schiller Park, IL: AEDC.

Blair, J.P. (1995) *Local Economic Development: Analysis and Practice*, Thousand Oaks, CA: Sage.

Boothroyd, P. and Davis, H.C. (1993) "Community Economic Development: Three Approaches," *Journal of Planning Education and Research*, 12: 230–240.

Christenson, J.A. and Robinson J.N. (1989) *Community Development in Perspective*, Ames, IA: Iowa University Press.

Committee for Economic Development, Research and Policy Committee (1995) *Inner-City Communities: A New Approach to the Nation's Urban Crisis*, New York: Committee for Economic Development.

Dunbar, J. (1972) "Community Development in North America," *Journal of the Community Development Society*, 7(1): 10–40.

Entergy Arkansas Corporation. (2005) *Business Retention and Expansion Guide*, Little Rock, Arkansas. Available online at http://www.entergy-arkansas.com/economic development/ (accessed April 13, 2007).

Green, P.G. and Haines, A. (2002) *Asset Building and Community Development*, Thousand Oaks, CA: Sage.

Grisham, V.L. (1999) *Tupelo: The Evolution of a Community*, Washington, DC: Kettering Foundation Press.

Haughton, G. (1999) *Community Economic Development*, London: The Stationery Office.

Hautekeur, G. (2005) "Community Development in Europe," *Community Development Journal*, 40(4): 385–398.

Huie, J. (1976) "What Do We Do About it? – A Challenge to the Community Development Profession," *Journal of the Community Development Society*, 6(2): 14–21.

IEDC (2006) *Organizing and Managing Economic Development*, Washington, DC: International Economic Development Council.

Long, H. (1975) "State Government: A Challenge for Community Developers," *Journal of the Community Development Society*, 6(1): 27–36.

Malizia, E.E. and Feser, E.J. (1999) *Understanding Local Economic Development*, New Brunswick, NJ: Rutgers University Center for Urban Policy Research.

Martin, H.A. (1996) "There's a 'Magic' in Tupelo-Lee County, Northeast Mississippi," *Practicing Economic Development*, Schiller Park, IL: American Economic Development Foundation.

Mattessich, P. and Monsey, M. (2004) *Community Building: What Makes It Work*, St. Paul, MN: Wilder Foundation.

Meridian Star (2007) "Why Not Follow the Tupelo Model?" Available online at http://www.meridianstar.com/editorials/local_story_070002240.html (accessed March 11, 2007).

National Research Council (1975) *Toward an Understand-*

*ing of Metropolitan America*, San Francisco, CA: Canfield Press.

Pittman, R.H. (2006) "Location, Location, Location: Winning Site Selection Proposals," *Management Quarterly*, 47(1): 2–26.

Pittman, R.H. (2007) "Business Retention and Expansion: An Important Activity for Power Suppliers," *Management Quarterly*, 48(1): 14–29.

Ploch, L. (1976) "Community Development in Action: A Case Study," *Journal of the Community Development Society*, 7(1): 5–16.

Roberts, R.T. (2006) "Retention and Expansion of Existing Businesses," *Community Development Handbook*, Atlanta, GA: Community Development Council.

Shaffer, R., Deller, S. and Marcouiller, D. (2006)

"Rethinking Community Economic Development," *Economic Development Quarterly*, 20(1): 59–74.

Shirouzu, N. (2006) "As Detroit Slashes Car Jobs, Southern Towns Pick Up Slack." *The Wall Street Journal*, February 1, p. A1.

Southern Economic Development Council (SEDC) (2006) *Member Profile Survey*, Atlanta, GA: SEDC.

Walker, C. (2002) *Community Development Corporations and Their Changing Support Systems*, Washington, DC: Urban Institute.

Warren, R.L. (1963) *The Community In America*, Chicago, IL: Rand McNally Press.

Wise, G. (1998) *Definitions: Community Development and Community-Based Education*, Madison, WI: University of Wisconsin Extension Service.

# 2 Seven theories for seven community developers[1]

## Ronald J. Hustedde

Community developers need theories to help guide and frame the complexity of their work. However, the field is girded with so many theories from various disciplines that it is difficult for practitioners to sort through them. Although many undergraduate and graduate community development programs have emerged in North America and throughout the world, there is no fixed theoretical canon in the discipline. This chapter focuses on the purpose of theory and the seven theories essential to community development practice.

Why seven theories? In Western cultures, seven implies a sense of near-completeness. There are seven days in a week, seven seas, seven climate zones, and seven ancient and modern wonders of the world. Rome was built on seven hills. While seven may or may not be a lucky number, seven theories are offered as a theoretical core for those who approach community development from at least seven contextual perspectives: organizations; power relationships; shared meanings; relationship building; choice making; conflicts; and integration of the paradoxes that pervade the field. Hence the chapter's title: "Seven theories for seven community developers."

## Introduction: why theory?

Theories are explanations that can provide help in understanding people's behavior and a framework from which community developers can explain and comprehend events. A good theory may be stated in abstract terms and help create strategies and tools for effective practice. Whether community developers want others to conduct relevant research or they want to participate in the research themselves, it is important that they have theoretical grounding. Theory is the major guide to understanding the complexity of community life and social and economic change (Collins 1998; Ritzer 1996).

The starting point is to offer a definition of community development that is both distinctive and universal and may be applied to all types of societies from postindustrial to preindustrial. Bhattacharyya (2004) met these conditions when he defined community development as the process of creating or increasing solidarity and agency. He asserts that solidarity is about building a deep sense of shared identity and a code of conduct for community developers. The developers need that solidarity as they sort through conflicting visions and definitions of problems among ethnically and ideologically plural populations. It may occur in the context of a "community of place" such as a neighborhood, city, or town. It may also occur in the context of a "community of interest" such as a breast cancer survivors' group, an environmental organization, or any group that wants to address a particular issue. Bhattacharyya contends that creating agency gives people the capacity to order their world. According to Giddens, agency is "the capacity to intervene in the world, or to refrain from intervention, with the effect of influencing a process or the state of affairs" (Giddens 1984: 14). There are complex forces that work against agency. However, community development is intended to build capacity, which makes it different from other helping professions. Community developers build the capacity of a people when they encourage or teach others to create their own dreams, to learn new skills and knowledge. Agency or capacity building occurs when practitioners assist or initiate community reflection on the lessons its members have learned through their actions. Agency is about building the capacity to understand, create and act, and reflect.

## Seven key concerns in the community development field

Following this definition of community development, there are seven major concerns involving solidarity and agency building: (1) relationships, (2) structure, (3) power, (4) shared meaning, (5) communication for change, (6) motivations for decision making, and (7) integration of these disparate concerns and paradoxes within the field. Horton (1992) shared similar concerns about African-American approaches to community development. He emphasized historic power differences and the influence of culture and black community institutions in his black community development model. Chaskin et al. (2001) focused on neighborhood and other structures and networks in their work on capacity building. Littrell and Littrell (2006), Green and Haines (2002), and Pigg (2002) all wove concerns about relationships, communicating for change, full participation, rational decision making, and integrating micro and macro forces into their community development insights.

Relationships are linked to a sense of solidarity. How critical are trust and reciprocity in the community development process? What is essential to know about relationship building? Structure refers to social practices, organizations, or groups that play a role in solidarity and capacity building. It also refers to the relationships among them. Some of these social practices and organizations may have a limited role. Therefore, to establish solidarity, new organizations may need to be built and/or existing ones could expand their missions.

Power refers to relationships with those who control resources, such as land, labor, capital, and knowledge, or those who have greater access to those resources than others. Since community development is about building the capacity for social and economic change, the concept of power is essential. Shared meaning refers to social meaning, especially symbols, that people give to a place, physical things, behavior, events, or action. In essence, solidarity must be built within a cultural context. Individuals and groups give different meanings to objects, deeds, and matters. For example, one community might see the construction of an industrial plant as an excellent way to bring prosperity to their town, while another community might see a similar construction as the destruction of their quality of life. Community developers need to pay attention to these meanings if they wish to build a sense of solidarity in a particular community or between communities.

Communication for change is linked to the concept of full participation, a consistent value in the community development literature. Within a framework often dominated by technicians, the corporate sector, or national political constraints, practitioners raise questions about how the voice of citizens can be heard at all. Motivation can influence many aspects of community development. It helps us understand whether people will or will not become involved in a community initiative. It also affects making difficult public choices, a process which usually involves thinking through all the policies to decide which will maximize individual and collective needs. Who is more likely to win or lose if a public policy is implemented? What are the potential consequences on other aspects of life if the policy is carried out? Essentially, the process of making rational choices can be nurtured as a form of capacity building. The integration of paradox and disparate macro and micro concerns are part of community development practice. How does one reconcile concerns about relationships, power, structure, shared meaning, communication for change, and motivational decision making? Is there a theory that ties some of these economic, political, and sociological concerns together?

These seven concerns form the basis for essential community development theory: social capital theory, functionalism, conflict theory, symbolic

■ **Table 2.1** *Concerns and related theories*

| *Concern:* | *Related theory* |
|---|---|
| 1 Relationships | Social capital theory |
| 2 Structure | Functionalism |
| 3 Power | Conflict theory |
| 4 Shared meaning | Symbolic interactionism |
| 5 Communication for change | Communicative action |
| 6 Motivations for decision making | Rational choice theory |
| 7 Integration of disparate concerns/paradoxes | Giddens' structuration |

interactionism, communicative action theory, rational choice theory, and Giddens's structuration theory. Table 2.1 lists these concerns and theories. Each of these seven theoretical perspectives will be examined and considered as to how they may be applied to community development practice.

## 1 Concerns about relationships: social capital theory

Community developers know inherently that the quality of social relationships is essential for solidarity building and successful community initiatives. Friendships, trust, and the willingness to share some resources are integral to collective action. Community developers build intuitively on these relationships. Social scientists view these relationships as a form of capital. *Social capital* is that set of resources intrinsic to social relations and includes trust, norms, and networks. It is often correlated with confidence in public institutions, civic engagement, self-reliant economic development, and overall community well-being and happiness.

Trust is part of everyday relationships. Most people trust that banks will not steal their accounts or that when they purchase a pound of meat from the grocer, it will not actually weigh less. Life can be richer if there is trust among neighbors and others in the public and private sectors. Think of settings where corruption, indifference, and open distrust might inhibit common transactions and the sense of

the common good. Equality is considered to be an important cultural norm that is high in social capital because it reaches across political, economic, and cultural divisions. Reciprocity is another cultural norm that is viewed as part of social capital. It should not be confused with a *quid pro quo* economic transaction; it is much broader than the concept of "I'll scratch your back if you'll scratch mine." When individuals, organizations, or communities provide food banks, scholarship funds, low-cost homes – or other forms of self-help, mutual aid, or emotional support – it stimulates a climate of reciprocity in which the recipients are more likely to give back to the community in some form. A culture with high levels of reciprocity encourages more pluralistic politics and compromise which can make it easier for community development initiatives to emerge.

Putnam (1993, 2000) has argued that social capital has declined in the United States since the 1990s. Social capital indicators have included voter turnout, participation in local organizations, concert attendance, or hosting others for dinner at one's home. Suburban sprawl, increased mobility, increased participation of women in the labor force, and television are among the reasons given for this decline. Some critics claim the indicators are linked too closely with "communities of place" because memberships in organizations such as the Sierra Club and other groups have increased significantly. They have also asserted that communities with strong social capital can also breed intolerance and smugness. They have distinguished between "bonding social capital" and "bridging social capital." They contend a mafia group or the Klu Klux Klan may have strong bonding social capital, but it does not build any new bridges that can expand horizons, provide new ideas, or generate wealth. They suggest to focus more on "bridging social capital" – the formation of new social ties and relationships to expand networks and to provide a broader set of new leaders with fresh ideas and information. For example, some communities have created stronger links between African-American and Caucasian faith-based communities or established leadership programs that nurture emerging and diverse groups of leaders. These activities both create new community linkages to broader resource bases and build new levels of trust, reciprocity, and other shared norms.

*How can social capital theory serve as a guide for community development practice?*

Community developers can integrate social capital theory into their initiatives. In some cases, they will find communities which have relatively low levels of social capital. In such cases, they may have to begin by nurturing "bonding social capital" through sharing food and drink, celebrations, storytelling, dance, or public art. They will have to create opportunities for people to get to know each other and build new levels of trust through shared interests including music, book clubs, games, or other pursuits.

In other cases, communities may have strong bonding social capital but really need "bridging social capital" if they are going to prosper and increase their quality of life. Take the case of tobacco-dependent counties in rural Kentucky that have limited communications with sister counties to build new regional initiatives such as agricultural and ecological tourism. The Kentucky Entrepreneurial Coaches Institute was created to build a new team of entrepreneurial leaders through a mutually supportive network and linkages with the "best and brightest in rural entrepreneurship" from around the world, nation, and region (Hustedde 2006). Social

capital was built through the mutual support of multi-county mini-grant ventures consisting of international and domestic travel seminars in which participants shared rooms, buses, seminars, and programs. These activities led to new forms of bonding and bridging social capital which stimulated not only entrepreneurship but an entrepreneurial culture.

## 2 Concerns about structure: functionalism

Second, it is important to look at structure, which underlies organizational and group capacity to bring about or stop change. In essence, structure is related to Giddens's concept of agency or capacity building. The theoretical concept concerned with structure is known as *structural functionalism*. It is also called *systems theory, equilibrium theory*, or simply *functionalism*. According to this theoretical framework, societies contain certain interdependent structures, each of which performs certain functions for societal maintenance. Structures refer to organizations and institutions such as health care, educational entities, business and nonprofits, or informal groups. Functions refer to their purposes, missions, and what they do in society. These structures form the basis of a social system. Talcott Parsons and Robert K. Merton are the specialists most often associated with this theory. According to Merton (1968), social systems have manifest and latent functions. Manifest functions are intentional and recognized. In contrast, latent functions may be unintentional and unrecognized. For example, it could be argued that the manifest function of urban planning is to assure well-organized and efficiently functioning cities, whereas the latent function is to allocate advantages to certain interests such as those involved with the growth machine or real estate developers.

Functionalists such as Parsons argue that structures often contribute to their own maintenance, not particularly to a greater societal good. Concern for order and stability also leads functionalists to focus on social change and its sources. They view conflict and stability as two sides of the same coin. If a community development practitioner wants to build community capacity, he or she will have to pay attention to the organizational capacity for stimulating or inhibiting change. Structural functionalism helps one to understand how the status quo is maintained. Some critics claim that the theory fails to offer much insight into change, social dynamics, or existing structures (Collins 1988; Ritzer 1996; Turner 1998).

*How can structural functionalism guide community development practice?*

Structural functionalism is a useful tool for practitioners. Looking at the case of an inner city neighborhood that is struggling to create a micro-enterprise business that will benefit local people, if one applied structural functionalism to community development practice, one would help the community analyze which organizations are committed to training, nurturing, and financing micro-enterprise development and what their latent or hidden functions might be. A functionalist-oriented practitioner is more likely to notice dysfunctions in organizations. If existing organizations are not meeting local needs in this area, the functionalist would build community capacity by transforming an existing organization to meet the same concerns. A functionalist would also want to build links with broader social systems, such as external organizations, that could help the community's micro-entrepreneurs to flourish. In essence, a functionalist would see structures as important components of capacity building. While structural functionalism is an important tool for community development, it is limited because it does not fully explore the issue of power that may be found in other theories.

## 3 Concerns about power: conflict theory

Power is the third key issue for community development. Power is control or access to resources (land, labor, capital, and knowledge). Since community

development builds capacity, concerns about power are pivotal. Insights into power tend to be found in political science or political sociology. More contemporary theorists have added to the richness of the literature. In his later writings, Foucault (1985) argued that where there is power there is resistance. He examines the struggles against the power of men over women, administration over the ways people live, and of psychiatry over the mentally ill. He sees power as a feature of all human relations (Foucault 1965, 1975, 1979, 1980, 1985; Nash 2000). Power has fluidity in the sense that it can be reversed and exists in different degrees. Beyond conventional politics at the state level, Foucault's focus extends to the organizations and institutions of civil society and to interpersonal relations.

Wallerstein (1984) applied Marxist theory to understand the expansion of capitalism to a globalized system which needs to continually expand its boundaries. "Political states," such as Japan, the UK, the European Union and the U.S., are among the core developed states based on higher level skills and capitalization. These states dominate the peripheral areas such that weak states are economically dependent on the "core." The low-technology states form a buffer zone to prevent outright conflict between the core and the periphery. Some have applied Wallerstein's world system theory to regional economics, with places like Appalachia serving as a "periphery" to global market forces. Mills (1959), one of the earliest American conflict theorists, examined some of the key themes in post-World War II American politics. He argued that a small handful of individuals from major corporations, federal government, and the military were influencing major decisions. He believed this triumvirate shared similar interests and often acted in unison. Mills' research on power and authority still influences theories on power and politics today. However, Mills also had critics such as Dahl (1971), who believed that power was more diffused among contending interest groups. Galbraith (1971) asserted that technical bureaucrats behind the scenes had more power than those in official positions. Neo-Marxists argued that Mills and Dahl focused too much on the role of individual actors. They believed that institutions permit the exploitation of one class by another. They also posited that the state intervenes to correct the flaws of capitalism and preserve the status quo, both of which are in the institutions' interests.

In summary, conflict theory suggests that conflict is an integral part of social life. There are conflicts between economic classes, ethnic groups, young and old, male and female, or among races. There are conflicts among developed "core" countries and regions and those that are less developed. It is argued that these conflicts result because power, wealth, and prestige are not available to everyone. Some groups are excluded from dominant discourse. It is assumed that those who hold or control desirable goods and services or who dominate culture will protect their own interests at the expense of others. Conflict theorists such as Coser (1956), Dahrendorf (1959), and Simmel (cited in Schellenberg 1996) have looked at the integrative aspects of conflict and its value as a contributing force to order and stability. Conflict can be constructive when it forces people with common interests to make gains to benefit them all. Racial inequalities or other social problems would never be resolved to any degree without conflict to disturb the status quo. Simmel discusses how conflict can be resolved in a variety of ways including disappearance of the conflict, victory for one of the parties, compromise, conciliation, and irreconcilability (Schellenberg 1996).

This theoretical framework that underlies both the power of one party over another and the potential for conflict is not intended to be exhaustive. Instead, it points to some of the major concerns that can guide community development practice.

*How can conflict theory serve as a guide*
*for community development practice?*

Community organizers tend to more readily embrace conflict theory as a pivotal component of their work. However, it may be argued that community developers also need conflict theory if their goal is to build capacity. Power differences are a reality of community life and need to be considered as devel-

opment occurs. Take the case of an Appalachian community near a major state forest. The state Department of Transportation (DOT) wanted to build a highway through the state forest. They claimed it would lead to more jobs and economic development. A group of local citizens questioned this assumption. They believed the highway would pull businesses away from the prosperous downtown area to the edge of town, lead to sprawling development that would detract from the quality of life, destroy a popular fishing hole, and harm the integrity of the forest. The DOT refused to converse with the community; they claimed the proposed highway's economic benefits were irrefutable.

Conflict theory served as a reference point for moving the community's interests further. At first glance, it appeared that the DOT was in charge of making the major decisions about the highway. However, the community developer put conflict theory into practice. Community residents were encouraged to analyze the power of the DOT as well as their own political, technical, economic, and social power. Through its analysis, the group was expanded to include downtown businesspeople, hunters, environmental, and religious groups. In this particular case, the community decided it needed more technical power. They were able to secure the services of university researchers, such as economists, foresters, sociologists, and planners, who had the credentials to write an alternative impact assessment of the proposed highway. This report was widely circulated by the community to the media and prominent state legislators. Gradually, external support (power) emerged to help the community and the DOT decided to postpone the project.

In a similar situation, the use of conflict theory took another twist. The opponents of a DOT-proposed road sought a mediator/facilitator to help them negotiate with the DOT and other stakeholders. They believed a neutral third party could create a safe climate for discussion, and that during such discussions power differences would be minimized. In this particular case, their use of conflict theory paid off because the dispute was settled to everyone's satisfaction.

In summary, community developers need conflict theory because it helps them gain insight into why specific differences and competition have developed among groups and organizations in a community. It can help them to understand why some people are silent or have internalized the values of elites even to their own disadvantage. Practitioners and researchers can use Simmel's theory to see how people resolve their differences. Alternately, they can borrow from Marx and the neo-Marxists to consider the sharp differences between and among class economic interests, gender, race, and other concerns.

Conflict theory can help communities understand the kind and extent of competing interests among groups. It also can shed light on the distribution of power, whether concentrated in the hands of a few or more broadly distributed. Communities can also explore the use of conflict to upset the status quo – whether through protests, economic boycotts, peaceful resistance, or other ranges of possibilities – especially if competing groups or institutions refuse to change positions or negotiate.

While conflict theory is an essential tool for capacity building, it should be noted that critics claim it is limited because it ignores the less controversial and more orderly parts of society and does not help in understanding the role of symbols in building solidarity (Collins 1998; Ritzer 1996; Turner 1998). This leads to another theoretical framework about shared meaning.

## 4 Concerns about shared meaning: symbolic interactionism

Shared meaning is the fourth key concern in community development. If the field is committed to building or strengthening solidarity, then practitioners must be concerned about the meaning people give to places, people and events. Herbert Blumer (1969) named the theory "symbolic interactionism" because it emphasizes the symbolic nature of human interaction rather than a mechanical pattern of stimulus and interaction. For sym-

bolic interactionists, the meaning of a situation is not fixed but is constructed by participants as they anticipate the responses of others. Mead (1992) explored the importance of symbols, especially language, in shaping the meaning of the one who makes the gesture as well as the one who receives it.

Goffman (1959) argued that individuals "give" and "give off" signs that provide information to others on how to respond. There may be a "front" such as social status, clothing, gestures, or a physical setting. Individuals may conceal elements of themselves that contradict general social values and present themselves to exemplify accredited values. Such encounters may be viewed as a form of drama in which the "audience" and "team players" interact. In his last work, Goffman (1986) examined how individuals frame or interpret events. His premise involves group or individual rules about what should be "pictured in the frame" and what should be excluded. For example, a community developer's framework of a community event may exclude ideas such as "citizens are apathetic." It will probably include shared "rules" such as "participation is important." The emphasis is on the active, interpretive, and constructive capacities of individuals in the creation of social reality. It assumes that social life is possible because people communicate through symbols. For example, when the traffic light is red, it means stop; when the thumb is up, it means everything is fine. Flora, Flora and Tapp (2000) investigated how two opposing community narratives moved through the stages of frustration, confrontation, negotiation, and reconciliation. Their case study could be viewed as the employment of social interactionism. They concluded that, among the symbols that humans use, language seems to be the most important because it allows people to communicate and construct their version of reality. Symbolic interactionists contend that people interpret the world through symbols but stand back and think of themselves as objects.

For example, a group of Native Americans view a mountain as a sacred place for prayer and healing, and react negatively when someone tries to develop or alter access to it. Developers, foresters, tourism leaders, and others are likely to have other meanings for the mountain. Different individuals or groups attach a different meaning to a particular event. These interpretations are likely to be viewed by others as a form of deviance which may be accepted, rejected, or fought over. Social interactionists argue that one way people build meaning is by observing what other people do, by imitating them, and following their guidance.

*How can symbolic interactionism serve as a tool for community development practice?*

Symbolic interactionism is essential for community development because it provides insight into the ways people develop a sense of shared meaning, an essential ingredient for solidarity. When a community developer helps a community develop a shared vision of their future, she is helping them build a sense of unity. A community-owned vision comes about through the interaction of people and is related through pictorial, verbal, or musical symbols. A symbolic interactionist would be keen on bringing people together to develop a shared understanding.

For example, take a case where some citizens have expressed an interest in preserving the farmland adjacent to the city and have asked a community developer for assistance. If one employed a symbolic interactionist perspective, one would ask them what the presence of farmland means to them. One would link them with farmers and others to see if there were a different or competing meaning. Participants would be asked how they developed their meaning of farmland. A symbolic interactionist would not ignore the concept of power. Participants would be asked questions as to whose concept of farmland dominates public policy. Through the employment of symbolic interaction theory, a sense of solidarity could be gradually established in a community.

A symbolic interactionist would identify groups that deviate from the dominant meaning of something and would engage them with other groups in order to move the community toward solidarity. Symbolic interactionists would also use symbols to build capacity. For example, a community might choose to preserve a historic structure because they believed it was beautiful, or explain its importance in a labor, class, racial, or gender struggle or some other interests. A community developer could augment their meaning with data about the historical and architectural significance that external agents see in the structure. Community capacity could be built in other ways such as providing information about tax credits for historic structures or how to locate grants for preservation. Increasingly, community development researchers and practitioners are asked to help citizens reflect and understand the meaning of their work. The symbolic interactionist concepts may be used to aid in collective evaluations. Essentially, it all boils down to what it means and who gives it meaning.

Symbolic interactionists probe into the factors that help people understand what they say and do by looking at the origins of symbolic meanings and how meanings persist. Symbolic interactionists are interested in the circumstances in which people question, challenge, criticize, or reconstruct meanings. Critics argue that symbolic interactionists do not have an established systematic framework for predicting *which* meanings will be generated, for determining *how* meanings persist or understanding how they change. For example, say a group of Mexican workers and a poultry processing firm move into a poor rural community that was historically dominated by Anglo-Saxon Protestants. The events may trigger cooperation, goodwill, ambivalence, anger, fear, or defensiveness. The cast of characters involved in these events may be endless. What has really happened and whose interpretation captures the reality of the situation? Symbolic interactionists have limited methodologies for answering such questions. In spite of these limitations, it is hoped that a strong case has been made as to why symbolic interactionism is an essential theory for community development practice.

## 5 Communication for change: communicative action theory

It is safe to assume that community development occurs within the context of democracy that is deliberative and participatory. Public talk is not simply talk; it is essential for democratic participation. It is about thinking through public policy choices. Deliberation occurs when the public examines the impacts of potential choices and tries them on, just as one might try on clothing in a department store before making a choice. In such settings, public talk involves rich discussions among a variety of networks. From the community development perspective, participation occurs in a setting where a diversity of voices are heard in order to explore problems, test solutions, and make changes to policies when the community finds flaws. Communities with robust democratic networks may be viewed as *communicatively integrated* (Friedland 2001). This type of integration involves the communicative activities that link individuals, networks, and institutions into a community of place or interest.

Habermas argues that communicative action is shaped at the seam of a system and *lifeworld.* Systems involve macro-economic and political forces that shape housing, employment, racial, and class divisions in a particular community. Local politics are also influenced by federal and state laws, national party politics, and regulations. Although the system is embedded in language, it is self-producing. Power and markets can be relatively detached from community, family, and group values. At the same time, there is the world of everyday life or the *lifeworld.* Habermas views the lifeworld as constituted of language and culture:

> The lifeworld, is, so to speak, the transcendental site where speaker and hearer meet, where they reciprocally raise claim that their utterances fit the world ... and where they can criticize and confirm those validity claims, settle their disagreements and arrive at agreements.
>
> (Habermas 1987: 126)

Habermas is concerned about the domination and rationalization of the lifeworld, in which science and technology are the *modi operandi* to address complex public issues. He believes that science and technology maintain the illusion of being value-free and inherently rational. In practical terms, citizens find it difficult to engage in dialogue with "more rational" scientists, engineers, or political and corporate elites. The problem is compounded when there is technical arrogance or limited receptivity to local voices. For example, many local newspapers and television stations are corporately owned. It is therefore difficult to hear local voices since they are filtered through more dominant perspectives. Habermas is concerned about the colonization of the lifeworld of culture and language, a colonization that reduces people to the status of things. He also argues that technical knowledge is not sufficient for democratic settings in which community developers work. It must be balanced by hermeneutic knowledge which he calls "practical interests." Hermeneutics deals with the interpretation of technical knowledge and what it means for an individual, his or her family, or community. It is action oriented and involves mutual self-understanding.

The third dimension of knowledge is emancipatory. It regards the liberation of the self-conscious and transcends and synthesizes the other two dimensions of knowledge. While science and technology may help liberation, they can also suffocate it. Emancipatory knowledge incorporates both technical and hermeneutic knowledge into a fresh perspective and outlook that leads to action.

In essence, Habermas's theory of communicative action is that it builds a linkage between the "rational" system and the lifeworld. His communicative action theory and political objective are based on free, open, and unlimited communication. It should be noted that Habermas grew up in Nazi Germany and his focus on reason could be viewed as a response to the unreason of Holocaust. At the same time, unlimited public talk could be seen as reaction to the curtailment of intellectual freedom and public dialogue during the Hitler years. Habermas's insights about communicative action theory, and his emphasis on reason and unrestrained public talk are viewed by some critics as utopian liberal ideals in which people talk their ideas to death. Others assert that universal principles of justice and democracy have been replaced by relativistic and egocentric perspectives. They assert that "reason" is a rationale for the powerful to suppress others. While Habermas emphasizes the potential to reach common ground, his detractors claim that common ground is not possible and that there is nothing wrong with competition between groups. They say he is merely moralizing and that communicative action theory is a hotchpotch of ideas gathered from the Enlightenment, Karl Marx, Max Weber, and others.

On the other hand, it should also be asserted that Habermas is continually expanding his perceptions and that, in spite of these criticisms, he is one of the world's leading public intellectuals. He has been a powerful influence on the formation of social democracies in Germany and the rest of Europe. "Communicative action" describes the seam where monetary and bureaucratic structures meet the lifeworld. This emphasis on reason, unfettered public discussion, and the potential for common ground provides an essential theory for community development practice in its concern for process.

*How can communicative action theory
guide community development practice?*

By its very nature, community development involves the participation of networks, groups, and individuals whose voices are part of the lifeworld. While this lifeworld operates within the context of technical, political, and market realities, it should be noted that the principles of community development entail participation of citizens in defining their own problems and dreams. If technicians or political and corporate interests dominate discussions, citizen involvement and participation becomes a mere afterthought. If technical knowledge is discarded or minimized, community development efforts may not be successful. Habermas's communicative action theory is guided by the intersection of technical and corporate knowledge with local and practical knowledge.

Combined, they can lead to a new kind of "emancipatory knowledge" that offers fresh ideas and action.

There are many ways for community developers to carry out Habermas's communicative action theory. For example, the National Issues Forums are held in many communities wherein individuals, networks, and groups explore public issues through the perspective of several public policy choices. Rather than choose sides, these forums are designed for the participants to examine the applicability, strengths, limitations, and values of each choice. National Issues Forums are conscious acts of deliberation that make it easier for the system and the lifeworld to interact.

In another community development case, an Appalachian Cancer Network was developed by homemakers and health care professionals to deal with high rates of breast and cervical cancer in that region. The health care leaders were tempted to tell the homemakers what to do. However, the community developers who guided this initiative did not begin with technical knowledge. They started with storytelling in which technical and lay participants responded to the questions: *Have you or a family member ever been touched by cancer or another serious illness? If so, what happened?* The stories that emerged told of triumph, heartache, loss, and anger. The next set of questions was: *What do our stories have in common? What should we do, if anything, about our common issues?*

Eventually, the community development principles of full participation were carried out. The network acted in ways that brought out technical, practical, and emancipatory knowledge. That is, new ideas and action emerged from this initiative that would have been impossible if technicians or lay leaders had acted independently.

# 6 Motivation for decision making: rational choice theory

The rational economic man model was proposed by Alfred Marshall (1895). He believed that humans were interested in maximizing their utility, happiness, or profits. The rational man would investigate each alternative and choose that which would best suit his individual needs. While Marshall recognized that irrational decisions were made, he believed that the overwhelming number of decision makers would operate in a maximizing fashion and cancel out irrational actions. Marshall assumed all the relevant information was available to the economic man and that he could understand the consequences of his choices. The focus was on the individual rather than the collective. Rational choice theory has several embellishments and spinoffs from various social scientists. For example, Mancur Olson (1965) explored whether rational calculation would lead a few individuals to pursue collective action as a way to obtain public goods because they could pursue these goods whether they were active or not. He believed that collective behavior could be expected under two conditions: (1) selective incentives – such as increased stature in the community, tax breaks, or other benefits – could increase the rewards of those engaging in collective action, and (2) the threat of sanctions against those who fail to participate.

In recent years, social scientists have explored how four structural factors relate to individual participation in collective activities. One is prior contact with a group member because it is easier to recruit through interpersonal channels. A second is prior membership in organizations due to the likelihood that those who are already active may join other groups and, conversely, isolated individuals may perceive joining as a type of risk. The second is a history of prior activism because those with previous experience are more likely to reinforce their identity through new forms of activism. The fourth factor is biographical availability, which pulls people toward and away from social movements. For example, full-time employment, marriage, and family responsibilities may increase the risks and costs of becoming involved. Conversely, those who are free of personal constraints may be more likely to join. There is some empirical evidence that students and autonomous professionals may be more likely to join social movements (McAdam 1988).

Critics of rational choice theory have argued that actors do not have equal access to information or that information is distorted. Others assert that many people's choices are limited by social, political, and economic interests and values, which limits their participation in rational choice making.

*How can rational choice theory serve as a guide for community development practice?*

Community developers know that while people may have altruistic concerns, they also have their own needs and make choices about how to invest their time. There have been many creative responses to rational choice theory. For example, the Cooperative Extension Service Master Gardener Program offers free horticultural training but participants must volunteer hours back to the community in order to receive the training. Leadership programs have popped up in many communities where participants gain the advantage of expanding their network and knowledge bases. Their positive experience in meeting and working with others in collective settings leads to a greater openness and involvement.

When applied to community development, rational choice theory is concerned with finding appropriate rewards and minimizing risks to individuals who become involved in community initiatives. Such rewards might be as simple as free babysitting services or an awards and recognition banquet. Both examples would facilitate people's choices to invest their time or money in community development efforts. In other situations, there is a tendency toward misinformation, misunderstanding, competing sets of data, or different interpretations of the same data. Any or all of these make it difficult to reach common ground and establish solidarity. In such cases, community developers can find new ways to gather data, interpret information, or glean new information from mutually respected third party sources. It should be asserted that in many settings universities are no longer viewed as neutral or objective. They may be perceived as instruments of the state, the corporate sector, or a particular political or economic interest. One of the limitations of rational choice theory is that it can be implemented by technicians, the corporate sector, and bureaucracies in ways that can overwhelm and silence citizens who may not understand such knowledge. Habermas's theory of communicative action can provide a counterbalance to such shortcomings.

## 7 Integration of disparate concerns and paradigms: Giddens's structuration theory

The classical theories of structural functionalism, conflict theory, and rational choice theory are essential concepts for building community capacity. The fluid contemporary theories of social capital, communicative action, and the classical theory of symbolic interactionism are important for creating or strengthening solidarity. There are obvious tensions inherent in these theories. The dualism of macro versus micro characterizes much of the theoretical thinking in sociology. Sharing the same goal of picturing social reality, these schools choose to proceed from opposite directions. The macro-thinkers attempt to draw a holistic picture and lay down the works of society, whereas the micro-theorists hope to arrive at the same results by scrutinizing what happens "in" and "between" individual people. Neither approach is entirely successful in producing a complete and exhaustive picture for community development practice. In a more recent development, efforts have been made at a "micro-translation," which seeks to visualize social reality as composed of individuals interacting with one another to form "larger interaction ritual chains" (Collins 1988).

However, recent theory also recognizes that social agency itself, pointed out above as a key concern for community development, needs to be theoretically addressed. This must be done in a way that transcends both the established orientations in modern social theory and the whole macro–micro split. In his structuration theory, Anthony Giddens (1984, 1989) offers a perspective that is more fluid and

process-oriented. He introduces a third dimension, or an "in-between" level of analysis, which is neither macro nor micro. It has to do with the cultural traditions, beliefs, societal norms, and how actors draw upon those in their behavior (Collins 1988: 399). For Giddens, those normative patterns of society exist "outside of time and space" (Collins 1988: 398–399), meaning they are neither properties of the empirical social system nor of the individual actors. Their actuality consists in the moments when individuals' behaviors rise to that level of society's traditions and norms. People also draw and act upon thought patterns or cultural "molds"; for example, the classical notion of reciprocity – getting one thing in return for something else. Cultural traditions and patterns become modalities by virtue of placing them on Giddens's analytical scheme. They represent a third level, that between individualistic behavior and the macro-structures. Even though the reality of modalities may be only momentary, when people actually rise to them in their behavior, then the social process and the role of culture and normative patterns can be better visualized. "Actors draw upon the modalities of structuration in reproduction of systems of interaction" (Giddens 1984: 28). Social structure is upheld and existing divisions of society carry on through these "mental molds."

The laying out of society on the six above-mentioned levels – social capital theory, functionalism, conflict, symbolic interactionism, communicative action theory, and rational choice theory – reflects a fluid process in which all levels interact. Individuals represent the agency whereby interaction among levels takes place. Coming back to the community development profession and its key concerns, Giddens's model is perhaps best suited to grasp how social agency is exercised and solidarity established amid and often against the existing structural divisions of society. Behavior is neither haphazard nor merely a reflection of the existing social structure and its divisions. Modalities represent the levels at which people establish solidarity by following the symbolic norms and patterns of their cultures and traditions.

Similarly, new rules of behavior also occur through the medium of modalities, in this instance their creative redefinition. This is how the existing divisions can be overcome and new bonds between people forged. For this to take place, genuine social creativity is necessary. This means that people come up with solutions and ideas that simultaneously draw on the common reference point of their cultural traditions, and transcend those traditions to establish new bonds and patterns of solidarity. Modalities serve not only as the rules for the reproduction of the social system, but for its transformation (Turner 1998: 494). Giddens's concept of modalities is the link between macro- and micro-theories. Modalities are part of the analytical scheme in a particular place. For example, individualism in the United States is a strong modality and can keep citizens from united action. The notion of the common good is another American modality which may be used to transform a divided community into one with a greater sense of solidarity. Modalities may be used to influence the macro- or micro-level of social change. There are several substantive analyses looking at cultural patterns and systems of ideas and how they mediate the social process. In these analyses, social processing and the dynamics of social transformation are at least partly carried out on the level of modalities. Gaventa (1980) examines the modalities of Appalachia with a focus on rebellion and quiescence. He analyzes how power is used in the region to prevent or implement decisions. The use of force and threat of sanctions are discussed along with less intrusive aspects such as attitudes that are infused into the dominant culture by elites and internalized by non-elites. For example, there are perspectives such as "you can't change anything around here" or "you don't have to be poor if you want to really work." Gaventa argues that there are other modalities in which Appalachian culture has resisted the penetration of dominant social values. Those with less power can develop their own resources for analyzing issues and can explore their grievances openly. He views the "myth of American democracy" as another modality that can set the stage for greater openness and transparency in local government.

Staniszkis (1984) provides further insights about modalities through her ideas about how workers' solidarity emerged in Poland. She saw the working class under the communist regime as a unified bloc, both in a positive hegemonic way and negatively, as subject to the party's control and manipulation. Solidarity and its charismatic leader Lech Walesa transformed these modalities with references to workers' common identity, as opposed to their identity with the Communist Party apparatus. To further create a sense of solidarity and unity in opposition to the Communist Party and the system, Walesa incorporated Polish workers' strong Christian identification into helping define their new self-understanding and self-image. In her work on the change in workers' collective identity, Staniszkis's consistent attention to symbolic meanings and their interplay with the social structure aptly demonstrates how modalities can be transformed.

Analytically, Giddens's structuration theory stands as a middle ground between the micro- and the macro-theories as well as the issue of agency and solidarity. Giddens's structuration theory suggests that the micro-theories associated with symbolic interactionism can influence cultural and traditional norms and patterns (modalities) and vice versa. While the symbolic interactionists tend to ignore structure, Giddens' mid-level theory about modalities is a crucial link among symbolic interactionism, rational choice theory, social capital, the micro–macro conflict, communicative action, and structural functionalist theories (Giddens 1984).

Max Weber's social action theory was originally cast at an "in-between level." If his theory was not explicit, his intentions were at least implicit. Weber attempted to view society as a fluid process, dissecting it into various components for analytical purposes (Turner 1998: 17) much like Giddens did. Although Weber never attempted an analytical model of society along micro-theoretical lines, some observers have categorized Weber as a micro-theorist because of his subjective interpretation of behavior and its meaning to the actor. Others argue that Weber is a strong macro-theorist since his intentions may lie closer to Giddens' perspective. This was especially obvious in his attempts to explain the rise of modern capitalism through the interplay of social structural conditions and the religious beliefs of Protestantism. He followed similar analyses for non-Western societies in his volumes on the sociology of religion. What Giddens delineated in theory Weber actually performed in his works, bridging the macro and the micro dimensions in his attention to society's traditions and norms. He observed how people, independent of the macro-structural forces of society, transform these traditions and norms by interpreting and reinterpreting them. Similarly, Gaventa and Staniszkis demonstrated how one can connect communities or groups to structure them in a way that is not fixed or mechanical.

In contrast to debates on whether structure shapes action to determine social phenomena or the reverse, Giddens believes that structure exists in and through the activities of human agents. He views it as a form of "dualism" in which neither can exist without the other. When humans express themselves as actors and monitor the ongoing flow of activities, they contribute to structure and their own agency. He contends that social systems are often the results of human action's unanticipated outcome. Giddens views time and space as crucial variables. Many interactions are face-to-face, and hence are rooted in the same space and time. However, with the advent of new technologies, there can be interaction across different times and spaces. Community developers are likely to feel some kinship with Giddens because he has a dynamic rather than static concept of the world. He recognizes the interplay of humans and structure in shaping and being shaped. Critics are likely to argue that he has oversubscribed to the concept of the power of human agency. The space of this chapter limits a response to those critiques and a fuller exploration of Giddens's theoretical insights.

*How can Giddens's structuration theory guide community development practice?*

Structuration theory provides many theoretical insights (Ritzer 1996: 433) for those engaged in community development because it links disparate macro-theories about structure and conflict with micro-theories about individual and group behavior

such as social capital, rational choice, and symbols or symbolic interactionism. Giddens's concept of modalities is essential for community development practice.

Revisiting the case of the Appalachian community group that opposed the construction of a road through a nearby state forest, as addressed under the heading, "How Can Conflict Theory Serve as a Guide for Community Development Practice?," the group believed they were overpowered by the Department of Transportation (DOT) that wanted to build the road. The community found it difficult to argue against the DOT report, which contained sophisticated economic, social, and natural resource information. Here is what the community development practitioner did. First, the practitioner asked community residents to identify the strengths of their local traditions – particularly storytelling and the arts – as a venue for building solidarity regarding the integrity of the forest. Together, the community and the practitioner examined the modalities of storytelling and the arts to see if they could use the media to make an impact on the public and local legislators. The community's strong respect for the local Cooperative Extension Service was identified as another modality to mobilize the broader information resources of the land grant university. Without spending much money, the community developer was able to draw upon the services of professional economists, sociologists, foresters, and others. These professionals developed an alternative to the DOT report that was widely disseminated. Storytelling, the local arts, and links with the local Extension Service influenced broader structures and led to fewer power imbalances. Eventually, the DOT decided to permanently "postpone" the development of the road. Because the community developer understood the power of modalities (local cultural traditions and patterns), the community was able to develop a sense of shared meaning. This led to greater influence on structure and resolved the conflict.

How do Giddens's structuration theory and the concept of modalities relate to some of the theories discussed earlier, particularly the classical theories of structural functionalism, conflict theory, rational choice theory, and symbolic interactionism?

When one looks at functionalism through a Giddens lens, one sees how structures shape and can be shaped by modalities. From a Giddens perspective, community change agents are not powerless when faced with powerful structures. Cultural patterns can be transformed to influence or break down structural constraints that inhibit solidarity and capacity building. Giddens's structuration theory illuminates conflict theory because it suggests that communities can influence power imbalances through cultural norms and patterns. It also suggests that external power can shape behavior.

Based on a Giddens perspective, the micro-theories associated with symbolic interactionism and making rational choices can influence cultural and traditional norms and patterns (modalities) and vice versa. While the symbolic interactions and rational choice theorists tend to ignore structure, Giddens's mid-level theory about modalities is a crucial link among symbolic interactionism, rational choice making, the macro "conflict" theory, and structural functionalism. The fluid theories associated with Habermas's communicative action and social capital may be viewed as mid-level theories, as part of structuration theory. They also address the intersection of modalities and structure.

However, there are several limitations to Giddens's theories. His writing is analytical and abstract to the point of being vague and imprecise. He rarely gives concrete examples, which can be frustrating to those community developers who are more empirically grounded. Giddens's analysis is also difficult because it involves constant movement among the levels of modalities, societal institutions, and the actions of individuals. In spite of these limitations, structuration theory is especially useful for community developers because of the potent role of symbolic norms and cultural patterns (modalities) in creating new structures, influencing power differences, and infusing individual behavior with a sense of solidarity.

## Conclusion

Community development is often thought of as intention to build solidarity and agency (capacity building). Theory is essential for community development practice because it provides explanations of individual and group behavior. It also provides frameworks so that community developers may comprehend and explain events. There are seven theories that should be part of a community development canon, or knowledge: (1) social capital; (2) structural functionalism; (3) conflict; (4) symbolic interactionism; (5) communicative action; (6) rational choice; and (7) structuration theory. Each theory should be explored along with its limitations and applicability for community development practice.

This chapter is about reaching across the conceptual divide between theory and action. It should stimulate dialogue and further discussion on essential theory for community development practice. The classical theories of structural functionalism, conflict, symbolic interactionism, and rational choice can be balanced by the more fluid and synthesizing theories of social capital, communicative action, and structuration. These theoretical camps may be linked in novel ways to help community developers become more effective.

## Note

1  This chapter is an expansion of the article: Hustedde, R.J. and Ganowicz, J. (2002) "The Basics: What's Essential about Theory for Community Development Practice?," *Journal of the Community Development Society*, 33(1): 1–19. The editor of the journal granted permission to duplicate and integrate parts of the article into this chapter.

## Keywords

Solidarity, agency building, structure, power, shared meaning, social capital theory, structural functionalism, conflict theory.

## CASE STUDY: COMMUNITY DEVELOPMENT AND INTERNATIONAL CONFLICT RESOLUTION

Arguably the most pressing international issue of this and future generations is the relationship between Islamic and Western countries, as evidenced by the wars in Iraq and Afghanistan and the ongoing conflict between the Israelis and Palestinians. One scholar believes that community development could serve as a valuable tool to improve Islamic–Western relations and help ease conflicts across the globe. In a series of articles, Jason Ben-Meir states his belief that participatory, grass-roots community development in conflict areas will empower local residents and encourage them to reject religious extremism, engage in community and nation building, and appreciate the foreign aid efforts of Western countries. Ben-Meir is President of the High Atlas Foundation, a U.S. nonprofit organization that assists community development in Morocco.

According to Ben-Meir, the billions spent on foreign aid reconstruction in Iraq and Afghanistan typically channeled through third party contractors and national governments often fosters resentment toward Western countries because input from the communities where the projects take place is not obtained and local residents feel they are not in control of rebuilding their own economic and social life. Ben-Meir argues that sustained development and genuine reconstruction require funding local projects designed by the entire community. The community's priorities would be established by facilitated interactive dialogue where all local residents have a right to express their opinions and collective priorities are developed in a true inclusive and participatory community development process. He believes this will encourage community residents to actively support local rebuilding and economic development efforts. As they feel empowered, develop hope for the future, and see tangible signs of progress of their own design, they will be less likely to embrace extremism born of frustration and alienation. Ben-Meir also believes that successful community and economic development outcomes fostered by this approach will engender goodwill toward Western countries funding these local projects and helping with the community development capacity-building process. Furthermore, progress will be sustainable, since citizens in communities throughout turbulent regions will have learned community- and nation-building skills and local infrastructure will be improved.

In the case of Iraq, Ben-Meir believes the national government should:

- Train local schoolteachers and other community members in group facilitation methods and begin the community development process in all communities with inclusive, participatory meetings to establish local priorities.
- Create community reconstruction planning and training centers in all communities to help implement local priorities and redevelopment projects. The centers would also provide further training in facilitation, conflict management, modern agricultural techniques, health care, and other development topics.

Encouraging community development and funding local priority projects will also help alleviate the Israeli–Palestinian conflict, according to Ben-Meir. He points out that the Palestinian economy is almost totally dependent on Israel's, and when political tensions rise, economic links and flows of people and goods are severely restricted, causing huge hardship for the Palestinians. He argues that Israel and the West can generate tremendous goodwill and help make the Palestinian people economically self-reliant by promoting the community development process and investing in projects designed and managed by local residents.

Whether or not the community development can help achieve these lofty goals is an open question, but there is no doubt that its principles of conflict resolution, group decision making, inclusiveness, and fairness are certainly relevant to international affairs and foreign policy. Community development is germane to countries all over the world and its principles transcend geo-political boundaries.

The Editors

**Sources**

Ben-Meir, J. (2005) "Iraq's Reconstruction: A Community Responsibility," *The Humanist*, 65(3): 6.

Ben-Meir, J. (2004) "Create a New Era of Islamic–Western Relations by Supporting Community Development," *The International Journal of Sociology and Social Policy*, 24(12): 25–41.

## Review questions

1 What are the seven concerns of community development discussed?

2 What are the seven theories of community development related to the concerns?

3 What can be learned from theory for community development practice? Give an example of an application.

## Bibliography and additional resources

Bhattacharyya, J. (2004) "Theorizing Community Development," *Journal of the Community Development Society*, 34(2): 5–34.

Biddle, W. and Biddle, L. (1965) *The Community Development Process*, New York: Holt, Rhinehart & Winston.

Blumer, H. (1969) *Symbolic Interactionism: Perspective and Method*, New York: Prentice Hall.

Chaskin, R.J., Brown, P., Venkatesh, S. and Vidal, A. (2001) *Building Community Capacity*, Hawthorne, NY: Aldine de Gruyter.

Christenson, J. and Robinson, J. (eds) (1989) *Community Development in Perspective*, Iowa City, IA: University of Iowa Press.

Collins, R. (1988) *Theoretical Sociology*, New York: Harcourt Brace Jovanovich.

Coser, L. (1956) *The Functions of Social Conflict*, New York: The Free Press.

Dahl, R.A. (1971) *Polyarchy: Participation and Opposition*, New Haven, CT: Yale University Press.

Dahrendorf, R. (1959) *Class and Class Conflict in Industrial Society*, Stanford, CA: Stanford University Press.

Flora, C.B., Flora, J.L. and Tapp, R.J. (2000) "Meat, Meth and Mexicans: Community Responses to Increasing Ethnic Diversity," *Journal of the Community Development Society*, 31(2): 277–299.

Foucault, M. (1965) *Madness and Civilization: A History of Insanity in the Age of Reason*, New York: Vintage.

—— (1975) *The Birth of the Clinic: An Archeology of Medical Perception*, New York: Vintage.

—— (1979) *Discipline and Punish: The Birth of Prison*, New York: Vintage.

—— (1980) *The History of Sexuality, Volume 1, An Introduction*, New York: Vintage.

—— (1985) *The History of Sexuality, Volume 2, The Use of Pleasure*, New York: Pantheon.

Friedland, L.A. (2001) "Communication, Community and Democracy," *Communication Research*, 28(4): 358–391.

Fussell, W. (1996) "The Value of Local Knowledge and the Importance of Shifting Beliefs in the Process of Social Change," *Community Development Journal*, 31(1): 44–53.

Galbraith, J.K. (1971) *The New Industrial State*, Boston, MA: Houghton Mifflin.

Gaventa, J.L. (1980) *Power and Politics: Quiescence and Rebellion in an Appalachian Valley*, Champaign, IL: University of Illinois Press.

Giddens, A. (1984) *The Constitution of Society*, Berkeley, CA: University of California Press.

—— (1989) "A Reply to My Critics," in D. Held and J.B. Thompson (eds) *Social Theory of Modern Societies: Anthony Giddens and His Critics*, Cambridge, UK: Cambridge University Press.

Goffman, E. (1959) *The Presentation of Self in Everyday Life*, Garden City, NY: Anchor.

—— (1986) *Frame Analysis: An Essay on the Organization of Experience*, Boston, MA: Northeastern University Press.

Green, G.P. and Haines, A. (2002) *Asset Building & Community Development*, Thousand Oaks, CA: Sage.

Habermas, J. (1987) *The Theory of Communicative Action, Vol. 2, Lifeworld and System: A Critique of Functionalist Reason*, Boston, MA: Beacon Press.

Horton, J.D. (1992) "A Sociological Approach to Black Community Development: Presentation of the Black Organizational Autonomy Model," *Journal of the Community Development Society*, 23(1): 1–19.

Hustedde, R.J. (2006) "Kentucky Leadership Program Coaches Entrepreneurs," *Economic Development America*, winter: 28–29.

Hustedde, R.J. and Ganowicz, J. (2002) "The Basics: What's Essential about Theory for Community Development Practice?," *Journal of the Community Development Society*, 33(1): 1–19.

Jeffries, A. (2000) "Promoting Participation: A Conceptual Framework for Strategic Practice, with Case Studies from Plymouth, UK and Ottawa, Canada," *The Scottish Journal of Community Work and Development*, Special Issue 6 (autumn): 5–14.

Littrell, D.W. and Littrell, D.P. (2006) *Practicing Community Development*, Columbia, MO: University of Missouri-Extension.

Marshall, A. (1895) *Principles of Economics, Third Edition*, London: Macmillan.

McAdam, D., McCarthy, J. and Zald, M. (1988) "Social Movements," in N.J. Smelser (ed.) *Handbook of Sociology*, Newbury Park, CA: Sage.

Mead, G.H. (1992) *The Individual and the Social Self: Unpublished Work of George Herbert Mead*, Chicago, IL: University of Chicago Press.

Merton, R.K. (1968) *Social Theory and Social Structure*, revised edn, New York: The Free Press.

Mills, C.W. (1959) *The Sociological Imagination*, New York: Oxford University Press.

Nash, K. (2000) *Contemporary Political Sociology: Globalization, Politics, and Power*, Malden, MA: Blackwell.

Olson, M., Jr. (1965) *The Logic of Collective Action*, Cambridge, MA: Harvard University Press.

Parsons, T. (ed.) (1960) "Some Reflections on the Institutional Framework of Economic Development," in *Structure and Process in Modern Societies*, Glencoe, IL: Free Press.

Parsons, T. and Shils, E.A. (eds) (1951) *Toward a General Theory of Action*, New York: Harper & Row.

Perkins, D.D. (1995) "Speaking Truth to Power: Empowerment Ideology as Social Intervention and Policy," *American Journal of Community Psychology*, 23(5): 569–579.

Pigg, K.E. (2002) "Three Faces of Empowerment: Expanding the Theory of Empowerment in Community Development," *Journal of Community Development Society*, 33(1): 107–123.

Putnam, R.D. (1993) *Making Democracy Work: Civic Traditions in Modern Italy*, Princeton, NJ: Princeton University Press.

—— (2000) *Bowling Alone: The Collapse and Revival of American Community*, New York: Simon & Schuster.

Ritzer, G. (1996) *Sociological Theory*. 4th edn, New York: McGraw-Hill.

Rothman, J. and Gant, L.M. (1987) "Approaches and Models of Community Intervention," in D.E. Johnson, L.R. Meiller, L.C. Miller and G.F. Summers (eds), *Needs Assessment: Theory and Methods*, Ames, IA: Iowa State University Press.

Schellenberg, J.A. (1996) *Conflict Resolution: Theory, Research and Practice*, Albany, NY: State University of New York Press.

Shaffer, R.E. (1989) *Community Economics: Economic Structure and Change in Smaller Communities*, Ames, IA: Iowa State University Press.

Staniszkis, J. (1984) *Poland's Self-limiting Revolution*, Princeton, NJ: Princeton University Press.

Turner, J.H. (1998) *The Structure of Sociological Theory*, 6th edn, Belmont, CA: Wadsworth Publishing.

Wallerstein, I. (1984) "The Development of the Concept of Development," *Sociological Theory*, 2: 102–116.

Weber, M. (1947) *The Theory of Social and Economic Organization*, trans. A.M. Henderson and T. Parsons, New York: Oxford University Press.

# Asset-based community development

## Anna Haines

Building on a community's assets rather than focusing on its needs for future development is the basic approach of asset-based community development. By focusing on successes and small triumphs instead of looking at what is missing or negative about a place, a positive community outlook and vision for the future can be fostered. This approach also focuses on a sustainable approach to development. This chapter outlines the process and the major steps in identifying individual, organizational, and community asset development.

## Introduction

Chapter 1 focused on the definition and overall scope of community development and its relationship to economic development. This chapter discusses community development from the perspective of concentrating and building on community assets rather than focusing on needs and problems. This approach leads to a more sustainable approach to development. The term "community" is used throughout this chapter to refer to a place. A place may be a governmental entity, such as a city, or it may be a neighborhood that has no specific or official boundaries. Finally, this chapter outlines the major steps in planning for an asset-based community development strategy.

## Definitions of community development

As noted in Chapter 1, community development is both a process and an outcome, as seen in the various definitions from the literature. Let us consider yet a few more definitions, in the context of assets, to further characterize community development.

- "Community building in all of these efforts consists of actions to strengthen the capacity of communities to identify priorities and opportunities and to foster and sustain positive neighborhood change" (Chaskin 2001: 291).
- "Community development is asset building that improves the quality of life among residents of low- to moderate-income communities, where communities are defined as neighborhoods or multi-neighborhood areas" (Ferguson and Dickens 1999: 5).
- "Community development is defined as a planned effort to produce assets that increase the capacity of residents to improve their quality of life" (Green and Haines 2007: vii).
- "Community development is a place-based approach: it concentrates on creating assets that benefit people in poor neighborhoods, largely by building and tapping links to external resources" (Vidal and Keating 2004: 126).

Critical components of these definitions include:

- *A place-based focus*
  Communities may be thought of as the neighborhoods, towns, villages, suburbs, or cities in which people live. These are places that are rooted in a physical environment. In contrast, communities may also be interest based. Many people identify with groups of

people who share similar interests, for example, professional associations, sports teams, religious affiliations, service clubs.

- *The building up or creation of assets*

    The next section of this chapter will spend time discussing asset-based community development. For now, the definition of an asset is: a resource or advantage within a community (of place).

- *The improvement of quality of life*

    Quality of life is a vague notion, and, therefore, each community must define indicators in order to be able to monitor whether or not improvement is occurring. Quality of life can refer to economic, social, psychological, physical, and political aspects of a community. Examples of indicators include: number of violent crimes within a neighborhood; hours of work at the median wage required to support basic needs; percentage of employment concentrated in the top 10 employers; percentage of the population that gardens; and tons of solid waste generated and recycled per person.

Also stated in the above definitions of community development are the following aspects:

- *Financial, economic, environmental, and social sustainability*

    More and more, the idea behind community development is to build up resources and advantages in a community so that the community and the individuals within it can be sustained over time.

- *The approach, while universally applicable, is particularly relevant to non-wealthy communities*

    This is the notion that, unlike wealthy communities which not only have assets but recognize these assets and use them in the formal economy, many lower income communities do not.

# Needs-based community development

There are two primary methods of approaching community development. The conventional or traditional approach is to identify the issues, problems, and needs of a community. In many low-income neighborhoods, it is easy to point to problems – vacant and abandoned houses, boarded-up store fronts, empty lots filled with trash, and countless others. By focusing on problems, community residents tend to concentrate only on what is missing in a community. For example, a neighborhood may point to problems such as high unemployment rates or lack of shopping opportunities and identify the need for more jobs and businesses. If community residents focus only on trying to fix the problems that they see, they may miss or ignore the causes of these problems.

Many of the problems identified, like poverty or unemployment, are issues too large for one community to solve by itself. By focusing on the causes of problems, community residents may end up wringing their hands or giving up because of the overwhelming nature of the causes. This approach can create unreasonable expectations that may lead to disappointment and failure over time. In addition, this approach can point to so many problems and needs that people feel overwhelmed, and, therefore, nothing is done. Figure 3.1 provides an example of a community needs map which outlines problems within a community. This map illustrates numerous problems, many of which are difficult to resolve by any one community, neighborhood, or organization.

# Asset-based community development

An alternative approach is asset-based community development. One could argue that this approach is the reverse of the conventional approach. The idea is to build capacity within a community – to build and strengthen a community's assets. In contrast to focusing on problems and needs, this alternative

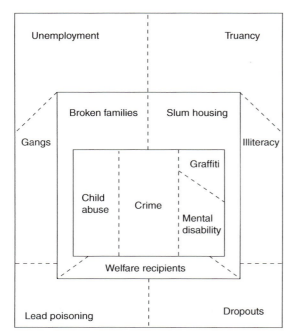

**Figure 3.1 Community needs map**

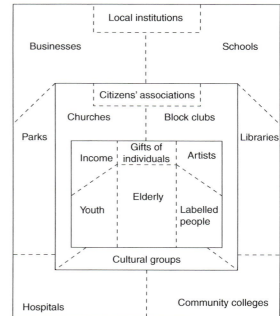

**Figure 3.2 Community assets map**

approach focuses on a community's strengths and assets.

This asset-based approach is focused on a community's capacity rather than on its deficits. For instance, rather than focusing on missing small businesses, this approach would focus on existing small businesses and their success. Further, by focusing on its assets, the community as a whole will see its positive aspects (such as community gardens, a mentoring program, and the many skills of its residents) and can then work on developing these assets even more. By implication, concentrating on community assets will create a snowball effect that will influence other areas within a community such as its needs and problems. This alternative approach does not ignore the problems within a community, but focuses first on its strengths and small triumphs in order to provide a positive perspective of the community rather than a discouraging one. Figure 3.2 shows an example of the "mapped" assets within a community and the capacities of its individuals, associations, and

institutions. The assets map underscores the potential for community development because it is starting from a positive base rather than from a base rooted in problems. Chapter 10 provides additional information about the process of asset mapping.

## Assets defined

Before moving forward in this discussion, it is critical to define the term assets. Assets are the stock of wealth in a household or other unit (Sherraden 1991: 96). Another definition is that assets are "a useful or valuable quality, person or thing; an advantage or resource" (Dictionary.com). Thus, individuals, associations, local institutions, and organizations are useful and valuable within the asset-based community development framework. Kretzmann and McKnight (1993) defined assets as the "gifts, skills and capacities" of "individuals, associations and institutions" (p. 25). The idea that individuals within a community are assets is important. What would a community be without its residents?

Within an economic context, assets can be forms of capital such as property, stocks and bonds, and cash. Within a community context, assets may be seen as various forms of capital as well. Assets take a variety of forms within a community. Ferguson and Dickens (1999) talk about five forms of community capital: physical, human, social, financial, and political. Green and Haines (2007) identify seven forms of community capital: physical, human, social, financial, environmental, cultural, and political. Rainey et al. (2003) present three forms of capital that they see as essential: human, public (physical), and social. While there can be debate about the forms of community capital and which forms are more essential than others, the important point here is that a community can identify its own assets – its own capital.

For the purpose of this chapter, three types of capital – physical, human, and social – will be defined and discussed. Why only these three forms of capital? These forms of capital may be subdivided into other forms of capital. For example, physical capital may comprise the built environment and natural resources. Thus, environmental capital, which comprises the natural resources within a community, is part of a community's physical characteristics and, thus, its assets. Natural resources may have shaped the community in the past either through its physical shape – constraining where and how a place grew – or by influencing its economy. The built environment – buildings, infrastructure – is a critical component of a community and clearly is part of its physical capital.

*Physical capital* comprises the roads, buildings, infrastructure, and natural resources within a community. When thinking about a specific community, what is often considered are its physical attributes and key features. These attributes and features may include roads, rail, water and sewer, a downtown, residential neighborhoods, parks, a riverfront, industrial areas, strip development, schools, government buildings, a university or college, a museum, a prison, and many others. In contrast to

the other forms of capital, physical capital is largely immobile. Although redevelopment of buildings and infrastructure occurs, physical capital endures over a long time period and is rooted in place. Another quality of physical capital is the degree of both public and private investment – public investment into infrastructure (roads, sewer, water) and private investment into structures (residential, commercial, and industrial) – with the expectation of a return on that investment.

*Human capital* is defined as the skills, talents, and knowledge of community members. It is important to recognize that not only are adults part of the human capital equation, but children and youth also contribute. It may include labor market skills, leadership skills, general education background, artistic development and appreciation, health, and other skills and experience (Green and Haines 2007: 81). In contrast to physical capital, human capital is mobile. People move in and out of communities, and, thus, over time, human capital can change. In addition, skills, talents, and knowledge change due to many kinds of cultural, societal, and institutional mechanisms.

*Social capital* often refers to the social relationships within a community and may refer to the trust, norms, and social networks that are established (Green and Haines 2007). "Social capital consists of the stock of active connections among people: the trust, mutual understanding, and shared values and behaviors that bind the members of human networks and communities and make cooperative action possible" (Cohen and Prusak 2001: 4).

In the community development context, the importance of social relationships is critical to mobilizing residents and is often a critical component for the success of a project or program. Social capital comprises the formal and informal institutions and organizations, networks, and ties that bind community members together. There are many forms of social capital – formal and informal, strong and weak, bonding and bridging – to name the more well-defined types. Formal ties or networks are those ties that are established through organizations, such as service clubs, and are seen as weak ties. Informal ties are those established through personal relationships. Often these ties are strong, and time and energy are involved in maintaining them. Bonding capital refers to bringing together people who already have established relationships or ties. In contrast, bridging capital refers to the idea of widening individuals' networks and ties. By establishing new networks or ties, people will have access to new information and more networks for sharing and using information.

In addition, social capital may be subdivided into various forms such as financial, political, and cultural. While these three forms can easily be separated from social capital and each other, social capital is central to these forms of community capital. Financial capital refers to access to credit markets and other sources of funds. Poor and minority communities often lack access to credit markets. Without sources of financing for home ownership, business start-ups and expansions, these communities are unable to put underused resources to work (Green and Haines 2007). Political capital is the capacity of a community to "exert political influence" (Ferguson and Dickens 1999: 5). Weir (1988) discusses three categories of the relationship between community-based organizations (CBOs) and local political systems. These categories include elite domination with weak ties to CBOs, a political patronage system, and a more inclusionary system where community and political leaders overlap. Based on Weir's categories, the exertion of political influence would differ between communities and the organizations within those communities.

One important concept to take from this discussion is that all forms of community capital are intricately linked together and are necessary for sustaining communities and achieving a better quality of life.

### The process of asset-based community development

Many community development professionals and others have, or are moving toward, an asset-based approach to community development. This next section introduces a general outline of this approach and describes its major steps. Figure 3.3 illustrates a community development process with four main steps. It is shown moving from community organizing to visioning to planning to implementation and evaluation and back to organizing. While the illustration moves from one step to another and creates a feedback loop, community development is far more messy and non-linear in practice. Many of these steps continue throughout the process. In addition, one step may be given more emphasis than the others in specific time periods. Every community is different, and the actual time it takes for each step will differ as well. Some communities may be fairly organized and cohesive and can move through organizing, visioning, and planning in a short amount of time and spend the bulk of their time and effort on implementation. However, other communities may find they are spending a great deal of time on organizing. Another aspect to consider in this illustration is the absence of a time frame. All the steps in the process appear to take a similar amount of time, but, as indicated above, the amount of time spent on any one step will depend on the community's residents and what they are trying to accomplish. The illustration includes a step for implementation – the action phase from which outcomes will be felt and measured. This step is a crucial part of the process – it is not separate from it. Chapter 6 continues the discussion of the community development process, focusing on visioning and strategic planning.

### Community organizing

Community organizing focuses on mobilizing people within a specific neighborhood or community. It is distinct from other forms of organizing because of its focus on communities of place rather than communities of interest. Community organizing does not need to be conceived of as a task for getting everyone in a community mobilized for doing something. In fact, community organizing may be thought of as a way to mobilize small groups of people to accomplish a particular task. Mobilizing community residents involves direct action ranging

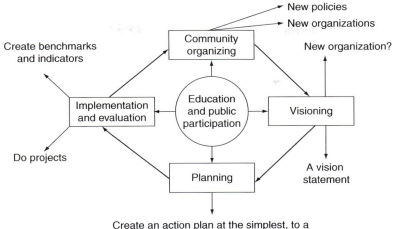

Create benchmarks
and indicators

Do projects

Create an action plan at the simplest, to a
comprehensive plan at the most ambitious

■ *Figure 3.3* **A community development process**

from writing letters to the editor to organizing a protest outside the offices of the school district. Often community organizing uses a problem-oriented approach rather than an asset-based approach. Community residents are mobilized to "solve" a particular problem recognized in their neighborhood. There are two strategies for mobilizing residents: social action campaigns and the development model. Social action campaigns are those direct actions, like the examples above, that aim to change decisions, societal structures, and cultural beliefs.

Another form of organizing occurs through the development model and is more prevalent at the community level. The development model is a way to organize communities of place to accomplish a variety of community goals. There are several different community organizing models (Rubin and Rubin 1992). The Alinsky model is probably the most popular and involves a professional organizer. The organizer works with existing organizations in a community to identify common issues. In contrast, the Boston model contacts welfare clients individually at their residences and relies heavily on appeals to the self-interest of each person. The Association of Community Organization for Reform Now

(ACORN) has mixed the Alinsky and Boston models. The Industrial Areas Foundation (IAF) model emphasizes the importance of intensive training of organizers. This model is a direct descendant of the Alinsky model but emphasizes the importance of maintaining close ties with existing community organizations. Community development corporations (CDC) use these development models to achieve community development goals. CDCs often represent the type of organization that provides economic and social services in low-income neighborhoods and communities (Rubin and Rubin 1992).

*Visioning*

Visioning is one method among many, such as future search, to establish a long-range view of a community. Box 3.1 provides a vision statement from one community. The term became popular in the 1990s, and many communities have used the technique to guide their future. While it is often used in the context of community planning, it has also been used to focus on specific topical areas, such as housing, transportation, and education. Many communities have found it useful to create multiple topical visions that can be more detailed and focused

---

**BOX 3.1 VISION STATEMENT FROM RUIDOSO, NEW MEXICO**

**A vision statement for Ruidoso**
**We treasure...**

- The **serene natural environment** – cool pines, high mountains, the Rio Ruidoso, comfortable weather, and clear skies.
- A **sense of community**. People are friendly; we prize the easy lifestyle where people know each other and where kids are safe riding their bikes.
- A **small-town atmosphere**, even during the summer and winter when the village serves an influx of part-time residents and tourists.

We like where we live, take pride in "our place" and we are willing to volunteer our time for community betterment.

Source: http://www.ruidoso.net/local/vision_statement.htm (accessed June 16, 2004).

---

rather than creating one broad vision, which many people view as too vague and broad to bring meaning to the necessary actions. The basic idea is to bring together a wide range of individuals, associations, and institutions within a community to arrive at, often through some form of consensus, a written statement – the vision – of the future and to prepare an action plan to move that community toward the vision.

There are at least three critical components of a visioning exercise. The first component is inviting a broad spectrum of the community so that many opinions and perspectives are represented. The second component is preparing a process that is meaningful, effective, and efficient. The process must be meaningful to the participants so that the time they are volunteering will have appropriate and useful results. The process must be effective in that it fulfills the purpose defined for it. Finally, the process must be efficient in terms of people's time, energy, and funds expended. This third component, which is closely related to the second one, involves choosing public participation techniques to accomplish a vision or multiple visions for a community. While visioning is thought of as a public participation technique, it must use techniques such as brain-

storming, SWOT analysis, and charrettes to accomplish its purpose.

*Planning*

During the planning phase there are at least three tasks in preparing an action plan: data collection and analysis, asset mapping, and a community survey. Data collection and analysis is important to understand current circumstances, changes occurring within a community over time, and the implications of the data collected.

Asset mapping is a process of learning what resources are available in a community. Kretzmann and McKnight (1993) provide the most hands-on and thorough asset mapping process. Their process "maps" or inventories the assets or capacities of:

- individuals including youth, seniors, people with disabilities, local artists, and others;
- local associations and organizations including business organizations, charitable groups, ethnic associations, political organizations, service clubs, sports leagues, veterans' groups, religious institutions, cultural organizations, and many others;
- local institutions for community building includ-

ing parks, libraries, schools, community colleges, police, hospitals, and any other institution that is part of the fabric of a community.

Asset mapping is an ongoing exercise. The purpose is to recognize the skills, knowledge, and resources within a community. It is a good first step in beginning to understand the assets of a community.

Community surveys can be useful in identifying issues at the beginning stages of a planning process and/or to refine particular ideas or policies as a community begins to think about its goals or its action plans. A community survey will allow various organizations in a community to:

- Gather information about public attitudes and opinions regarding precisely defined issues, problems, or opportunities.
- Determine how the public ranks issues, problems, and opportunities in order of importance and urgency.
- Give the public a voice in determining policy, goals, and priorities.
- Determine public support for initiatives.
- Evaluate current programs and policies.
- End speculation about "what people are thinking" or "what people really want."

(Laboratory of Community and Economic Development (LCED) 2002)

While community surveys can communicate important information about public attitudes and opinions, this public participation technique is focused only on input, not on shared decision making. Nevertheless, if carried out well, this technique allows for a much broader range of residents to participate than many other public participation techniques that call for face-to-face interaction.

*Public participation*

Determining the future of a community and how that community will get from Point A to Point B are important endeavors. Often residents leave community goals and decisions to others – a consul-

tant, a local government, a state or federal agency, a private developer, a business owner, or corporation. However, residents of a community need to participate in and get actively involved in determining the future course of their community. If they do not, others will determine their future for them. Thus, public participation is critical to the entire community development process. Figure 3.3 illustrates its importance by situating it at the center of the process.

Using public participation effectively and meaningfully is a difficult task for individuals, groups, or organizations that are trying to determine which techniques are most appropriate to use, when to use them, and who should be involved. Effective public participation needs to be both functional for the specific goal and meaningful to the public. Participation is functional when it helps to create better decisions and a more thoughtful community plan or some other document that can help organizations, institutions, and individuals understand how their community is moving forward. Participation is meaningful when it creates opportunities for the public to exercise influence over decisions and feel a sense of ownership toward the product.

*Implementation and evaluation*

Actions in community development are where change occurs and where people can see tangible results. This phase in the community development process is the point at which the rubber meets the road. It is the phase where individuals, groups, and organizations are active rather than passive participants in their community. Up until the implementation phase of a community development process, individuals and organizations have made a concerted effort to understand their assets, community attitudes, and opinions; have arrived at a shared understanding of the future; and have agreed upon initial actions, and possibly broader strategies to take that will lead to specific actions in the future. Action plans generally identify specific projects, deadlines, responsible parties, funding mechanisms, and other tasks that will accomplish specific goals. An action plan describes a set of activities that need to be accomplished to move the community toward its future vision and/or goals.

Another part of implementation is to consider the regulatory context within which development occurs, in particular the physical capital of a community. It is likely the action plan has identified areas that need changes through local government, such as zoning changes. It is important for the regulations to integrate with the plan.

An often overlooked but important component to community development is monitoring and evaluation. Monitoring is the act of assessing the community development process as it is taking place. Monitoring functions as a way to take the pulse of a community effort. It allows for adjustments to be made in the process rather than letting issues or situations get beyond the control of facilitators, a steering committee, and/or a project team to manage the process (Green et al. 2001).

Evaluation, in contrast, usually occurs after a project or a plan is considered completed. At least two types of accomplishments can be measured: outputs — the direct and short-term results of a project or plan such as the number of people trained, the number of affordable houses built, or the number of jobs created; and outcomes — the long-term results of a project or plan. Outcomes are much more difficult to measure and to link directly to a specific action. Outcomes that many community development plans aim for are long-term goals such as decreased levels of poverty, a better quality of life in a community, or increased levels of personal income (ibid.).

By establishing benchmarks and indicators to more easily track the accomplishments of specific actions, communities are able to more usefully conduct monitoring and evaluation. "Sustainable Seattle" offers a good example of a nonprofit organization that has used regional and neighborhood indicators to monitor the environmental health of the Seattle region in addition to using public participation to accomplish their goals (Sustainable Seattle 2004).

## Challenges of the community development process

An important caveat to any community development process is that they can be difficult, time-consuming, and costly. The difficulty can occur, for example, when many diverse interests cannot or will not find common ground about either specific actions or even the general direction the community should take. Thus, finding consensus and making compromises is not only difficult but can be time-consuming as well. Nevertheless, creating a forum where diverse interests can discuss issues is critical for improvement of a community. In addition to stumbling blocks and obstacles, the process itself may take time, which can be frustrating for those individuals and organizations that like taking action. It is difficult to maintain interest and commitment to the effort if participants cannot point to successes. It is important for motivation and trust to create a balance between initial actions and completing the process that was set out so that residents can see that something is occurring.

The cost of the process depends on the amount and types of public participation that are used within a process in addition to the amount and types of data collected and analyzed. While some forms of public participation are inexpensive, such as developing a website, they may only act to inform residents rather than provide a way to gather input and create partnerships. Forms of public participation that call for facilitators may be more expensive; however, residents can be trained as facilitators to reduce costs. Creating a process that is manageable, fits the community, and fits the budget is imperative; the process of community development can be as important and as valuable as its products. While there continues to be a debate over the importance of process versus outcomes in community development, it is clear that the ultimate goal is to improve the quality of life for the residents in the community. In the long run, both process and outcomes are essential parts of community development.

## Conclusion

Asset-based community development is a promising approach to achieving a better quality of life and sustaining communities not only over time or in an economic sense, but through the development of all forms of capital that are necessary for a community to thrive. Kretzmann and McKnight (1993) believe that the key to community revitalization is "to locate all of the available local assets, to begin connecting them with one another in ways that multiply their power and effectiveness, and to begin harnessing

---

**CASE STUDY: ASSET BUILDING ON THE SHORES OF LAKE SUPERIOR**

Place-based communities across the U.S. are making changes towards a more sustainable future – reducing greenhouse gases, becoming less reliant on fossil fuels, promoting local food systems, encouraging green buildings, higher density and more active lifestyles. In the past two years, approximately eleven communities in Wisconsin have declared themselves "eco-municipalities" to indicate their commitment to move toward a sustainable future. In a small region of Wisconsin comprised of about 32,000 people living along the southern shore of Lake Superior, residents and community leaders joined together in creating a strategic plan for sustainable development. Its three small cities and one township have passed "eco-municipality" resolutions, and a series of groups called "green teams" are taking a variety of actions initially identified in the strategic plan as well as other actions that were not.

One of the region's strongest regional assets is its natural capital. The region is rich in natural resources which underlies the current economy that is based in tourism, wood products, and farming. Within the two counties are over 1000 lakes, many miles of rivers and streams (including over 100 miles of Lake Superior shoreline), and thousands of acres of wetlands and forests. The major tourist draw is the Apostle Island National Lakeshore comprised 22 islands in Lake Superior within a short ferry and boating distance to the City of Bayfield. Other key environmental assets are the Kakagon and Bad River sloughs – 16,000 acres on and around the Bad River Band of the Lake Superior Tribe of Chippewa Indians Reservation – that represent the largest undeveloped wetland complex in the upper Great Lakes. These sloughs have been called Wisconsin's Everglades and were designated a National Natural Landmark by the U.S. Department of the Interior in 1983 (Nature Conservancy 2000). This area is considered one of the healthiest wetland ecosystems in the area and produces wild rice for members of the Bad River Tribe.

In the first half of 2007, the area cities and several organizations have each created a "Green Team" to review each organization's operations to determine specific ways in which sustainable practices may continue to occur. For example, one of the first initiatives the City of Ashland Green Team will be completing is an energy audit to determine where its energy consumption is occurring and where it can make changes. The City of Washburn has installed energy-efficient compact fluorescent bulbs and tubes in the civic center and library; studied how lighting and heating can be improved for the city garage; replaced a hot water boiler at a local park's shower building with a tankless coil system that operates on demand; and installed geothermal heating and cooling in public housing designed for low-income and elderly citizens (Boyd 2007).

While the environmental capital is the motivation for the current efforts to create sustainable communities in this small region, it is the social, financial, and political forms of capital that have functioned to bring about the results described above in the first paragraph. Local politicians, wealthy residents, faculty, and staff at Northland College (an environmental liberal arts college), local environmental groups, and the League of Women Voters worked together in multiple ways over many years to get to where they are today.

those local institutions that are not yet available for local development purposes" (p. 5).

This chapter has defined asset-based community development in contrast to a needs-based approach and has touched upon the steps in a community development process.

## Keywords

Asset-based community development, asset identification, community organizing, community participation, community revitalization, visioning, sustainable development, physical capital, social capital.

## Review questions

1 What is asset-based development, and why is it different from past approaches?
2 Define the types of assets that a community may have.
3 Name some of the steps in the process of community development and discuss their importance.
4 What is the importance of social relationships in community development?
5 What are some challenges to community development?

## Bibliography and additional sources

Boyd, D. (2007) "Chequamegon Bay Communities Support Sustainability Framework," MSA Professional Services Update, 18: 3.

Chaskin, R. (2001) "Building Community Capacity: A Definitional Framework and Case Studies from a Comprehensive Community Initiative," Urban Affairs Review, 36(3): 291–323.

Cohen, D. and Prusak, L. (2001) In Good Company: How Social Capital Makes Organizations Work, Boston, MA: Harvard Business School Press.

Ferguson, R.F. and Dickens, W.T. (eds) (1999) "Introduction," in Urban Problems and Community Development, Washington, DC: Brookings Institution Press, pp. 1–31.

Future Search Network. Available online at http://www.futuresearch.net/ (accessed June 16, 2004).

Green, G.P. and Haines, A. (2007) Asset Building and Community Development, 2nd edn, Thousand Oak, CA: Sage.

Green, G.P. et al. (2001) RRD182 Vision to Action: Take Charge Too, Ames, IA: North Central Regional Center for Rural Development. Available online at http://www.ag.iastate.edu/centers/rdev/pubs/contents/182.htm (accessed August 9, 2004).

Kretzmann, J.P. and McKnight, J.L. (1993) Building Communities from the Inside Out: A Path Toward Finding and Mobilizing a Community's Assets, Chicago, IL: ACTA Publications.

Kretzmann, J.P. and McKnight, J.L. (1996) "Assets-based Community Development," National Civic Review, 85(4): 23.

Laboratory of Community and Economic Development (LCED) (2002) Assessing and Developing: Your Community Resources, University of Illinois – Extension, winter 2002 Issue 1(3). Available online at http://www.communitydevelopment.uiuc.edu/toolbox/ (accessed August 9, 2004).

Nature Conservancy (2000) "Chequamegon Bay Watershed Site Conservation Plan." October 2000, unpublished manuscript.

Rainey, D.V., Robinson, K.L., Allen, I. and Christy, R.D. (2003) "Essential Forms of Capital for Sustainable Community Development," American Journal of Agricultural Economics, 85(3): 708–715.

Rubin, H.J. and Rubin, I.S. (1992) Community Organizing and Development, 2nd edition, Boston, MA: Allyn & Bacon.

Sherraden, M. (1991) Assets and the Poor: A New American Welfare Policy, Armonk, NY: M.E. Sharpe.

Shuman, M.H. (1998) Going Local: Creating Self-reliant Communities in a Global Age, New York: The Free Press.

Sustainable Seattle (2004) Available online at http://www.sustainableseattle.org/nd/programs/default.htm (accessed August 27, 2004).

Vidal, A.C. and Keating, W.D. (2004) "Community Development: Current Issues and Emerging Challenges," Journal of Urban Affairs, 26(2): 125–137.

Weir, M. (1988) "The Federal Government and Unemployment: The Frustration of Policy Innovation from the New Deal to the Great Society," in M. Weir, A.S. Orloff and T. Skocpol (eds) The Politics of Social Policy in the United States, Princeton, NJ: Princeton University Press, pp. 149–197.

# 4 Social capital and community building

## Paul W. Mattessich

Social capital or capacity lies at the heart of community development. If citizens cannot plan and work together effectively and inclusively, then substantial proactive community progress will be limited. This chapter discusses community social capacity and how it relates to community development. Based on an extensive review of previous studies, the chapter lists and categorizes factors that affect the likelihood that community building efforts can succeed to increase the social capacity of geographically defined communities.

## Introduction

The opening chapter distinguished social capital (sometimes called social capacity) from human, physical, financial, and environmental capital. All of these constitute resources that communities need to function. The extent to which communities have these forms of capital influences their ability to accomplish tasks and to develop themselves.

Development of a community includes, in part, the building of its social capacity. Conversely, the level of a community's social capacity both influences the way development evolves for that specific community and the pace at which its development efforts can occur. The opening chapter of this book defined social capacity and social capital. It explained how the social capital of communities relates to the process of community development, both as an antecedent that predisposes a community to further development in the "community development chain" and, as a consequence, a feature of the community that increases as a result of community development. The discussion in this chapter places emphasis on social relationships embedded within, <u>and</u> distributed throughout, a community. Individuals can amass social capital, as well as other forms of capital. They can have strong ties to social networks that enable them to do things for themselves. However, unless the social capital that they possess becomes a resource available to and used by the entire community, it has little or no direct effect upon that community's development.

## Social capital: what is it?

Chapter 1 provided some definitions of social capital. In the simplest sense, what comprises the core of these definitions are "social networks and the associated norms of reciprocity" (Clarke 2004). Analogous to other forms of capital (e.g., financial, human), social capital constitutes a resource. It provides value to individuals and can also benefit communities. It has effects, often called "externalities," in economics literature.

No single definition has achieved universal acceptance though eventually a common standard may emerge. Most definitions stress interconnections among people or social networks. Early definitions tended to place emphasis on how individuals could use social relationships as a resource to accomplish goals; they did not add a community dimension to the definition. More recent definitions tend to recognize the distinction between social capital at the individual level and social capital at the community level.

Bourdieu's (1986) early definition, for example, focused on individuals:

> Social capital is an attribute of an individual in a social context. One can acquire social capital through purposeful actions and can transform that capital into conventional economic gains. The ability to do so, however, depends on the nature of the social obligations, connections, and networks available to you.

In this tradition, Sobel (2002) stated that social capital "describes circumstances in which individuals can use membership in groups and networks to secure benefits."

Eventually, however, social scientists began to realize that social capital in the control of individuals, but not shared with others in the same community, did not fit into the "community development chain" illustrated in Chapter 1. It does not necessarily produce community social capacity that affects the ability of the community to develop. In fact, individuals who have such capital can even use it to the detriment of their own or other communities.

Consider, for example:

- A gang or a network of organized criminals that lives and operates within a community but uses its close interconnections to commit crimes and to engage in acts that detract from the livability of the community.
- Wealthy elite residents of a small city in a developing country who separate themselves from others in the population, maintain social and business connections with individuals and organizations in other locations, and channel their financial wealth to places outside the country.
- In a financially struggling United States city, a gated cluster of town homes for high-wealth residents who don't engage in ongoing relationships with the local population and who invest most of their money elsewhere.

In all the above cases, social capital, not to mention financial capital, exists within a geographically defined area. However, that capital produces few, if any, benefits for the community as a whole. In fact, those who have such capital may actually use it to exploit others in the same locality. A "map" of the social interconnections within any of the communities in the above examples would reveal one or both of the following:

- Individuals with no ties to other individuals in the community but to those outside the community. For example, many of the wealthy members of the developing society, such as residents of the gated community, may have strong ties with business and social associates in locations throughout the world.
- Highly cohesive social networks among one or more subsets of the community, with no strong ties between any of these networks. For example, members of criminal gangs may maintain strong ties with one another or wealthy elite residents in an area may do the same.

Recognition of the distinction between individual social capital and community social capital (or community social capacity) does not negate the importance of individual social capital. An individual's social capital provides an important resource for the individual; it has real effects. However, unless amalgamated with the social capital of others in the same community, it does not necessarily produce benefits for that community. As noted in Chapter 1, "bonding" capital ties individuals to others like themselves (race, economic status, nationality); "bridging" capital ties individuals to a diverse set of others, some like themselves, some not.

## Community social capacity: what is it?

Mattessich and Monsey (1997) define community social capacity as: "The extent to which members of a community can work together effectively." This includes the abilities to: develop and sustain strong relationships; solve problems and make group decisions; and collaborate effectively to identify goals and get work done. Communities with high community social capacity can identify their needs; establish priorities and goals; develop plans, of which the members of that community consider themselves "owners"; allocate resources to carry out those plans; and carry out the joint work necessary to achieve goals.

The term "community social capacity" applies holistically to an entire community. It is an attribute of a community, not of any specific members. The level of community social capacity depends upon the number and strength of ties or bonds that community members have with one another. Thus, it is a form of social capital – since it involves "social networks and norms of reciprocity" – but it distinctly involves interconnections among people who reside in the same community (defined by geographic location).

If people who live within the same geographic area do not know one another and have little contact with one another, the likelihood is low that they can get together to define community goals or respond productively with one voice to a community issue. Their community social capacity is low. On the other hand, if people who live within the same geographic area do know one another, share a large number of social ties, and feel a commitment to the place where they live, then community social capacity is high.

## How does community social capacity influence development?

Jane Jacobs recognized the importance of community social capacity for community vitality. She observed that deep and heterogeneous social relationships seemed to enable communities to thrive; barriers imposed upon these relationships seemed to lead to community deterioration (Jacobs 1961).

The level of community social capacity (or community social capital) influences community development in two broad ways: structural and cog-

nitive (see e.g., Uphoff 2000). Structurally, interconnections among people within a community create a web of social networks. These networks facilitate community development by enabling the flow of information, ideas, products, and services among residents. Cognitively, interconnections create a shared sense of purpose, increase commitment, promote mutual trust, and strengthen norms of reciprocity among community residents.

## Intentional action to increase social capacity

Communities often recognize the need to increase their social capacity and take steps to do so in a community-building process. Mattessich and Monsey (1997) offer a brief definition of community building: "Any identifiable set of activities pursued by a community in order to increase community social capacity."

The elements of community building come across in a longer definition from a review of comprehensive community initiatives by Kubisch et al. (1995):

> Fundamentally, community building concerns strengthening the capacity of neighborhood residents, associations, and organizations to work, individually and collectively, to foster and sustain positive neighborhood change. For individuals, community building focuses on both the capacity and "empowerment" of neighborhood residents to identify and access opportunities and effect change, as well as on the development of individual leadership. For associations, community building focuses on the nature, strength, and scope of relationships (both affective and instrumental) among individuals within the neighborhood and through them, connections to networks of association beyond the neighborhood. These are ties through kinship, acquaintance or other more formal means through which information, resources, and assistance can be received and delivered. Finally, for organizations, community

building centers on developing the capacity of formal and informal institutions within the neighborhood to provide goods and services effectively, and on the relationships among organizations both within and beyond the neighborhood to maximize resources and coordinate strategies.

Community organizing can support community building. It refers to the process of bringing community members together and providing them with the tools to help themselves. The process may include:

> identification of key local resources, the gathering of information about the community context, the development and training of local leaders to prepare them to serve effectively as representatives of the community and as full partners in an initiative, and the strengthening of the network of the various interests both internal and external to a community.
>
> (Joseph and Ogletree 1996)

## Factors that influence the success of community-building efforts

Communities often wish to improve themselves. They want to attract new businesses, improve housing stock, reduce crime, improve the education of their children, or accomplish any number of tasks that will better the quality of life for community residents. All of these goals, whether adopted individually or together by a community, constitute goals that fall under the umbrella of "community development."

After adopting goals, communities often attempt collaborative action involving individuals and organizations in order to improve themselves. Mattessich and Monsey (1997) synthesized the research literature on what makes one aspect of such efforts successful; that is, the aspect of "community building," a term used purposefully to distinguish it from the more inclusive concept of community development.

## What is community building?

Community building refers to activities pursued by a community in order to increase the social capacity of its members (the term capacity building is often used interchangeably as in Chapter 1). In the words of Gardner (1993), community building involves "the practice of building connections among residents, and establishing positive patterns of individual and community behavior based on mutual responsibility and ownership."

## What influences the success of community building efforts?

The review, *Community Building: What Makes It Work* (Mattessich and Monsey 1997) synthesized research on community building to identify factors that influence its success. The factors fall into three categories:

1 Characteristics of community – social, psychological, and geographic attributes of a community and its residents that contribute to the success of a community building effort.
2 Characteristics of a community-building process – components of the process by which people attempt to build community.
3 Characteristics of Community-building organizers – qualities of the people who organize and lead a community-building effort such as commitment, trust, understanding, and experience.

No community building effort has a 100 percent likelihood of success. Whether an effort will succeed depends upon many circumstances, some within and others outside the control of community residents. Based on the synthesis of research, it is reasonable to assume that the higher a community stands on the factors influencing the success of community building, the greater the likelihood that it will succeed in a community-building effort.

Building community is much like improving or maintaining one's health. If someone eats nutritious food, exercises, and has regular checkups, for example, that person maximizes the likelihood of good health. Good health is not guaranteed but its much more likely. It is similar to the factors that influence the success of community building. If present, they maximize the likelihood that an effort will succeed but there are no guarantees. Box 4.1 lists the success factors in community building as identified by the Mattessich and Monsey literature review.

## BOX 4.1 TWENTY-EIGHT FACTORS THAT INFLUENCE THE SUCCESS OF COMMUNITY BUILDING

### 1 Characteristics of the community

*A Community awareness of an issue*

Successful efforts more likely occur in communities where residents recognize the need for some type of initiative. A community-building effort must address an issue which is important enough to warrant attention, and which affects enough residents of a community to spark self-interest in participation. Residents must know that the problem or issue exists.

*B Motivation from within the community*

Successful efforts are more likely to occur in communities where the motivation to begin a community-building process is self-imposed rather than encouraged from the outside.

*C Small geographic area*

Successful efforts are more likely to occur in communities with smaller geographic areas, where planning and implementing activities are more manageable. Interaction is harder to achieve if individuals are separated from one another by a great distance.

*D Flexibility and adaptability*

Successful efforts are more likely to occur in communities where organized groups and individuals exhibit flexibility and adaptability in problem solving and task accomplishment.

*E Preexisting social cohesion*

The higher the existing level of social cohesion (that is, the strength of interrelationships among community residents), the more likely that a community building effort will be successful.

*F Ability to discuss, reach consensus, and cooperate*

Successful efforts tend to occur more easily in communities that have a spirit of cooperation and the ability to discuss their problems and needs openly.

*G Existing identifiable leadership*

Successful efforts are more likely to occur in communities with existing, identifiable leadership – that is, communities with at least some residents whom most community members will follow and listen to – who can motivate, act as spokespersons, and assume leadership roles in a community-building initiative.

*H Prior success with community building*

Communities with prior positive experience with community-building efforts are more likely to succeed with new ones.

### 2 Characteristics of the community-building process

*A Widespread participation*

Successful efforts occur more often in communities that promote widespread participation, which is:

- Representative – it includes members of all, or most, segments of the community at any specific point in time.
- Continuous – it recruits new members over time, as some members leave for one reason or another.

*B Good system of communication*

Successful efforts have well-developed systems of communication within the community itself and between the community and the rest of the world.

*C Minimal competition in pursuit of goals*

Successful efforts tend to occur in communities where existing community organizations do not perceive other organizations or the leaders of a community-building initiative as competitors.

*D Development of self-understanding*

Successful efforts are more likely to occur when the process includes developing a group identity, clarifying priorities, and agreeing on how to achieve goals.

*E Benefits to many residents*

Successful community-building efforts occur more often when community goals, tasks, and activities have visible benefits to many people in the community.

*F Concurrent focus on product and process*

Community-building initiatives are more likely to succeed when efforts to build relationships (the process focus) include tangible events and accomplishments (the product focus).

*G Linkage to organizations outside the community*

Successful efforts are more likely to occur when members have ties to organizations outside the community, producing at least the following benefits: financial input; political support; source of knowledge; source of technical support.

*H Progression from simple to complex activities*

Successful community-building efforts are more likely to occur when the process moves community members from simple to progressively more complex activities.

*I Systematic gathering of information and analysis of community issues*

Successful community-building efforts are more likely to occur when the process includes taking careful steps to measure and analyze the needs and problems of the community.

*J Training to gain community building skills*

Successful community-building efforts are more likely to occur when participants receive training to increase their community-building skills. Examples include group facilitation, organizational skills, human relations skills, and skills in how to analyze complex community issues.

*K Early involvement and support from existing indigenous organizations*

Successful community-building efforts tend to occur most often when community organizations of long tenure and solid reputation become involved early, bringing established contacts, legitimization, and access to resources.

*L Use of technical assistance*

Successful community-building efforts are more likely to occur when residents use technical assistance to gain necessary skills.

*M Continual emergence of leaders, as needed*

Successful community-building efforts are more likely to occur when the processes produce new leaders over time.

*N Community control over decision making*

Successful community-building efforts are more likely to occur when resident have control over decisions, particularly over how funds are used.

*O The right mix of resources*
Successful community building efforts occur when the process is not overwhelmed by too many resources or stifled by too few, and when there is a balance between internal and external resources.

**3   Characteristics of community-building organizers**
*A An understanding of the community*
Successful community-building efforts are more likely to occur when organizers understand the community they serve. This includes an understanding of the community's culture, social structure, demographics, political structures, and issues.
*B Sincerity of commitment*
Successful community-building efforts are more likely to occur when organized by individuals who convey a sincere commitment to the community's well-being; are interested in the community's long-term well-being; have a sustained attachment to community members; are honest; and act primarily to serve the interests of the community, not of an external group.
*C A relationship of trust*
Successful community-building efforts are more likely to occur when the organizers develop trusting relationships with community residents.
*D A high level of organizing experience*
Successful community-building efforts are more likely to occur when the organizers are experienced.
*E Flexibility and adaptability*
Successful community building efforts are more likely to occur when organizers are flexible and able to adapt to constantly changing situations and environments.

Source: Mattessich and Monsey 1997: 14–17

## Conclusion

As stated in the opening of this chapter, the development of a community in part includes the building of the community's social capacity. Conversely, the level of a community's social capacity influences the way community development evolves for that specific community; it also influences the pace at which community development efforts may occur.

It should be apparent that communities are never "built" in the finite sense of that word. Community building is a continual process, not a set of steps to a permanent conclusion. In the process of building community, a set of individuals who live in the same geographic region can develop social networks and community social capacity; however, sustaining those networks and the social capacity requires ongoing effort.

In this chapter, factors were presented that research has demonstrated as affecting the success of community building, both in initial efforts to build community and ongoing efforts to sustain community social capacity. The greater the extent to which those factors are in place, the greater the likelihood that community building can be successful.

## CASE STUDY: COMMUNITY DEVELOPMENT IN LITHUANIA: THE CHALLENGE OF BUILDING SOCIAL CAPITAL IN A POST-COMMUNIST SOCIETY

The centrally planned, top-down command-and-control approach to governing in communist countries is in many ways antithetical to the principles of community development which emphasize the grassroots, bottom-up approach with local cooperation and self-initiative. In rural areas, the Soviet-era collective farm policy for agriculture rendered community development and organizing not only irrelevant but in many cases illegal. Post-communist Eastern European villages therefore make interesting laboratories for studying how the seeds of social capital and community development can sprout on previously barren soil.

Juska, Poviliunas and Pozzuto (2005) provide an example of this phenomenon through a case study of Balninkai, a small village of 500 people in eastern Lithuania. The authors note that a form of "shock therapy" was carried out on villages like Balninkai when the state farms or *kolkhoz* were disbanded and their assets privatized with the fall of the Soviet Union. Independent family farms replaced collective large-scale production. This major social disruption and accompanying disputes over the distribution of land and assets led to unrest and conflicts among neighbors which further poisoned the atmosphere for community development. Few if any of this chapter's community development success factors were present in post-communist Eastern Europe and Balninkai. Preexisting social cohesion, experience in reaching consensus and community building, and experienced leadership were all almost non-existent.

However, by the late 1990s continuous declines in living standards and a chronic lack of economic opportunity led to the birth of community development efforts and organizations aimed at improving the quality of life. Necessity was the mother of invention. As the authors point out, "successful mobilizing usually occurs when favorable opportunities are seized by social entrepreneurs or ambitious leaders who possess skills in communicating a mission and inspiring followers." In Balninkai, four individuals proved to be critical in the drive to create a community center: the Mayor, a local teacher, a local priest and a rural sociologist who happened to live in the village. Immediately, these four individuals faced a leadership challenge: the villagers would not trust or accept any top-down form of local organizing because of their experience with communist rule. These nascent leaders knew they had to solicit widespread support and input for any community development effort, so they turned to the most trusted and neutral local institution for help in announcing and promoting the community development effort: the Catholic Church. The Church disseminated information on the initiative through printed materials and word of mouth, and sponsored many community meetings.

To ensure that all residents felt they were a part of the process, a door-to-door survey was conducted to solicit views on village problems, priorities, and suggestions for improvement. The survey data was analyzed and presented at numerous community meetings. Priority issues were employment opportunities, village beautification, cultural activities, and medical care. Of these, lack of employment opportunities represented the biggest challenge and the authors acknowledge that without long-term concrete results in this area, the overall effectiveness of community development in Balninkai will likely be limited. Still, community development progress has been made, with initial efforts focused on village beautification, including improvements on the main street, construction of staircases and decorative windmills, and planting of flowers and trees. These initial results attracted the attention of national foundations which provided grant funding to assist with the community-development process.

The authors report that the village is trying to transform the structure of its community development effort from one based on volunteers to a not-for-profit organization. This is causing some difficulty because many villagers still maintain the communist-era distrust of organizations, even local ones.

However, rural community organizations are growing throughout Lithuania: in 2003, there were 300 such groups and just one year later the number had increased to 481. While Balninkai itself lacked many of the community development success factors, it did embrace some of the key community-building process factors such as widespread participation, a good system of communication, and systematic gathering of information and analysis of community issues. The example of Bakninkai demonstrates that community development and social capital building can occur even under the harshest of circumstances.

The Editors

Source: Juska, A., Poviliunas, A. and Pozzuto, R. (2005) "Rural Grass-roots Organizing in Eastern Europe: the Experience from Lithuania," *Community Development Journal*, 41(2): 174–188.

## Keywords

Individual social capital (capacity), social capacity, social networks, community social capital, community (capacity), community capacity, community building, social cohesion, linkages, indigenous organizations.

## Review questions

1  What is social capital (capacity)?
2  What is the difference between individual social capital and community social capital?
3  What is the interrelationship between community social capacity and development?
4  What is community (capacity) building?
5  What are some key factors that influence the success of community building?

## Bibliography and additional resources

Bourdieu, P. (1986) "The forms of capital," in J.G. Richardson (ed.) *Handbook for Theory and Research for the Sociology of Education*, New York: Greenwood Press.

Clarke, R. (2004) "Bowling Together," *OECD Observer*, 242 (March) Available online at www.oecdobserver.org (accessed April 14, 2007).

Gardner J. (1993) *Community Building: An Overview Report and Case Profiles*, Washington, DC: Teamworks.

Jacobs, J. (1961) *The Death and Life of Great American Cities*, New York: Random House.

Joseph, M. and Ogletree, R. (1996) "Community Organizing and Comprehensive Community Initiatives," in R. Stone (ed.) *Core Issues in Comprehensive Community Building Initiatives*, Chicago, IL: Chapin Hall Center for Children.

Kubisch, A. et al. (1995) *Voices from the Field: Learning from Comprehensive Community Initiatives*, New York: The Aspen Institute.

Mattessich, P. and Monsey, B. (1997) *Community Building: What Makes It Work: A Review of Factors Influencing Successful Community Building*, St. Paul, MN: Wilder Foundation.

Sobel, J. (2002) "Can We Trust Social Capital?," *Journal of Economic Literature*, XL (March): 139–154.

Uphoff, N. (2000) "Understanding Social Capital: Learning from the Analysis and Experience of Participation," in P. Dasgupta and I. Serageldin (eds) *Social Capital: A Multifaceted Perspective*, Washington, DC: The World Bank.

# 5  Community development practice

## John W. (Jack) Vincent II

Community development is a wide-ranging discipline that encompasses economic development. Community development is a process whereby all citizens are involved in the process of community change and improvement. Success in community development leads to more success in economic development. A set of values and beliefs and ethical standards has been developed that should always guide the community development process.

## Introduction

As discussed in previous chapters, community development is a broad subject incorporating many different disciplines. As a process and an outcome, it is inextricably linked to economic development, which may also be defined as a process and an outcome. Because of its multidisciplinary nature, recent research on establishing a theoretical foundation for community development has drawn from different fields including sociology, economics, psychology, and many others. The focus of this chapter is on the practice of community development. It provides a broad overview of community development practice as a foundation for subsequent chapters dealing with specific aspects of community development.

## First step: define the community

As noted in Chapter 1, a community is not necessarily defined by geographical or legal boundaries. A community could involve interaction among people with common interests who live in a particular area. Or it could involve a collection of people with common social, economic, political, or other interests regardless of residency.

Professional community development (CD) practice could involve a community that is comprised of a single city, town, or village. On the other hand, it may involve working with a community comprised of several cities, towns, and villages, or a community of counties that decide to form a regional development organization.

When considering a single city (town or village), it is also important to realize that citizens may not be residents. Those who make investments in, work in, or operate businesses in a city but live outside of the city boundaries are still very much citizens and stakeholders of that community. Similarly, major corporations may also be corporate citizens who invest in a community and represent the interests of the many individuals who have invested in their company. As the city goes, so go the fortunes of the investing corporation and its shareholders. Therefore, when organizing a community development initiative, it is important to carefully define the community and who should be included in the process. The community should not be too narrowly defined.

Looking to the future, CD professionals may find themselves working in unique communities such as "virtual communities." With the great advances in computers and electronic communications, it is possible for individuals with common interests and concerns to have electronic communities without legal or physical boundaries. Their interactions, business dealings, and common concerns may result in an interdependency that requires support from a community developer. This brings a whole set of unique problems and opportunities to the CD professional.

## Practicing community development

As noted in Chapter 1, community development is both a process and an outcome. The practice of community development may be described as managing community change that involves citizens in a dialogue on issues to decide what must be done (their shared vision of the future) and then involves them in carrying it out.[1]

Critical issues addressed in the CD process include: jobs and economic development (business attraction, expansion and retention, and new business development); education and workforce development; infrastructure development and improvement; quality of life, culture, and recreation; social issues such as housing, crime, teen pregnancy, and substance abuse; leadership development; the quality of governmental services; community image and marketing; and tourism development.

## The professional community developer's role

The king (or queen) is dead! Plato's philosopher-king, described in Plato's Republic, may have been a good person, but his benevolent dictatorship is inappropriate in the community environment. Assuming control and telling people what they should do is a subtle form of dictatorship, regardless of the intent. Many consultants assume this role in such a sophisticated advisory manner that clients do not even realize that they are relegating their rights and personal responsibilities. Further, this type of consulting does little to foster local

leadership development and self-sufficiency. To be successful, CD initiatives must originate from the citizens themselves. The people who live in a community must have "ownership" of the process and the will to work toward bringing about necessary change. Otherwise, the process will either fail or result in the creation of a plan of action that is never fully carried out.

A professional community developer's primary role is to facilitate the CD process. He or she is an agent of change, an individual in an advisory role focused on helping citizens assess their current situation, identify critical issues and options for the future, weigh those options, create a shared vision of the future, and make informed decisions about what they will do to make that vision a reality. If a professional community developer finds dependent citizens, he or she works to empower them so that they come to realize that they are not helpless and that their actions can make a positive difference in their communities. The professional developer moves citizens from a state of helplessness and dependency to one of self-directed interdependence. For a community developer, success is being "no longer needed." The professional works him or herself out of a job.

Professional community developers as defined in this chapter would include consultants trained in the field, but also local residents whose jobs involve community development activities such as economic development, planning, community health, and a host of other fields. The community development professional could also be a volunteer who is taking a lead role in the community development process. People who spend a large portion of their time in community development activities – whether paid or volunteer – may be considered professional community developers. As discussed below, professional community developers can obtain national certification as an indication of their experience and knowledge in the field. Whether considered a professional community developer or not, the more local citizens that understand the principles and practice of community development, the more successful the process and outcome will be.

## Inclusiveness

CD is based on the idea that all people are important and should have a voice in community decisions, have potential to contribute, resources to share, and a responsibility for community action and outcomes. Citizens are entitled to make informed decisions about the factors influencing their community. The process is always open and transparent. Therefore, CD is an inclusive approach to working with people who participate in the process to the extent that they are capable and interested.

The professional community developer uses many techniques that offer citizens an opportunity to be part of the development process. These techniques may include responding to a survey, being interviewed as part of a community assessment, and participating in a public meeting, task force, subcommittee, or strategic planning committee. These actions greatly increase local citizen involvement and, thus, the potential to improve their quality of life.

CD is guided by a set of values and beliefs. These may be thought of as mental tools that professional community developers use to provide guidance and direction for their work (see Box 5.1).

Professional community developers believe in people being proactive within the framework outlined in Box 5.1. These values and beliefs help the process of decision making to be open, inclusive, interactive, and focused on the well-being of the entire community, not just one segment.

## Community development principles of practice

Over time, a set of principles has evolved that also act as guides to CD practice. These principles, based on the author's experience and observations from over 25 years as a CD consultant and researcher, are:

1  *Self-help and self-responsibility are required for successful development.* No one knows better about what must be done or is more committed to change than those who live in a community. It is impossible for CD consultants to do all that is necessary for a community to realize its full potential. They do not possess all the knowledge and skills needed by their clients' community. The CD consultant's role is to organize citizens so that they realize their power, capabilities, and potential during the process of change.

---

### BOX 5.1 COMMUNITY DEVELOPMENT VALUES AND BELIEFS

1  People have the right to participate in decisions that affect them.
2  People have the right to strive to create the environment they desire.
3  People have the right to make informed decisions and reject or modify externally imposed conditions.
4  Participatory democracy is the best method of conducting community business.
5  Maximizing human interaction in a community will increase the potential for positive development.
6  Creating a community dialogue and interaction among citizens will motivate citizens to work on behalf of their community.
7  Ownership of the process and commitment for action is created when people interact to create a strategic community development plan.
8  The focus of CD is cultivating people's ability to independently and effectively deal with the critical issues in their community.

2  *Participation in public decision making should be free and open.* While not everyone can attend or participate equally in every CD activity, citizens' opinions, ideas, and support can be sought throughout the process. Participation may come in many forms including the completion of surveys, volunteering for a particular project, attending a public meeting, or serving on a committee or task force.

3  *Broad representation and increased breadth of perspective and understanding are conditions that are conducive to effective CD.* The CD professional strives to organize a leadership group that is representative of the community and its stakeholders. Individuals representing all major groups within the community should be offered membership in this group, and all their constituents should be encouraged to participate to some extent in the many development activities created in support of their shared vision.

4  *Methods that produce accurate information should be used to assess the community, to identify critical issues, strengths, weaknesses, opportunities, and threats (SWOT analysis).* This assessment, or environmental scan, can result in the creation of a community profile that details all information collected about the community. It may involve a literature review, interviews with key informants, focus group meetings, citizen and business opinion surveys, and a review of multiple statistical and demographic data resources. Assessment results support the creation of a strategic development plan and, eventually, marketing material.

5  *Understanding and general agreement (consensus) is the basis for community change.* Citizens need to make informed decisions. Community dialogue and assessment feeds that educational process. When making decisions, consensus should be sought. There is an important distinction between consensus and compromise. Compromise may not always result in the best alternative being selected. Usually one group gives up something important to get one thing it wants, and another group does the same. This usually results in neither side getting what they really want, nor the best alternative for everyone. In compromise both sides have win-win and lose-lose outcomes.

Consensus seeking, on the other hand, supports creation of an alternative that everyone can support, one that was developed through an open dialogue on options and is focused on producing the greatest good and positive outcomes. The process involves having all opinions heard, options discussed (including positives and negatives of each option), and developing a course of action that best addresses the identified critical issues. It is problem/solution-focused versus group- or individual-focused. Consensus seeks win-win results that produce positive outcomes for all citizens.

6  *All individuals have the right to be heard in open discussion whether in agreement or disagreement with community norms.* Participants in community organizations owe it to themselves to hear and consider all ideas and opinions. When everyone's views and ideas are heard, it provides more information for citizens to consider when making decisions. Often ideas from different individuals can be synergistically combined to produce even better solutions. However, individual rights should be exercised with respect and not be abused to the point of disrupting the activities and decisions of the majority of citizens. Individuals do not have a right to filibuster.

7  *All citizens may participate in creating and re-creating their community.* The wonderful thing about community development is that there is a role for everyone. Young and old, rich and poor, highly educated and modestly to uneducated individuals, and citizens of all races and cultures can create or join a project in support of the shared community vision. Those with gardening as a hobby might get involved in community beautification. Others may have information technology skills that can be utilized in developing a website and creating community marketing material. Still others may have leadership and organizational skills that can be focused on organizing and leading groups. CD professionals and citizens

should always be looking for new individuals to involve in the process and align their experience, knowledge, skills, and interests to the work that needs to be done.

8 *With the right of participation comes the responsibility to respect others and their views.* Every citizen should be treated with respect and kindness. The CD professional must work diligently to facilitate the process so that individuals feel positive about their own participation while respecting the rights of others. It is possible to be both fair and firm. Personal attacks or personalizing issues and ideas is counterproductive. It forces individuals into defensive posturing and conflict that disrupts progress.

9 *Disagreement needs to be focused on issues and solutions, not on personalities or personal or political power.* Disagreement is not a bad thing. It is a normal result of human interaction. As thinking individuals, humans naturally have opinions and develop alternatives based on their own life experiences, education, values, culture, and beliefs. CD professionals focus the group on what can be done instead of what cannot be done, and on areas of agreement rather than on areas of disagreement. This focus avoids getting bogged down trying to resolve irreconcilable differences. The focus of CD activities needs to be on problem identification and solutions rather than on who did what to whom when, or whose opinion or idea is the most important. The importance of ideas is not in their origins but in their utility for addressing community concerns.

10 *Trust is essential for effective working relationships and must be developed within the community before it can reach its full potential.* An old cliché states that trust must be earned before it can be given. There is generally a lack of trust among the very community groups that must work together in order for the community to be successful. This is particularly true in communities where there are many problems. Trust will only be developed if the preceding principles are practiced and supported within the community. Development of trust will take some time as the group forms and moves through a normal developmental cycle.

These principles are very practical and, when used, will protect a working group from selective participation that can create mistrust, rumors, misinformation, destruction of worthwhile efforts, and loss of key participants. Increasing the breadth of representation helps ensure that many different points of view are heard. This leads to the ability to implement decisions. General agreement also leads to a commitment to implement changes and a positive working relationship that supports long-term development initiatives. By hearing everyone, even those who disagree, it is possible to assemble a broad base of information. Sometimes those who disagree with the majority or general opinion may be right; their "weird" ideas or opinions may foster creative thinking in the group and innovative approaches to solving long-term or difficult problems.

## The community development process

The central theme of CD is that people, in an open and free environment, can think and work together to fashion their own future. However, when communities face serious problems, there is a tendency for citizens to feel frustrated and helpless. Anger develops and is often focused on each other. Certainly someone must have caused these problems. Differences among residents are exaggerated and passions run high. The community tends to splinter (e.g., rich vs. poor, white vs. black, long-term residents vs. newer residents, residents vs. non-resident property owners and business operators, and voters vs. elected officials). The CD professional often begins his or her work in a very contentious environment.

One of the most difficult roles that a professional developer has is facilitating communication within the working group so that all views are heard and discussions do not degenerate into non-productive complaint sessions and personal arguments. There is a fine line between facilitating meetings and manipulating meetings. This is a tightrope that all CD professionals must learn to walk. This is a particularly difficult task for those professionals who live and work full-time in the communities where they are consultants. The professional community developer has to continually reflect on his or her actions to determine whether they are facilitating or directing, providing information or solutions.

Information and professional experiences should be shared carefully so as to provide examples, options, and possibilities for consideration versus telling citizens what they must or should do. If actions and ideas come primarily from the professional developer, they are not likely be supported or implemented.

It is important to realize that people support what they help develop. The opposite is also true. Many perfectly written plans created under the direction of knowledgeable professionals with good intentions, and at significant cost, lie unused in city hall bookshelves throughout the world. The consultants did part of their job. They were asked to create a plan and they did; but without citizen participation and ownership, the plan is useless. Participation promotes citizen ownership of and commitment to the actions that have been planned.

CD is a process through which people learn how they can help themselves. Self-help is the cornerstone of CD. Through self-help, people and communities become increasingly interdependent and independent rather than dependent on outsiders to make and implement decisions.

The CD process is a set of steps that guide the identification of a program of work and movement toward the ultimate CD goal. These steps require the involvement of community members and serve as a guide to problem solving, planning, and completion. The steps do not necessarily follow a sequential path. They may not follow the exact sequence below, and some may occur concurrently. The steps are as follows:

- *Establish an organizing group* – This might be a strategic planning committee or development task force. It could be a new, independent organization that has broad representation from many different community organizations and includes a broad cross-section of community leaders. Or it might be an inclusive group sponsored by a successful community organization such as a Chamber of Commerce or economic development group. The most successful development organizations are public–private partnerships that involve a blend of prominent citizens, religious and neighborhood leaders, major community stakeholders (such as major property owners and managers of major businesses owned by external investors), elected officials, and local business leaders.

- *Create a mission statement* – This may be a mission statement for a strategic planning committee detailing why it was formed and what they aim to accomplish. This statement is important because it lessens the threat to and helps prevent role conflict with other community organizations by communicating its unique mission and role. Further, it keeps the group focused and helps prevent "mission creep," namely a loss of focus and drifting away from the primary purpose for which a group was formed. The following is a mission statement for a development organization in Phillips County, Arkansas:

  > The Phillips County Strategic Planning Steering Committee will generate planned action that rebuilds the community on a shared vision of the future using a focused alliance of community groups, leaders, resource partners, and stakeholders.

- *Identify community stakeholders* – Who are the stakeholders that should be involved in the process? What roles should they play? When? For practical reasons, inviting all citizens to all meetings is not only an inefficient use of human resources, but prevents detailed analysis and discussion of critical issues and the development of strategies. Initially, a representative group of citizens has to be created. As its work progresses, more and more citizens are involved by serving on subcommittees, task forces, or project teams through which they provide information, opinions, ideas, and questions; challenges to the status quo, approval of the final plan, and help in implementing it.

- *Collect and analyze information* – Before beginning work, it is important to identify the current community environment. There are many methods of conducting this environmental scan, and many types of information that may be gathered in support of community development. It is often useful to assemble a community profile. This is a statistical overview of the current and past demographics of the community – income, population growth/decline, age of the population, community boundaries, population density, major employers, employment by sector, and so on. Other useful tools include surveys of various types. A business opinion survey is useful in identifying economic issues and conditions that are impacting the community. This survey is also useful in retention and expansion program efforts. It surveys a sampling of local businesses and seeks opinions about the quality of local government, infrastructure, workforce, and other issues impacting business growth. Citizen attitude surveys are also useful in identifying a variety of issues that impact both economic and quality of life factors. Other assessments include comprehensive studies, surveys, and leadership workshops that examine all aspects of the community in order to identify critical issues and the strengths, weaknesses, opportunities, and threats (SWOT analysis) impacting development.
- *Develop an effective communications process* – It is extremely important to keep an open line of communication with the public. This will ensure that the process is inclusive and that activities are transparent. Many development activities are conducted by citizen representatives without a large public attendance. It will be important for them to keep their constituents informed, but this should not be the sole source of communications. If possible, involve the local press in the CD process. In addi-

tion to newspapers, it may be possible to use local television stations or public access cable channels as well as Internet websites. In addition, strive to establish two-way communications by providing a phone number and an email address which the public can use to send ideas, comments, and questions about what is being done.

- *Expand the community organization* – Once an organizing group is established, additional organizations and citizens can be involved in addressing specific problems that are of direct interest to them. For example, an economic development subcommittee might be created, and its membership expanded to include additional business leaders; representatives from development boards, airport, and port commissions; vocational educational institutions; and economic specialists from the state or other governmental agencies. Similarly, a tourism development subcommittee might be created where additional hotel, restaurant, gift shop, and tourist attraction operators would be invited to participate in discussions to determine how tourism might be expanded and supported within a community. This expansion continues the process of conducting a community dialogue that began with the environmental scan.
- *Create a vision statement* – As soon as the development group have identified the critical issues and conducted a SWOT analysis, they can create a strategic vision. This forward-looking vision statement provides guidance and direction for the actions that will be taken to make improvements. It is usually one sentence that embodies the desired state of the community in the future. It must be a vision that can be realistically achieved within 15 to 20 years but should have enough "stretch" to challenge the community to achieve dramatic positive change. Below is a strategic vision for Kaniv, Ukraine, that was developed in 2005.

By 2020, Kaniv will be: The spiritual capital of Ukrainians, where tourist attractions, communal infrastructure, education, culture, and scientific, ecologically-clean, high-tech industries are brought together for the well-being of people.

- *Create a comprehensive strategic plan* – After the development group have created the vision they would like to achieve, they begin to create a strategic plan to support the achievement of that vision. This can be done by subcommittees or task forces, each working on one of the critical issues. Communication between the subcommittee chairs is important to avoid duplication of effort and to share ideas that may be useful for other groups. The typical structure of a strategic plan is a list several goals, each supported by numerous objectives.

  An effective plan will be realistic and credible. Objectives are written down so that they have specific completion dates and clear measurable outcomes. They are also supported by specific tasks and milestones that lead to achievement of the objective. Each objective should include the names of individuals responsible for its completion as well as the funding and resources to be used. If an objective does not have an individual responsible for managing it or funding or dedicated resources, it should not be included in the plan.

- *Identify the leadership and establish a plan management team* – From the very beginning of the process, the CD professional needs to identify those who will become leaders and champions of the process when he or she leaves. These individuals may come from the original organizing group; be the chairs of some of the committees, subcommittees, and task forces; or come from individuals included in the process as it expands to involve more citizens. Generally a group of seven to nine individuals should be identified and selected by the entire planning committee to become a plan management team.

  Since it is unrealistic to have a large group of individuals meet frequently, the plan management team is charged with acting on the planning committee's behalf. The team meets periodically to manage the ongoing process, and its membership should reflect the public–private nature of the group by including representatives from each of the critical issue/goal areas. The team's role is to manage the process, assisting those charged with implementing goals and objectives to move the process forward.

  The plan management team might initially meet twice a month to get an update on the overall progress of the plan. Later they might meet once every two months. They are also charged with making periodic written and verbal reports to the entire group, which might meet quarterly or semi-annually. Throughout the process, the work is continually being implemented by those responsible for goals and objectives.

- *Implement the plan* – Implementing the plan is when the "rubber meets the road." It is a crucial time that the plan management team needs to monitor very closely. Objective team leaders can be quickly discouraged if they do not see results. For that reason, it is important to build early successes into the planning process. When facilitating the process, the CD professional, plan management team, and committee chairs should include some objectives with the following characteristics:

  - short time frame for implementation
  - highly visible
  - money and resources are available
  - popular with the vast majority of residents
  - low risk of failure.

  Ensuring early success is important because it builds momentum, helps attract additional volunteers, and instills the belief that things are changing for the better.

- *Review and evaluate the planning outcomes* – One key aspect of total quality management is the Plan, Do, Check, and Adjust cycle. It is also important to realize that planning is a dynamic process and that the plan is a living document. Some objectives will be completed ahead of schedule. Others will be delayed, and their time schedule must be revised. Some will have to be eliminated because of changing environmental conditions or the loss or lack of anticipated funding. Most interestingly, however, is the fact that new objectives will be created and added to the plan as environmental conditions change and new challenges and opportunities emerge.

The plan management team should perform periodic evaluations of the plan, including a review of each objective. What is going well? What problems need to be addressed? What needs to be changed? Their primary mission is to keep the planning activities and the community's progress moving forward so that the shared vision can be realized.

- *Celebrate the successes* – Winning events need to be built into the planning process. In addition to an annual report and celebration of the community's accomplishments, it is important to have smaller ongoing celebrations that provide reward, recognition, and continued motivation for volunteers and citizens. A comprehensive reinforcement process could be developed and implemented and could include having publicity in the newspaper; drinks and snacks, gifts and discount certificates, and/or awards for volunteers; and special T-shirts or hats for those who do the work. Since motivation is unique to each individual, leaders are charged with structuring celebrations into their activities that continue to recognize, reward, and reinforce volunteer efforts. Simply put, participation in the planning process should be fun, and citizens should want to participate.

- *Create new goals and objectives as needed* – A comprehensive plan usually has several goals, each supported by a number of objectives. It contains a multitude of projects that will be completed at different times. Some objectives actually lay the foundation for what must be done next for the community to realize its shared vision. As stated above, the community and the plan operate in a dynamic environment. In order to remain relevant, the plan must be continually updated, and new goals and objectives added as others are achieved or completed.

## How does community development practice relate to economic development?

Traditional economic development (ED) focused on the attraction and location of businesses (industrial development) in a community. The focus was strictly on job creation and business investment. In recent years, economic development has been expanded to include the retention and expansion of existing businesses as well as the incubation of new businesses.

Attraction of new business development is a very competitive process. Each year thousands of communities in hundreds of countries compete on a continual basis for new business facilities. Location decisions are usually driven by economics, competition, and cost decisions. During the latter part of the twentieth century, much of the rural southern United States lost unskilled jobs to foreign workers. The same quality of labor for cut-and-sew operations, for example, could be found in Honduras, Vietnam, Sri Lanka, and China for a much lower cost.

An economic developer might promote a community to a business investor based on a very specific set of objective, but narrow criteria. However, at some point the investor will visit "short-listed" communities to select a location. It is at that point that the realities of the location are driven home.

Not every community can be successful in locating major industries. Often conditions other than available sites and buildings, labor costs, and location incentives impact location decisions. For the decision maker, these other factors may be very subjective and could be related to personal, cultural, sociological, and quality of life issues. Some of these factors are directly under the influence or control of local leaders and citizens.

A shortcoming of traditional ED activities is that they are focused on the "now" rather than on the future. What business can I locate here based on the current resources (e.g., labor, raw materials, location, infrastructure capacity, and available buildings or

## BOX 5.2 HOW COMMUNITY DEVELOPMENT CREATED ECONOMIC DEVELOPMENT IN SLOVAKIA

Representatives of a major Japanese electronics manufacturer were visiting a location in Trnava, western Slovakia. They were in Trnava during the Christmas season (St. Nicolas Day). At night, the historic town center was brightly decorated and lighted. Fresh snow had just fallen, and children and parents were on the street celebrating. Saint Nicholas was making his rounds visiting homes and delivering gifts to children. It was a totally unplanned coincidence, but one that demonstrated many aspects about the city's quality of life. After witnessing this event, the selection team told the economic developer, "We will visit the other cities, but we have already made our decision." It was obvious that the site selection team was greatly impressed with the spirit of the town and its people, which was very visible during the holiday event.

sites)? What is the best business development match for a specific community based upon its current strengths? Traditional ED activities were not focused on addressing problems such as social issues that, if solved, could result in making a community an attractive location for different industries. Initially, ED was asset marketing focused on "picking the low-hanging fruit."

On the other hand, community development tends to be a long-term process. It is holistic in that it sees all aspects of the community as related and as affecting development. Citizens and business investors want more than just jobs or a good investment location. They know that their managers and technical workers will move there to work and live. Therefore other desirable traits, "soft factors," drive the final selection process. These include:

- Effective community leadership.
- High-quality education for pre-school, primary, secondary, higher education, and workforce training and retraining.
- Economic development that creates a variety of quality job opportunities through the attraction of new businesses, the retention and expansion of existing businesses, and support of entrepreneurship within the community.
- Attractive community known for its "curb appeal."

- High-quality, affordable housing and a variety of housing for citizens at all income levels.
- Quality and affordable health care and emergency services.
- Recreational and cultural activities (golf courses, resorts, parks, festivals, theaters).
- Safe, crime-free community.
- Honest and effective government that delivers efficient services.
- Good infrastructure including roads, drainage, water and sewage services, solid waste disposal, and so on.
- Reliable and competitively priced utilities (electric, gas, and telecommunications).

## What do community developers do?

Positive change in communities is driven by many factors. Therefore, positive change will be difficult to bring about by working on only one or two projects. Community development has to take a holistic approach to organizing, planning, and implementing change. The CD process identifies and organizes local leadership, involves the public, identifies critical issues, creates a plan, and implements actions to solve problems across a broad spectrum of areas so that desired change occurs.

As part of their practice and by the very nature of the community environment, professional community developers perform "grass-roots" economic development. A community may have a very desirable building or green field site available. Initiating a marketing campaign to attract businesses to these sites may fail, however, if some underlying development issues are not addressed before or concurrently with the marketing initiative. The local workforces' skill level may need to be addressed. Social issues, such as a high percentage of the working-age population abusing illegal substances and alcohol, may prevent successful business attraction. The community's "curb appeal" may also detract from its competitiveness. The local infrastructure may be at or near full capacity or be in need of major renovations.

CD professionals draw on many knowledge and skill bases in order to be successful in their work. Some professionals are generalists, while others specialize in one or more areas of the CD practice. In their practice, professionals may be found to perform one or more of the following functions (while not an exhaustive list, the following provides an overview of the broad range of knowledge, skills, and abilities exercised by CD professionals):

- *Community assessment*
  - Perform statistical and demographic research.
  - Organize citizens to conduct assessments (environmental scans).
  - Design and conduct surveys.
  - Prepare and present assessment reports.

- *Strategic planning*
  - Organize a planning committee.
  - Facilitate the planning process.
  - Assemble and edit the plan.
  - Assist in plan implementation and management.

- *Organizational development*
  - Identify stakeholders.
  - Recruit volunteers.

  - Organize community groups.
  - Recruit and work with volunteers.

- *Leadership development*
  - Recruit leaders.
  - Conduct learning needs assessment; create and evaluate training and development activities.
  - Deliver training and development activities.

- *Economic development*
  - Create and manage business attraction programs.
  - Support expansion and retention of existing businesses.
  - Create and deliver entrepreneur training and business incubation activities.
  - Foster technology transfer to help businesses regain/retain their competitiveness.

- *Public and private development financing*
  - Identify governmental grant opportunities and write applications.
  - Identify foundation grant opportunities and write applications.
  - Seek venture capital sources and write applications.
  - Assist businesses in writing bank and financial institution loan applications.

- *Community land-use planning and research*
  - Plan industrial parks, commercial and residential developments.
  - Plan and design utility infrastructure.
  - Plan and design roads.
  - Plan and design seaports and airports.
  - Plan and design parks and recreational centers.

## Professional standards of ethical practice

Every profession has a set of ethical standards that it expects its members to follow. Certified Professional Community and Economic Developers (PCED) subscribe to the following ethical standards adopted by the Community Development Council (CDC – www.cdcouncil.com). These are based largely on the Standards of Ethical Practices of the International Community Development Society from which the CDC professional certification process originated. Failure to adhere to these principles and ethical practices may result in the Community Development Council rescinding a professional's certification.

## Professional values

The following values guide the professional practice of community development:
- Honesty
- Loyalty
- Fairness
- Courage
- Caring
- Respect
- Tolerance
- Duty
- Lifelong learning

## Professional principles

- The purpose of community development is to raise living standards and improve the quality of life for all citizens.
- Community development seeks to build initiatives around shared values and critical issues after identifying existing strengths, weaknesses, opportunities, and threats.
- Positive change begins with creating a shared vision that can be transformed into reality by the actions of citizens using goals, objectives, and action plans.

- Community development is inclusive and involves developing leaders and building teams across class, gender, racial, cultural, and religious lines.
- Community development is more than social service programs and "bricks and mortar" construction. It is a comprehensive initiative to improve all aspects of a community's interdependence: human infrastructure, social infrastructure, economic infrastructure, and physical infrastructure.
- Community development involves consensus building that looks for the best solutions to community problems rather than only those that are politically expedient or popular or that benefit a few citizens.
- Community development is directed toward increasing a community's leadership capacity for solving problems and moving citizens from dependence to interdependence.
- Leadership in community initiatives is shared so that responsibility and commitment are encouraged across a broad base of the population.
- Development leaders work to transform their communities for the better and inspire others to do the same.
- Community development is an educational process that helps citizens understand the economic, social, political, environmental, and psychological aspects of various solutions.
- Community development is focused on action that improves communities by transforming learning into performance.
- Community development includes economic development initiatives that help bring high-quality employment, career, and business opportunities to citizens.
- Successful initiatives at the community level lay the groundwork for regional alliances and cooperation directed toward solving common problems.

## Ethical standards

Ethical standards adopted by the Community Development Council are as follows:

- Establish and maintain a professional and objective relationship with the client community and its representatives, one that advances the ethical standards of practice.
- Always perform in a legal and ethical manner.
- Immediately disengage from activities when it becomes apparent that they may be illegal or unethical, reporting illegal activities to the appropriate authorities as required.
- Adhere to the professional principles outlined above.
- Clearly and accurately detail personal knowledge, experience, capabilities, and the outcomes of past consulting when requested.
- Clearly and accurately detail the scope of work to be performed (and its anticipated outcomes), and the fee for and terms of that work prior to engaging in consulting.
- Avoid conflicts of interest and dual relationships, especially those that could result in or appear to result in personal benefit (outside the scope of work) at the expense of a client community or its representatives.
- Disengage from activities that may result in one group or individual unethically or illegally benefiting at the expense of another.
- Adhere to all professional principles and practices regarding selecting, administering, interpreting, and reporting community assessment measures.
- Keep confidences, and only reveal confidential information at appropriate times and with proper authority.
- Maintain confidential records in a secure location and under controlled access.
- Discuss ethical dilemmas that arise with other Certified Professional Community and Economic Developers or the Community Development Council to solicit guidance and opinions regarding possible actions.
- When feasible, consult confidentially with professional colleagues whose behavior is in question or when the colleague requests assistance in resolving ethical or legal dilemmas at a personal level.
- Notify the Community Development Council when ethical and legal dilemmas are unable to be resolved at a personal level.

## Getting started

1 *Ground rules* – It is often helpful to make copies of the values, principles, and process of CD and to distribute them to participants in the process. They may be used as a checklist or reference for guiding the process. You may also help the group create a set of ground rules that will guide their group activities. This might involve such topics as confidentiality, conflict of interest, and resolving disagreements. It may also be useful to guide the group through the creation of a credo that lists their beliefs and values. These actions will help guide the CD process and serve as a tool for the facilitator to refocus the group when problems occur.
2 *Ice breaker* – Ask people to make a list of all the topics they know well enough to teach someone else. Either post the results on a flip chart page, whiteboard, or project the results on a wall or screen using a digital projector and a laptop so that all participants can see the results. This exercise can help citizens come to realize the wide range of resources within the group. You may want to record this information for future reference as a guide for involving individuals in later activities.

## Conclusion

This chapter has provided a broad overview of the practice of community development. It has discussed community development values and beliefs, practice principles, the community development process, community development tasks, and professional standards of ethical practice.

Professional community developers include consultants and those whose jobs include community and economic development activities. Volunteers who devote a significant portion of their time to community development and community activities might also be classified as professional community developers. People who practice community development should be aware of its underlying beliefs, values, and ethical standards. The more familiar all local citizens – not just professionals – are with the principles of community development practice, the more success a community is likely to enjoy.

---

## CASE STUDY: MAYVILLE AND LASSITER COUNTY

Mayville is a town with a population of 50,000 in the south central portion of the United States. It is the county seat of Lassiter County. In its early years, Mayville grew as an agricultural market, a transportation hub, and a central location for regional services and retail. German immigrants came into Mayville in the early twentieth century and established many of the retail businesses. To this day, the town still has several excellent restaurants founded by the German immigrants. On one side of the menu is traditional German fare and on the other side is traditional southern fare.

In the latter part of the twentieth century, Mayville attracted several manufacturing operations paying good wages and the town prospered. Most of these industries located within the city limits because of good municipal services. Despite the fact that people commuted from all over the county (and surrounding counties) to Mayville and these good jobs, county elected officials often seemed envious of the jobs and industries in Mayville. Some of them even began to believe that the city, which had a strong economic development program and staff, was actively steering prospects and jobs away from the county and into the city.

Despite this mistrust, the economies of Mayville and Lassiter County continued to prosper because of the good local industries. Amidst this prosperity, many county elected officials remained suspicious that the city did not have the county's best interests at heart. Some county officials and residents even came to believe that city residents looked upon them as uneducated "rubes" who did not deserve high-wage jobs.

One day both the city and the county received notice that the local U.S. military base was on the Base Realignment and Closing (BRAC) list to be considered for downsizing or closing. For the first time, the city and county faced a common imminent danger and were required to work together. They formed a Local Re-use Authority (LRA) to begin contingency planning in case the base did close, although the initial job of the LRA was to make the case to the Department of Defense that the local base should not be closed.

The LRA hired an outside consultant to help develop the contingency plan, which included a strategy for economic development in the city and county to replace the jobs that might be lost through base closure. Following the principles of community and economic development practice, the consultant conducted an assessment of the community identifying the strengths and weaknesses and performing a SWOT analysis. The consultant conducted interviews with business leaders, elected officials, and other community stakeholders; and public meetings were held to solicit feedback and ideas.

City officials forewarned the consultant about the negative attitude and lack of cooperation on the part of the county, saying they were at a loss to explain this behavior. Despite this warning, the consultant was shocked at the first meeting with city and county officials together in the same room. Not only would certain county officials not cooperate and help develop the strategic plan, they were outright hostile to city officials, essentially claiming that the city did not respect them. It was obvious that there was absolutely no mutual trust, and, as a result, county officials always assumed that the city had an ulterior motive for anything they suggested and the county's response was always "No!"

The consultant realized that to accomplish anything, his team had to restore some semblance of mutual trust and let the county officials know how much their lack of cooperation was jeopardizing the continued strength of the local economy. With the city and county openly disagreeing and even feuding, they would not be able to develop an effective case for keeping the military base open or an effective strategic plan for economic development in the future. In addition, as the consultant pointed out, it was unlikely that any businesses would choose to move to the area in the future with such open inter-governmental hostility.

Through one-on-one confidential interviews with city and county officials, the consultant attempted to dispel the rumors and misguided notions that the city was trying to undermine the county. In the written report, the consultant pointed out (as diplomatically as possible and without including names) the harm resulting from the city/county distrust and hostility, including the fact that it would jeopardize the military base outcome even more and interfere significantly with economic development in the future.

Following the confidential interviews and report, many of the county officials began to realize the consequences of their actions and pledged to work together with the city for the good of all local residents. They realized that they had been putting personal issues and disputes before the good of the city, the county, and the residents who had elected them. However, one county commissioner continued to point fingers at city officials and refused to change his hostile ways. In the next election, he was the only county commissioner not re-elected. The county residents and the other county commissioners had certainly gotten the community developer's message, even if the lone commissioner had not.

A year after the consultant's work, it was announced that the local military base would not be downsized or closed after all. The city and county had worked together and combined their resources to make the case that the military base was vital to national security and demonstrated that it was one of the most productive and efficient in the country. Not only did they not lose the military base, but the city and county gained a joint strategic plan to enhance future economic development and a strong working relationship with which to implement it.

The Editors

## Keywords

Community, community development, CD values and beliefs, economic development, ethical standards.

## Review questions

1  Define the practice of community development.
2  Describe the professional community developer's role.
3  What are the community development practice values and beliefs?
4  What are the community development practice principles?
5  List and describe the steps in community development practice. Do you agree with the order they are listed in the chapter?
6  List some of the tasks professional community developers perform.
7  What are some of the ethical standards to which professional community developers should adhere?

## Note

1 Circa 1989, a working definition developed by Bill Miller, retired Director of the Community Development Institute Central in Arkansas; George McFarland, retired Community Development Manager, State of Mississippi; and the author.

## Bibliography and additional resources

Block, P. (1981) *Flawless Consulting: A Guide to Getting Your Expertise Used*, San Francisco, CA: Jossey-Bass/Pfieffer.

Dyer, W. (1994) *Team Building: Current Issues and New Alternatives*, 3rd edn, Reading, MA: Addison-Wesley.

Schein, E. (1987) *Process Consultation: Lessons for Managers and Consultants*, Vol. 2 (paperback), Reading, MA: Addison-Wesley.

# PART II

# Preparation and planning

# 6 Community visioning and strategic planning

## Derek Okubo

## Introduction

There are many examples of communities that have faced highly complex issues and reached their goals through sheer determination and a collaborative spirit. These communities succeeded in large part because they were willing to convene different players and undergo an extensive planning process. Not allowing the plan to sit on the shelf, these communities continued on and persevered throughout the plan's implementation. All sectors – government, business, nonprofit, and the citizens themselves – participated in the development of a common agenda. In addition, the community at large received ample opportunity to provide input. Because all sectors of these communities were involved in the creation and ongoing development of programs for the future, such programs received widespread support and encountered minimal resistance.

Some communities allow the future to happen to them. Thriving communities recognize that the future is something they can create. These communities take the time to produce a shared vision of the future they desire and employ a process that helps them achieve their goals. Achieving the desired future is hard work. Yet successful communities understand that the things they dream about will only come true through great effort, determination, and teamwork.

One way of achieving these community goals is through a community-visioning project. Such a process brings together all sectors of a community to identify problems, evaluate changing conditions, and build collective approaches to improve the quality of life in the community. The participants must define the definition of a community. Some projects define their community as a neighborhood; others a whole city or town; many projects have focused on regions that include multiple cities, towns, and counties.

## Process principles

Collaboration and consensus are essential – successful community efforts focus on ways in which business, government, nonprofits, and citizens work together. In reviewing successful collaborative efforts around the country, it has been found that all possess the following ingredients:

- People with varied interests and perspectives participated throughout the entire process and contributed to the final outcomes, lending credibility to the results.
- Traditional "power brokers" viewed other participants as peers.
- Individual agendas and baggage were set aside so the focus remained on common issues and goals.
- Strong leadership came from all sectors and interests.
- All participants took personal responsibility for the process and its outcomes.
- The group produced detailed recommendations that specified responsible parties, timelines, and costs.
- Individuals broke down racial, economic, and sector barriers and developed effective working relationships based on trust, understanding, and respect.

- Participants expected difficulty at certain points and realized it was a natural part of the process. When these frustrating times arose, they stepped up their commitment and worked harder to overcome those barriers.
- Projects were well timed – they were launched when other options to achieve the objective did not exist or were not working.
- Participants took the time to learn from past efforts (both successful and unsuccessful) and applied that learning to subsequent efforts.
- The group used consensus to reach desired outcomes.

These ingredients make up the essence of collaboration itself. True collaboration brings together many organizations, agencies, and individuals to define problems, create options, develop strategies, and implement solutions. Because they typically involve larger groups, collaborative efforts help organizations rethink how they work, how they relate to the rest of the community, and what role they can play in implementing a common strategy. Many times it becomes clear that no single organization has the resources or mandate to effectively address a particular issue alone. A group effort can help mobilize the necessary resources and community will.

Effective collaboration requires that decisions be made by consensus. In *A World Waiting to Happen*, M. Scott Peck (1993) describes consensus as:

> A group decision (which some members may not feel is the best decision, but which they can all live with, support, and commit themselves not to undermine), arrived at without voting, through a process whereby the issues are fully aired, all members feel they have been adequately heard, in which everyone has equal power and responsibility, and different degrees of influence by virtue of individual stubbornness or charisma are avoided so that all are satisfied with the process.

Although a consensus-based decision-making process takes more time on the front end, it can save time during the back end of the implementation phase of a visioning project where blocking ordinarily occurs. If citizens are provided with a forum in which their ideas and opinions are heard, seriously considered, and incorporated into the action plan, they will be less inclined to resist or ignore new initiatives. Community "ownership" of a plan and willingness to assist in its implementation often corresponds directly with the public's level of participation in the plan's development. As a result, projects can be completed in timely fashion through the consensus-building process. In collaborative processes, the sharing of information and pooling of resources build understanding and lead to better decisions. Special interests are not as inclined to block implementation of the plan, since it reflects their own interests and efforts. While collaborative problem solving is not appropriate for every issue and situation, it is an absolute necessity for a community-visioning project. Collaborative problem solving should be used when:

- The issues are complex or can be negotiated.
- The resources to address the issues are limited.
- There are a number of interests involved.
- Individual and community actions are required to address the issue effectively.
- People are interested and willing to participate because of the importance of the issue.
- No single entity has jurisdiction over the problem or implementation of the solutions.

Community visioning is both a process and an outcome. Its success is most clearly visible in an improved quality of life, but it can also give individual citizens and the community as a whole a new approach to meeting challenges and solving problems. Citizens of all types who care about the future of their communities conduct community-visioning projects. These people are collectively called "stakeholders." The stakeholders in successful visioning efforts represent the community's diversity – politically, racially, geographically, ethnically, and economically – lending different "stakes," or personal and group interests, to the process. They form the core planning group for the visioning effort, perform community self-evaluation, set goals, and develop the action plan and implementation strategy. To ensure the success of the stakeholders' work, effective process design and structure are essential.

## Phase one: initiation

## Providing the groundwork for the process

The process of building a solid foundation for an effective community-visioning project includes a number of key tasks. The first is the selection of an initiating committee – a small group of 12 to 15 individuals who represent a slice of the community. Their job includes:

- Selecting a stakeholder group that reflects the community's interests and perspectives.
- Designing a process that will reach the desired outcomes of the community effort.
- Forming subcommittees that will play key roles throughout the project.
- Addressing key logistical issues such as staffing, siting, scheduling, fundraising, and the project name.

The initiating committee focuses on process, allowing the broader stakeholder group to work on content (identifying problem areas, formulating action plans). Preparation and completion of logistical tasks can send the visioning effort on its way toward success.

These individuals must be willing to invest a substantial amount of time over roughly three months in the development phase of the project. They may or may not wish to continue on as members of the stakeholder group for the planning effort itself. The initiating committee needs to reflect the community's diversity in terms of race, gender, economic sector, and place of residence and employment. Each member of the initiating committee should wear "multiple hats" or represent multiple interests. The initiating committee will make the first statements about the visioning initiative to the community, so it must be credible and well balanced. The two crucial attributes of the initiating committee are diversity and credibility. A good question to ask while forming the group is: "Will any community member be able to look at the initiating committee membership and say, 'Yes, my perspective was there from the beginning'?" If this isn't the case, then the missing perspective must be identified and a credible individual recruited to participate. The purpose of the initiating committee is to focus on the process and logistics required to move the project forward. The content of the community vision will be developed during the broader stakeholder-planning phase. The diverse voices on the initiating committee must create and agree to methods by which stakeholders can equitably address complex and controversial issues.

In order to create a safe environment for discussion of difficult issues, the initiating committee must complete a number of tasks. These tasks are made up of the following fifteen actions:

### 1 Identifying who must be at the table

Using a "stakeholder analysis," the initiating committee must identify a group of 100 to 150 individuals to serve as the core planning group. The stakeholder group must be as diverse as possible and represent every major interest and perspective in the community. Even more than the initiating committee, the stakeholder group must represent the community's demographic diversity in terms of age, race, gender, preferences, and places of residence and employment. In selecting stakeholders for the community visioning process, the initiating committee must consider the diverse sectors and various interests and perspectives of the community. The committee must avoid "rounding up the usual suspects" or forming a "blue ribbon panel" of the same community leaders and organizations that have always been involved in past community efforts.

These active people are valuable contributors, but this type of project must tap into populations and people that are traditionally excluded from community processes. A balance of the "old guard" and "new blood" is useful. Further, it is important that participants act as citizens with a stake in the quality of life in the *whole* community, not simply as representatives of a particular organization, part of town, or issue. In this process, stakeholders should be effective spokespersons for their interests and perspectives, but they should not simply serve as advocates for their agencies and organizations.

One of the most critical groups of stakeholders will be those who have a stake in the future of the community but have little political or financial power. It will also be important to include both "yes" people and "no" people in the stakeholder group. It is easy to pick positive people who have the power to get things done. It is harder, but no less important, to pick people who have the power to stop or delay a project. As with the initiating committee, it is useful to look for people who wear multiple hats or fall into a number of categories: for example, a single parent with kids who is a banker and lives in a northwest-quadrant neighborhood, or a small business owner who is on the planning commission and serves as a soccer coach to his child's team.

A sample of categories for identifying the stakeholders in the community may include:

- Pro-growth/no growth
- Business (small, corporate, industrial)
- Old/new resident
- Conservative/liberal/moderate
- Geographic location
- Age
- Ethnicity/race
- Service provider (seniors, different abilities, youth)
- Income level
- Education reform/back to basics
- Elected/appointed leadership
- Single-parent/dual-parent house
- Institution type (schools, police)
- Inside/outside city boundaries.

## 2 Designing the process

It is important to note two fundamental premises about the community-visioning process. First, key leaders and the community as a whole must empower the stakeholder group to make decisions. Citizens are too knowledgable to accept the role of only advising officials and community leaders, who may or may not choose to accept their advice. Although elected officials clearly have legal authority over issues such as taxes and the provision of services and corporate leaders are free to determine their own business development strategies, they must participate in this process as peers and agree to honor, while not necessarily rubber-stamping, the stakeholder group's conclusions. If the process works correctly, honoring the conclusions should not be a problem, since the "power" people were a part of building the same conclusions.

Second, the orientation of the entire process, from the very beginning, has to be proactive. Too many community task forces have been convened over the years with marginal results. The goal of this effort is not to conduct interesting discussions or forge new relationships, though these will certainly result. The goal must be to develop a broad, able to be implemented, community-owned action plan that will truly serve the whole community and then to put that plan into action. The process must be customized to fit the community's needs and desired outcomes. It must take into consideration local realities (budget, time constraints) and complement other useful community efforts. In addition, the outreach process of the project must take into account the community's political, social, cultural, and geographic characteristics and fit the specific language, literacy, and accessibility needs of the local population.

## 3 Setting the project timetable

The experience of successful efforts has shown that a comfortable schedule for the visioning project is to have the stakeholders meet once every three weeks over 10 to 12 months. Some extra time may be taken to work around major holidays or significant community activities. The initiating committee may choose to meet more frequently, such as once a week, in its preparatory work to speed up the process. The time frame will depend on the nature and needs of the community, local scheduling realities, and the urgency surrounding issues in the community. The timing of stakeholder meetings is an important factor. Successful visioning projects have made accessibility and participation in the project a priority. Therefore, stakeholder meetings often took place in the evenings to allow working people to participate on a regular basis.

### 4 Designing structure to coordinate the project

The project should have a project chairperson, at least three small subcommittees, and adequate staffing. Stakeholders, not those individuals staffing the project, must lead committees. Though initiating committee members may take leadership positions on subcommittees in the early phases of the project, new leaders may be available after the project kickoff once stakeholders are more involved and further recruitment can take place. The coordinating committee and the outreach committee are the best places to involve stakeholders who want to contribute.

### 5 Selecting a chairperson

All successful community projects have strong and fair leadership. Therefore, the selection of the project chairperson is critical. She or he must be (and must be perceived as) open, fair, neutral, and likeable. The chairperson's duties include:

- Formally opening and closing every stakeholder meeting.
- Chairing the meetings of the coordinating committee.
- Appointing the chairs of the other committees, representing the project in the press.
- Leading the fundraising effort.

- Being the spokesperson for the project to the broader community.
- Resolving any disputes within the group, and putting out any fires that may flare up during the course of the project.
- Working with the facilitation team to ensure that the meetings run effectively and a safe environment for discussion is maintained.

The chairperson also submits recommendations for the composition of the coordinating committee to the initiating committee. Every process goes through challenging periods, and heated discussions may take place during meetings. The chairperson has a crucial role to play during these periods. She or he must work closely with the project facilitator to remind stakeholders of the project purpose and goals and to keep the environment safe for discussion from all perspectives. Above all, the chairperson is a role model for the whole group and must have a strong commitment to the project and participants. If she or he is accountable, the entire group is more likely to be accountable. She or he must be willing to devote a substantial amount of time to the community-visioning project.

### 6 Forming a coordinating committee

The first subcommittee is the coordinating committee. This group of 10 to 15 stakeholders manages the process, but not the content, of the project. Its members guide the plan and schedule; serve as liaisons with the stakeholders; fundraise; supervise the other committees and the project staff; and generally keep the effort on track. They will also "own" the project on behalf of the entire community to ensure that the visioning process does not become merely an advisory effort. The coordinating committee will need to hold a planning/debriefing meeting for each meeting of the larger stakeholder group. Work will often have to be done between sessions, and the coordinating committee, with the support of staff, will need to ensure its completion. Some members of this committee, which begins its service at the kickoff and continues into the implementation phase, will likely have served on the initiating committee and, in some cases, may continue on into the implementation committee.

## 7  Forming an outreach committee

The second subcommittee is the outreach committee. This group of 10 to 12 stakeholders will take ownership of the community-outreach process, ensuring an active exchange of information between the stakeholders and the community at large. If the outreach strategies are successful, the community as a whole will have played a large role in developing the vision and action plans. All individuals will have had opportunities to provide input, and their interests, perspectives, and concerns will have been represented within the stakeholder group.

## 8  Forming a research committee

The third subcommittee is the research committee. Its purpose is to provide the stakeholders with information to help them determine current assets and challenges the community faces. This group of three to five individuals joins project research staff to develop at least two sets of documents:

- Preliminary materials for the external environmental scan on global, national, and regional trends that influence community quality of life.
- Local indicators and a profile of where the community is today (e.g., growth, population, crime rates, employment rates).

This information may also be used to educate the general public. Outreach committees in some projects have used the information to provide the public with a rationale for certain strategies.

It is important to make the distinction between primary and secondary research. Primary research involves the collection of raw data in the field. Such research should only be conducted if the desired information is not already available from other sources (i.e., through secondary research). Most information can be gathered from local health

departments, census data, government agencies, nonprofit organizations, chambers of commerce, local colleges and universities, and so forth. The research committee's work must begin with the initiating committee to assure availability of appropriate materials for the presentations during the stakeholder-planning phase.

## 9  Staffing the project

Administrative staff play a crucial role in the visioning process. The staff's ability to coordinate and complete the many logistical tasks involved often makes or breaks the overall effort. Administrative staff handle the following types of tasks:

- General communications (phone and written correspondence with stakeholders, committee members, and the community).
- Coordination of mailings and meeting reminder postcards.
- Coordination of speaker and information requests.
- Preparation of meeting room and other meeting logistics (refreshments, supplies).
- Taking of attendance at stakeholder sessions.
- Preparation of meeting materials.
- Taking of meeting notes.
- Copying and other general administrative tasks.

Should staff members come from the Chamber of Commerce, city government, or other influential body, it is critical that citizens, not staff, direct the stakeholder planning and outreach effort to avoid accusations that individuals with a hidden agenda developed the recommendations.

## 10  Selecting a neutral, outside facilitator

In visioning projects, it is helpful to have an experienced outside facilitator run the community-visioning meetings. Such a facilitator or facilitation team can assist in several ways including:

- Helping to design the process.
- Keeping the effort true to its purposes and values.

- Ensuring that the process stays on track and on schedule.
- Helping to identify experts from around the state and nation on various issues of priority importance to the community.
- Facilitating the large group stakeholder meetings – including encouraging wide participation and discouraging any personal attacks or group domination.

It is essential that the facilitator(s) be a neutral third party not connected with any organization in the process and possessing no specific stake in the outcome. As the project progresses, stakeholders can facilitate the small groups and task forces.

## 11  Identifying funding sources

Visioning projects require financial resources and in-kind contribution of other resources where possible to cover administrative, logistical, research, outreach, and facilitation costs. Successful visioning efforts have made a point of gathering financial and other resources in a cooperative fashion from throughout the community to ensure broad ownership of the project. Developing these resources early can help ensure success in the planning phase and guarantee the availability of adequate funding for those portions of the action plan requiring financial investment and other resources. Community-wide visioning projects usually range from $75,000 to $200,000 when all costs are taken into account. In developing a project budget, a community must consider the following questions:

- What types of resources are required (and in what amounts) for the successful completion of the planning phase of this project? Costs may include:
  - staffing ($20,000–$30,000);
  - facilitation costs ($30,000–$75,000);
  - food ($7,500–$15,000);
  - printing, copying, and office/ administrative costs ($4,000–$7,500);
  - travel ($1,500–$7,500);

  - community meeting-related costs ($4,000–$7,500);
  - outreach-related costs ($3,000–$10,000);
  - research-related costs ($1,000–$10,000);
  - equipment and meeting materials ($2,500–$7,500);
  - the final report ($3,500–$15,000);
  - the community celebration ($3,500–$10,000).
- What money and in-kind resources can be raised from within and outside of the community for implementation of the various action plans determined by the stakeholder group?
- Who will take the lead on resource development?

## 12  Creating a project name

Giving the visioning process a name is an early way to develop project identity and a following for the project. Some names of visioning projects from around the country include:

- Out of the Blue and Into the Future – Blue Springs, Missouri.
- Renew the Blue – Blue Springs, Missouri (update to the original Out of the Blue plan).
- Project Tomorrow: Creating Our Community's Future – Fargo, North Dakota/Moorhead, Minnesota.
- Lee's Summit: 21st Century – Lee's Summit, Missouri.
- Our Future By Design: A Greater Winter Haven Community – Winter Haven, Florida.
- Invent Tomorrow – Fort Wayne, Indiana.
- Liberty for All – Liberty, Missouri.

Project names should give the stakeholders a sense of ownership and enable the general public to identify with the effort.

## 13  Selecting a neutral meeting site

An accessible and neutral meeting site with a large and open layout, adequate parking, and supporting facilities is a must. If possible, avoid governmental and organizational facilities to prevent the perception that the effort is being driven by that entity. The site should have quality lighting, good acoustics, and no pillars to block the sight of participants. The room should have adequate wall space for the hanging of flip charts. The building should have adequate parking, restrooms, airconditioning, tables, chairs, a kitchen, and separate rooms for childcare needs. Community centers, schools, or churches typically serve as good neutral sites for meetings. In considering a site, room layout considerations must be taken into account. The 100 to 150 stakeholders typically sit at moveable round tables arranged comfortably around the room. One end of the room should be reserved for the facilitators, flip charts, screen, and an overhead projector. With large groups, two or three wireless microphones are crucial to aid people whose voices do not carry well.

*14  Recruiting the stakeholders*

A broad-based community visioning effort should start with an initial list of 300 to 400 prospective stakeholders. This list will be whittled down to a committed stakeholder group of 100 to 150 individuals who will attend all regular planning sessions. Past visioning projects have regularly shown that 50 to 70 percent of prospective stakeholders initially agree to participate in the effort. Of these, 5 to 10 percent never attend stakeholder meetings. An average of 15 percent of those invited turn down the request because they are unable to attend a regular session at any given time.

*15  Planning For the project kickoff*

The final tasks of the initiating committee are to ensure that all logistical details are covered and that significant public awareness of the community planning effort exists leading up to the kickoff. All staff and committees — especially the research and outreach committees — should be in place and carrying out their tasks by that time. Composition of the stakeholder group and committee may require fine-tuning through the first one or two stakeholder meetings. In addition, the stakeholders will be strongly encouraged to assist in the outreach effort by spreading the word to other community members and through other strategies developed by the outreach committee. The initiating committee must devise a plan to bring early attention to the project and focus media and public attention to the kickoff. Press conferences, public events, and other communication means have proven to be effective in building community awareness.

## Creating a parallel outreach process

An essential key to the success of the community visioning process is an active community outreach effort. Despite all efforts to recruit a stakeholder group that is representative of the community's diversity, there will be some gaps. For a variety of reasons, certain groups cannot or will not participate in stakeholder meetings. If certain groups cannot come to the stakeholder meetings, then the means must be developed to go out to them. Different strategies must be employed simultaneously to ensure that all sectors and segments of the community's population are kept informed throughout the life of the project. An effective two-way dialogue between the stakeholders and the community is a critical component in creating a relevant, widely supported, and effectively implemented action plan. An outreach effort running parallel to the stakeholder planning process, with activities at several key steps along the way, is necessary to test current thinking within the community and allow citizens to have input on an ongoing basis. An outreach committee of 10 to 20 stakeholders coordinates the effort. To attain its goals and objectives, the outreach committee will need the active support of project staff and the stakeholder group as a whole. The community outreach continues the principles ingrained in this community-based planning model by emphasizing an all-inclusive approach. An indication of a thorough outreach program is the absence of surprise and backlash when the action plan is released to the public. This is because people are already knowledgable of the plan's content due to the ongoing information loops established by the outreach committee.

A primary contact on the outreach committee should be designated for each major task area. One press-oriented person should serve as the media contact. Another should be recruited to liaise with community residents who may have questions regarding any of the activities or strategies. Others should be made responsible for the town meetings and speakers' bureau. In addition, the outreach committee should have strong contact with the coordinating committee, whose assistance it will need from time to time. It should be well organized, and develop and use a plan of action to cover both regular and special needs. Outreach is no different from any other community-based effort to the extent that one strategy alone is not enough to ensure success. A multiple-strategy approach is the only one that works consistently. A creative, highly prepared, hard-working outreach committee can attract positive attention and useful input to the community visioning process. The success of the community's initiative depends on it. Approaches to community outreach include the following actions.

### Project kickoff

The project kickoff has two primary audiences. The first consists of the stakeholders, who hold their first regular session and become familiarized with the project purpose, the planning process, and their colleagues. The second audience is the community as a whole. The kickoff can be the most effective way of introducing the visioning initiative to the media and the citizens whose support will be required throughout the project. Visioning teams often hold a public event/press conference prior to the kickoff to generate publicity. They may invite media representatives, key community leaders, and the public to a 30- to 45-minute presentation on the project given by three or four spokespersons (perhaps including the chair and/or a member of the initiating committee or

coordinating committee). The guests would be able to learn about the work completed to date and receive detailed information about the participants and the planning and implementation effort. The presentation might be followed by a 10- to 15-minute press conference wherein reporters would be able to ask additional questions. It is essential to prepare project fact sheets and media kits in advance.

### Surveys

At certain points throughout the community visioning process, the stakeholders will need specific feedback from the community in order to direct or refine their planning actions. Surveys and focus groups are common instruments for gathering such information. An entire industry centers on the effective use of these very powerful research tools. In this limited space, therefore, the subject can only be introduced and participants encouraged to seek professional assistance or to read further about these tools before using them to enrich the community project. There are many types of surveys, and any number of them may be used depending on the information needed. Standard surveys characterize a given problem after it has been identified but before a solution has been selected and implemented. Surveys should contain specific questions about individual topics, although multiple topics may be addressed in a single survey. Survey data may provide guidance on the most appropriate methods to use in addressing a given issue. In addition, surveys may be applied during any phase of the process to monitor the effectiveness of approaches being used.

Survey questions must be specific and designed to minimize the chances of misinterpretation by respondents (something that can skew the results). Moreover, questions must be relevant to the target population or, again, the results will be inconsistent. Finally, the analysis of the survey results will be invalid if it does not take historical patterns into account. Surveys may be administered in person, over the phone, or through forms filled out anonymously by large numbers of people. It is often effective to code the forms by the respondent's area of residence, income group, organization, and/or other characteristics.

Citizens from all walks of life in Mobile County, Alabama, came together to begin Mobile 2000. Businesspeople completed surveys. Citizens filled out forms that arrived with their Alabama Power bills. Students discussed and reached consensus about their rights to a quality education and the conditions under which they should receive it. Parents completed slightly longer surveys about changes they would like to see in the schools to improve their children's chances for academic success. After reading thousands of comments about what their community expected of its educational system, Mobile 2000 stakeholders created a vision. They described an educational system that would produce, as an end result, a professionally competitive population of critical thinkers engaged in the life of the community. Outreach committees in other communities took the time to identify certain segments of the population that are ordinarily overlooked in community processes. In Sioux Falls, South Dakota, the outreach committee went to public assistance offices and had front-line workers survey individuals. In Atlanta, Georgia, outreach committee members went to homeless shelters and held focus groups with individuals to get their input on the issues directly affecting them. Input from these individuals was brought back to the larger stakeholder group and incorporated into the action planning.

## Focus groups

Focus groups are a form of survey designed to identify and solve problems. Surveys help communities determine a course of action once a problem/issue has been identified; focus groups help communities find what problems/issues actually exist and how they should be defined. Focus groups are in-depth, specific interviews with people representing a cross-section of the community based on ethnicity, race, age, socio-economic status, perspective, and so forth. Focus groups are time-consuming, usually requiring a minimum of one month to

assemble and conduct. It is critical to ask the right questions of the right people and then base the conclusions on historical trends and community background. The focus group leader must make sure the respondent pool reflects the demographics of the community to ensure a valid sampling of perspectives. Because focus groups (and surveys) must be designed carefully if they are to achieve sound results, it is advisable to look carefully at your group's capacity before undertaking such projects without guidance. If no one on the planning team has extensive experience with surveys, college research departments or outside professionals should be consulted or even hired to do the job. A few tips for surveys/focus groups include keeping the language on the survey simple to allow participation by people of all levels of literacy/language proficiency. In addition, surveys should be translated into the language of non-English-speaking residents; focus groups for non-English-speaking residents will need a translator. Finally, allow sufficient lead-time for each method to give the designers a quality sampling of the community.

## Town meetings

Town meetings are large gatherings at which the stakeholders and planners can inform the public about the project and receive valuable feedback from community residents. Anyone may attend to listen, learn, and voice his or her opinions, interests, and concerns. An effective town meeting includes presentations by the planners, but most importantly it allows for public input. Individuals from all sectors of the community are encouraged to attend through carefully planned, highly proactive recruitment strategies. People tend not to come to meetings without a strong sense of their importance — especially the types of folks whose input is most critically needed. It is precisely the most marginalized community members who typically do not participate in such activities. It is recommended that at least three major town meetings be held during the planning phase of the visioning project. The first meeting should take place after the "current realities and trends" stakeholder session, immediately prior to the first visioning session. This meeting is intended to get the word out about the purpose and nature of the project and to solicit ideas from citizens on their visions for the future. The second town meeting should come after the visioning sessions and prior to the key performance areas (KPAs) sessions. At this gathering, stakeholders present their consensus on the vision and receive community input on KPAs and ideas for "trend-bending" action strategies. The third meeting takes place after the stakeholders have reached a rough consensus on the action plan and implementation strategy but before they have finalized that work. The community has the opportunity to give suggestions and help fine-tune the strategies prior to final consensus.

A strong turnout by community members and interested parties is crucial for town meetings. The outreach committee can employ various strategies to ensure adequate participation representative of the many sectors of the population. To begin with, the stakeholders themselves can spread the word. In addition, the outreach committee can send press releases to print, radio, and television media; mail flyers to key contacts or place them in conspicuous places; translate written materials for non-English-speaking populations; offer assistance with transportation and day care, and so forth. Neighborhood meetings are a variation on the town meeting theme. Such gatherings can target specific parts of a community whose residents might not attend larger meetings in other parts of a city.

## Press releases

Communities must enlist the aid of local experts in working with the media. Their knowledge of how to approach and follow up with news organizations can be crucial in effectively getting the word out about the community effort. A first step in publicizing a town meeting, the kickoff, or any major part of the visioning process is to maintain regular contact with the media. The most common tool in this effort is the press release, a very specific document announcing an event or major benchmark. Press releases will

frequently need to be drafted and sent to pre-developed contacts at each print, radio, and television news organization in the region. This mailing should always be followed up by a phone call to answer any questions and to lobby for coverage of the news item in question. The press release should be accurate and succinct. Media will cover events that are well supported (i.e., those with large attendance numbers, community leader participation, and so on). Press releases should be delivered to local, regional, and even statewide news organizations if appropriate. The papers that print the announcement will sometimes translate the text for specific non-English-speaking populations.

## Flyers

Flyers advertise upcoming events on single, brightly colored sheets of paper that give the group's name, date and time of the event, location, nature of the event, a contact phone number, and specifics regarding refreshments, transportation, and childcare. Flyers may be posted in public places and/or handed out to individuals on busy street corners. Flyers that catch the eye, are positive, and evoke an atmosphere of importance and fun are most effective.

## Speakers' bureau

Stakeholders can utilize their public speaking talents to spread specific messages to the community about the progress of the community visioning project. This is an effective way to receive input, share information, and promote visioning efforts in the community. Within certain pockets of the population, such as communities of color, it may be best to have a face-to-face meeting with elders or other community leaders to explain the program. Once they buy in, they may be able to inform their community and recruit new participants more effectively than "outsiders" could on their own. If these individuals do not have time to assist the outreach committee, ask for names of other people within the community who may be available. It is important to train members of the speakers' bureau together and provide them with good fact sheets and an overview of frequently asked questions so that they will deliver a consistent message to the public. In addition, preparation will assist in effectively reaching the targeted population. Consider the following:

- Accountability and follow-up plans should be addressed. The group needs to ask itself the following questions: "How can we ensure that people will show up for the meetings?" "How can we keep their attention once they are present?"
- A sign-in sheet for attendees should be used. The individual's name, address, and phone number may be valuable as the group attempts to recruit new members and to keep the community updated. Such information also supplements the record of the meeting itself.
- Finally, a contact person or persons should be designated so that those who did not offer feedback at the town meeting may do so at a later date, if desired.

## Op-ed articles

Opposite-editorial articles ("op-ed" stands for "opposite editorial," as in "opposite the editorial page") are written by non-journalists, usually community leaders and citizens, and are printed periodically by newspapers. They offer insight into local happenings, express grass-roots perspectives and interests, and update ongoing community programs. A newspaper's ultimate goal is to sell papers; for this reason, publishers want articles that are of high quality, are timely and in the public interest, and are positive in nature. They want to produce something the public will want to read. To get an op-ed article published, begin with a query letter to the editor. This letter should be short and to the point, including such facts as what inspired the effort, who is involved, how the project arrived at its current stage, where it is headed, and how specific plans will be implemented. This correspondence needs to be well written – the editor will look upon the letter as a sample of the author's writing ability. The language of the article itself should be positive,

focusing on the action the group is taking. Write about specifics – the obstacles the project has overcome, how breakthroughs were achieved, changes in team members' thinking. Focus on what the project is about, what it has accomplished, and what it will accomplish in the future. The writing should be inspiring for readers and leave them wanting to be a part of the effort, or, at the very least, highly supportive of and informed about its progress. Finally, the article must be concise and should not be more than 1000 words long (the newspaper will probably shorten the text of the article anyway). High-quality writing is critical to acceptance of the op-ed article. If the writing is good, the editor may ask for more articles to print in the future.

## Public service announcements

Radio and television stations were once required by the Federal Communications Commission (FCC) to provide public service announcements (PSAs). Many stations still do so as a community service. A PSA is a 30- or 60-second spot, provided free of charge, that informs the public about a cause, issue, program, service, or opinion. When contacting a broadcast outlet, ask for the individual in charge of PSAs, ask what the station's preferred PSA format is, and follow it carefully. Many PSAs are not broadcast because they do not follow station format. When working with the media, always strive to minimize the amount of work they must do.

## Websites/project home pages

In the technology age, an effective way to get the information out to the community is through a project home page. Most projects have stakeholders or other professionals who will gladly donate their time to create a project home page. These home pages can provide "surfers" with background information and the work to date. In addition, the home page can give users the opportunity to add their input through online surveys or feedback boxes on the website that the outreach committee or staff access and distribute to the relevant committees.

## Phase two: the stakeholder process

## Community visioning

Many communities begin their visioning project by determining the vision or desired future. Others look at where the community currently finds itself before identifying the desired future. Both approaches have produced quality results in visioning projects around the country. However, starting with the vision statement is preferred because it sets a positive tone for the process from the very start. This process convenes residents holding very diverse perspectives who come into the process with personal agendas. By starting the process with the development of vision themes, participants recognize early on that despite the different views, there are many areas on which they all agree. Experiencing such a "win" early on in the process sets the tone for participants to work toward agreement throughout the process.

### A *vision is a "stretch"*

In spring 1961, President John F. Kennedy, seeking increased funding for space exploration, described a most ambitious vision: to land a man on the Moon before the end of the decade and return him safely to Earth. At the time, the United States had only launched an astronaut into "sub-orbital" space, let alone going to the Moon. The vision, in the midst of the space race, was inspiring and motivating. The country vowed to move ahead on the vision and the ambitious timeline. Achieving the vision had its costs. In 1967, three Apollo 1 astronauts perished during a launch practice session because, some say, the timeline was too demanding. Staff within the space program learned from the tragedy, changed their approach, and continued working toward Kennedy's goal. On July 20, 1969, Neil Armstrong and Edwin Aldrin walked on the Moon and returned safely to Earth with fellow astronaut Michael Collins. Kennedy's clear vision with specific outcomes, the timing of the space race, the program's

ability to bounce back from loss, the enthusiastic commitment of the masses, and a number of other variables produced a technological achievement for the ages. On a summer day in 1963, Dr. Martin Luther King, Jr. addressed the masses at the Lincoln Memorial in Washington, DC. His "I Have A Dream" speech stirs as many souls today as it did on that memorable afternoon. Communities continue to struggle toward the future he described for all of the country's children and people.

### *Picture the desired future*

A community vision is an expression of possibility, an ideal future state that the community hopes to attain. The entire community must share such a vision so that it is truly owned in the inclusive sense. The vision provides the basis from which the community determines priorities and establishes targets for performance. It sets the stage for what is desired in the broadest sense, where the community wants to go as a whole. It serves as a foundation underlying goals, plans, and policies that can direct future action by the various sectors. Only after a clear vision is established is it feasible to effectively begin the difficult work of outlining and developing a clear plan of action. A vision may be communicated through a statement, a series of descriptions, or even a graphic depiction of how the community would look in the target year. Communities have used a number of methods and media to create and express their visions, their desired futures. The following ingredients are crucial to generating an exciting community vision.

### A INCLUDE A HEFTY DOSE OF POSITIVE THINKING

In developing a community vision, it is important not to be constrained by either political or economic realities. For many people who think negatively, it is challenging to focus their energy on how things *can*, rather than on why things cannot, happen. People who have been through successful visioning projects have challenged themselves to move beyond the constraints and to dream about what their ideal community would be like. In developing the action plans, they focus their thinking on what must happen to ensure that the vision becomes a reality. It is always better to aim too high than too low. The positive thinking will be reflected in the vision statement itself. The statement should be entirely in positive terms and in the present tense – as if it were a current statement of fact. The vision and its components should be stated in clear, easily understood language that anyone in the community could understand. The vision statement must be reached by consensus and encourage the commitment of diverse community members. It is the vision that will drive the entire planning process – every action plan will be designed such that, when implemented, it will help bring about the desired future.

### B STRONG VISUAL DESCRIPTIONS

In the visioning process itself, stakeholders can literally ask and answer such questions as:

- What words do you want your grandchildren to use to describe the health of the community?
- If the very best quality of life existed in the community, what would be happening?
- What common values exist across all perspectives and interests within the community, and how do they manifest themselves?
- How are people interacting with one another in this desired future? How are decisions being made?
- What is unique to our community that no other community has, and what does it look like 20 years from now?

### C A LONG TIME FRAME

The stakeholders select the time frame of the vision project. It is probably more useful to set a vision for a point at least 10 years into the future. Although communities would like to be able to achieve a desired future in the short term, the reality is that many changes will take a great deal of time to bring about.

An effective vision typically addresses a period stretching 15 to 25 years into the future. A quality vision statement has these important ingredients:

- positive, present tense language;
- qualities that provide the reader with a feeling for the region's uniqueness;
- inclusiveness of the region's diverse population;
- a depiction of the highest standards of excellence and achievement;
- a focus on people and quality of life;
- addresses a time period 15 to 20 years into the future;
- language that is easily understood by all.

## Step one: developing the vision statement

The process of refining the vision statement and its component points can be lengthy and arduous. There is no shortcut to working through the process as a group. Although groups often get caught up in "word-smithing" the statement, it is more important to reach agreement on the themes of the vision. The stakeholders may have to be reminded that the vision is the "end state," the final result. They will determine the specifics of how the vision will be reached later in the process, during the action-planning phase. The time required to generate a clear vision statement that expresses explicit themes can vary widely from one community to the next. It is unlikely that a broad group of citizens would complete the process in fewer than eight hours of working time, but they should not require more than 15. One effective format is a weekend visioning retreat. Typically, however, stakeholders work on vision statements over two nonconsecutive evenings. Through the visioning process, people draw heavily on the values that are important to them. The process translates these individual and collective values into a set of important issues that the community wants to address. With a clear vision statement articulated and the component points serving as a beacon for the future, the stakeholders can shift to determining their priorities.

## Step two: understanding trends, forces, and pressures

A community scan is a brief but important step in the community visioning process. It enables stakeholders to develop a shared understanding of the major events, trends, technologies, issues, and forces that affect their community and/or will do so in the future. National and global realities often have a significant impact on a community's ability to meet its challenges. It is not necessary for the group to reach true consensus on these observations, but all participants should recognize how their community relates to the world around it and how broader issues affect local choices. The research committee presents its first piece of work during this phase by providing to the stakeholder group a preliminary list of key current and future trends. The factors might include:

- The influence of population growth, age, and funding trends on the educational system.
- The affect of in- or out-migration on housing quality and affordability.
- New technologies, their costs, and the impact on jobs and the community's quality of life.
- Changes in funding and/or policies of national, state, and local government programs.
- Global trends regarding trade, the environment, and labor.

At an early initiating committee (IC) meeting, the research committee members, with the assistance of the IC, should generate a list of issues for a preliminary scan. However, this should only be considered as a first step; the preliminary scan should spark further discussion of the influence of these factors on the community's current and future quality of life. The final environmental scan must reflect more than merely the "experts' view." Community knowledge and perceptions of these larger issues must be considered during the stakeholder process. Following the presentation by the research committee, the stakeholders can discuss the issues in a large group format and then work in small groups to encourage greater participation. The small groups then report back to the larger group, discussing priority areas in greater detail. This step combines the findings of the research committee and community perceptions in general. While some of the issues raised during the environmental scan are beyond local control, their influence must be addressed if the community is truly to move to a new level. The discussion of these regional, national, and global forces sets the stage for identification of *local* realities and trends.

### A Looking at local realities and trends

During the late 1980s, the state of North Dakota entered into a major evaluation of the strengths and weaknesses of its economy through a series of public town hall meetings. The city of Fargo conducted an extensive assessment of the local development corporation and its previous management and leadership. As a result, the corporation's broad-based board of directors (representing labor, government, education, and business) redirected its goals and objectives based upon how Fargo compared to other major growth centers around the country. Through this evaluation, it was quickly determined that the development agency was not focusing on "primary sector" job development. The directors realized they had to move their focus in this direction or Fargo would not realize its potential in the twenty-first century. The evaluation generated a great deal of publicity and citizen involvement. The discussions were very candid and sometimes showed intense emotion; most importantly, they left no stone unturned. The success of Fargo's primary sector marketing strategy has already led the city to extend its growth initiative from 1997 out to the year 2000. This success gives meaning to daily problem solving because everyone now accepts the fact that the small details add up to create the big picture. Fargo's successful internal analysis provides an excellent example of self-examination as part of an ongoing strategic plan. Successful community efforts must begin with agreement on how the community is doing today. As in the Fargo example, such an inquiry can begin with the question, "What are our greatest strengths and most significant weaknesses?"

*B  Scanning the community*

The community scan consists of local indicators of how well the community is doing at a variety of levels. Developed from secondary data, the profile depicts assets (those areas/programs in which the community is doing well) and challenges (those areas in which the community is struggling). In many visioning projects, the research committee conducts a survey asking for residents' perceptions of their community's assets and challenges. These survey findings are combined with stakeholder perceptions to assist in identifying areas of focus during the action planning phase. The research committee collects community indicators from secondary sources. For instance, crime statistics may be obtained from local sheriff and police departments. Real estate, business, and other economic indicators may be collected from the local Chamber of Commerce. Health-related figures (e.g., teen pregnancy, sexually transmitted diseases, immunization rates) may be collected through social service and health departments. Effective research presentations have compared the latest data with baseline data from a number of consecutive years to display annual changes and to illustrate local trends. The community scan combines the survey results, the research data, and stakeholder perceptions into a single, powerful tool that the stakeholders can use in their own discussions and decision making. The combination allows stakeholders to base their deliberations on information and perceptions, a scenario that is both common and healthy in visioning processes. For instance, the community may perceive that violent crime has increased, but statistics may show that the opposite is true. Such occurrences build understanding of both the issues and the perspectives that exist within the community. In addition, the data and discussions provide the stakeholders and the community with a "likely future" – that is, the probable outcome of current trends and pressures if the community does not intervene. Stakeholders can identify areas of strength and those needing improvement by breaking into small groups and asking the following questions:

- What is the "likely future" of the community?
- Which elements of that direction are good or bad?
- Which aspects of it do we wish to maintain, and which should be altered?
- What are our most important opportunities and dangerous threats?

Once the likely future has been evaluated, new scenarios may be considered under different starting assumptions.

*C  Sample indicators of a community*

In 1992, Jacksonville, Florida, conducted a Quality of Life Project to monitor annual progress within the community. These indicators included the following:

- Public high school graduation rate.
- Affordability of single-family homes.
- Cost of 1000 KWH of electricity.
- Index crimes per 100,000 population.
- Percentage reporting feeling safe walking alone in their neighborhood at night.
- Compliance in tributary streams with water standards for dissolved oxygen.
- Resident infant deaths per 1000 live births.
- Percentage reporting racism to be a local problem.
- Public library book circulation per capita.
- Students in free/reduced lunch program.
- Tourism/bed-tax revenues.
- Taxable real estate value.
- Tons per capita of solid waste.
- People reporting having no health insurance.
- Employment discrimination complaints filed.
- People accurately naming two city council members.
- Percentage registered who vote.
- Bookings of major city facilities.
- Average public transit ridership per 1000 population.

Jacksonville set goals for each indicator and had mechanisms in place to measure progress for each year. Included with the indicators were community action steps for achieving the target goals. (For more information on Jacksonville's Quality of Life Project or specific indicators, contact Jacksonville Community Council Incorporated at (904) 396–3052.) A key factor in strategic planning is good information based on data and research available throughout the community. Information will be interpreted in different ways, therefore it is important to have sessions where facts and perceptions may be studied, analyzed, and discussed. Having a common perception about how the community is currently doing will assist in developing the desired future and identifying key areas in which to focus the action planning. There are several ways in which the community scan can be developed with each way requiring a different investment of time and resources. Many of these decisions can be resolved by the initiating committee during the early phases of the project. Whatever approach is decided on, the desired outcome of this phase of the project is to leave the stakeholders and the participants with a shared understanding of:

- The community strengths/assets.
- The difficult challenges the community faces.
- The realities the community faces that are both within and outside of its control.
- The community's likely future should no interventions take place.

## Step three: gauging the community's civic infrastructure

Scholars and practitioners of urban and community affairs are beginning to sense that associations and traditions play an integral role in the health of the communities, whatever their size. The National Civic League refers to the formal and informal processes and networks through which communities make decisions and solve problems as "civic infrastructure." Successful communities honor and nurture their civic infrastructures. They do not look primarily to Washington for money or program guidance. Rather, leaders in America's most vital communities recognize the interdependence of business, government, nonprofit organizations, and individual citizens. In particular, these communities recognize that solving problems and seizing opportunities is not the exclusive province of government. They carry on an ongoing struggle through formal and informal processes to identify common goals and meet individual and community needs and aspirations.

Examples of civic infrastructure include:

- What once was an impossible dream became reality when citizens of Broomfield, Colorado, embarked on a collaborative visioning process that led to Broomfield becoming its own city and county, the first change on the state map in over 94 years.
- In an effort to give citizens input in government decision making, the city of Fort Wayne developed community-oriented government where citizens take issues to one of their 227 different neighborhood organizations and work directly with city staff.
- Santa Maria, California, divided by racial and cultural barriers, instituted in 1997 a first-ever Peace Week designed to help erase violence and prejudice and bring Santa Maria residents together.
- After the closing of a major air force base in Denver, Colorado, the city created an unprecedented economic development partnership with the neighboring jurisdiction which was most affected by the closing.

What accounts for the different experiences of these four communities in addressing problems? In each case the strength or weakness of the civic infrastructure, the invisible structures and processes through which the social contract is written and rewritten in communities, determined success or failure. The Civic Index (see Box 6.1) is one way to analyze civic infrastructure. A comprehensive look at civic infrastructure is a fine point of embarkation — both for that conversation itself and for a reframing of the social contract.

## Step four: selecting and evaluating key performance areas

By this point in the process, the stakeholders will have discussed and reached consensus on where their community is today, where it is likely to be heading, and where they would like it to go. The next step in this results-oriented process is to decide how the community can get from where it is today to where stakeholders want it to be in the future. This step involves the selection and development of key perfor-

mance areas (KPAs). KPAs are highly leveraged priority areas for which specific actions will be developed to redirect the future of the community. Implementation of the strategies developed for the KPAs will bend the trend from the likely future (as determined by the community profile) toward the desired future (as articulated by the stakeholder group). Successful community visioning projects have prioritized their visions into four or five KPAs. They reasoned that only some issues are of high-level priority; moreover, not everything can be done at once. Choices must be made. Secondary priorities can be tackled later. The

---

### BOX 6.1 THE CIVIC INDEX: A TOOL TO QUALITATIVELY ASSESS CIVIC INFRASTRUCTURE

The National Civic League developed the Civic Index to help communities evaluate and improve their civic infrastructure. The 12 components of the Civic Index measure the skills and processes that a community must possess to deal with its unique concerns. Whether the specific issue is a struggling school system, an air pollution problem, or a lack of adequate low-income housing, the need for effective problem-solving and leadership skills is the same. A community must have strong leaders from all sectors who can work together with informed, involved citizens to reach consensus on strategic issues that face the community and the region around it. Committed individuals give communities the capacity to solve the problems they face. Communities must resolve to increase their capacity to address problems. Outside consultants can make recommendations, but action is unlikely without local ownership of a strategy and an implementation plan. The Civic Index provides a framework with which communities can increase their problem-solving capacity. It provides a method and a process for first identifying strengths and weaknesses and then structuring collaborative solutions to problems. It offers an environment within which communities can undertake a self-evaluation of their civic infrastructure. Creating civic infrastructure is not an end in itself. Rather, it is a community's first step toward building its capacity to deal with critical issues.

#### Incorporating the Civic Index into your visioning process

Some communities have used the Civic Index as a "stand-alone" project to enhance the community's civic infrastructure. Others have incorporated it into their visioning process. The steps taken to build the community's problem-solving capacity can easily be integrated into the action plans developed during the visioning process, particularly in areas where networks and communication mechanisms are identified as need areas from the Civic Index. These must be in place before longer term action steps can be implemented. The Civic Index's results are greatly enhanced when a large, diverse group such as the stakeholder group uses the Civic Index. Perceptions of the 12 components will vary among group members, and the discussion provides a great opportunity to build understanding and trust – the key ingredients of civic infrastructure. Discussion of the components takes place both in small groups (to ensure participation) and in large groups (to enhance the small group findings). Once the 12 component areas have been assessed collectively, it is important to focus on each component individually. Stakeholders then develop benchmarks that indicate progress toward the desired level. The benchmarks, with steps to reach them, are incorporated into the action plan during a later phase.

### Civic Index components

Stakeholders assess the community's current performance in each of the four sections listed below and consider how that performance affects the profile of the community.

#### Section one: what is our desired future?

Communities that deal successfully with the challenges they face have a clear sense of their past and have also developed a shared picture of where they want to go.

- Is there a shared sense among residents of what they want the community to become?
- Has the community completed a strategic plan to implement a community-wide vision?
- If a community vision exists, how is the vision being used?
- What are some examples of the community's positive self-image?
- What makes the community special and unique from other locales?

#### Section two: how are we fulfilling our roles in community governance?

In successful communities today, government is no longer the sole owner of the public agenda. Instead, citizens, businesses, nonprofit organizations, and government jointly hold the public interest.

*New roles for citizens.* The role for today's and tomorrow's citizens means that individuals must be willing to take responsibility for their community by stepping forward to share the burden of difficult decision making and challenging problem solving.

- What are some good examples of citizen participation in your community?
- What is the current nature of citizen participation? Is it confrontational or collaborative?
- Are there strong neighborhood and civic organizations? How so?
- Are citizens actively involved in major community projects? Why or why not?
- What opportunities exist for participation in community decision making? Are these opportunities the same for all people? If not, why not?
- Do citizens volunteer to serve on local boards and commissions? If so, do these citizens represent the diversity of the community? Why or why not?
- Is it difficult to find people to run for public office? Why or why not?

*New roles for local government.* One of the most powerful roles emerging for local government is that of the convener. By bringing together different sectors of a community for collaborative decision making and joint action, local government creates a greater sense of legitimacy and ownership for the solutions developed.

- Do the government and community share a common vision?
- How often does the government share community problem solving with private and nonprofit organizations?
- Does the government share decision making with the average citizen?
- Does the government listen to the community?
- Does the government provide services equitably to all people in the community?

*New roles for nonprofits.* Today's role for nonprofits continues to be one of service deliverer and change agent, but cutbacks in funding are compelling nonprofits to partner with each other, local government, and the private sector in order to meet increasing demand.

- What are the issues in your community that require collaboration among nonprofits? What collaborative efforts exist among nonprofits on these issues?
- How do nonprofits collaborate when resources are at stake?
- How do nonprofits collaborate with government, citizens, and businesses?
- Are nonprofits including their clientele in decision making?
- How do nonprofits know they are effective in the community?

*New roles for businesses.* The role of the private sector in a community's civic infrastructure is fundamental, yet often overlooked. Businesses must be willing to create cross-sector partnerships with government, nonprofits, and the community.

- How does the Chamber of Commerce participate in the community? What other things might the Chamber do?
- To what extent do businesses play a philanthropic role in the community?
- To what extent do businesses work with local nonprofits and schools?
- To what extent is corporate leadership a part of broad community improvement efforts? What about small business leaders?
- How well do businesses collaborate with government and nonprofits?
- Do businesses regularly encourage volunteerism among their employees?
- How do large and small businesses participate in the community in different ways? How can this improve?

## Section three: how do we work together as a community?

Successful communities understand that: (1) they are and will become increasingly diverse; (2) citizens demand a role in the decisions that affect their lives; (3) information is readily available through a variety of means; and (4) complex issues regularly cross regional boundaries.

*Bridging diversity.* Positive inter-group relations happens when groups of people (identified by race, ethnicity, age, gender, sexual orientation, income level, interest) are able to acknowledge differences while still being able to work toward common goals.

- What types of diversity exist within the community?
- How does the community view diversity? With distress, tolerance, or does it embrace diversity? On what do you base your assessment?
- How does the community promote communication among diverse populations?
- Do diverse groups cooperate in resolving conflicts before they escalate into major problems?
- How are all diverse groups involved and included into community-wide problem solving?
- Are non-U.S. citizens involved in community activities?
- Do schools have programs that deal with increasing community diversity?
- How does the community respond to discrimination, racism, and racist acts?

*Reaching consensus.* NCL defines consensus as being able to live with a decision to the point of supporting, and not blocking, its implementation. The process of consensus building across viewpoints and interests is a skill that sets successful communities apart from those that struggle.

- Are approaches to problem solving proactive or reactive? How so?
- Are there leaders in the community who are willing to set aside their own interests to help build consensus?
- Do citizens, government, nonprofits, and businesses work together to set common goals?
- Do leaders convene citizens to neutral forums where all opinions may be shared?
- Are there neutral forums and processes where all opinions are heard?

*Sharing information.* Shared information and a "safe space" for dialogue greatly enhances a community's ability to work toward cooperation and consensus, make balanced judgments, and head off disputes.

- How informed is the community of the plans and goals of its governing body?
- Where does the community get information about public issues? Do schools, libraries, and the government all provide public information?
- Do community leaders have regular opportunities to share ideas?
- Does the media play an active and supportive role in the community?
- Does the way the media frames the issues make it easier or more difficult for communities to solve its problems?
- Are issues in the media framed around conflicts or solutions? Are all sides presented?
- Do citizens have the information they need to make good decisions?
- Do all community members have access to current information technology?

*How do we work across jurisdictional lines?* Issues such as economic development, transportation, growth management, environmental protection, and recreation move beyond the boundaries of singular jurisdictions. Thus successful jurisdictions are working with neighboring municipalities in order to be effective in today's world.

- What issues should be addressed regionally? What collaborative efforts exist among communities in the region on these issues?
- How do local governments work together on regional issues?
- To what extent do institutions across the region collaborate with one another?
- Are services provided regionally?
- Is there a regional governance structure in place?

*Section four: how are we strengthening our community's ability to solve problems?*

All communities face the challenge of building and sustaining efforts. Key to building capacity is developing skills in individuals throughout the entire community, continually building networks by linking and convening people and organizations, and nurturing those relationships on an ongoing basis.

*Educating Citizens.* Strong citizen education must teach residents both what they can do to make a difference and how to apply their learning through actual participation in the community.

- Are there cradle-to-grave opportunities to learn about citizenship?
- Are there opportunities for all community members to learn about citizenship?
- Do a wide variety of organizations and institutions provide citizen education opportunities?
- Do citizen education programs develop the knowledge and skills necessary to participate in community governance?
- How are traditional power leaders supporting citizen education?

*Building leadership.* Today's complex times reveal that in order for communities to work more effectively, quality leadership must come from all parts of the community. These leaders must reflect the diversity of the community as well as possess the skill of convening different interests to share in decision making.

- What qualities do you want your community leaders to have? What kind currently exists?
- Are there leadership development opportunities for both formal (elected officials) and informal (neighborhood) leaders?
- Are current community leaders willing to adapt and modify the way they lead as times change?
- What kinds of leadership exist? Collaborative? Confrontational?

*Ongoing learning.* Many communities conducting initiatives experience varying degrees of success over the years. Participants in more successful efforts have taken the time to learn from past experiences and have incorporated that learning into subsequent efforts. Some people have said that the definition of insanity is "doing things as they have always been done but expecting different results." In such cases, realizing that a different result will require a new approach is a key to success.

- How is the learning generated from community projects and processes used to enhance future efforts?
- When conflict is managed or overcome successfully, how is that learning documented and incorporated into other settings?
- How can the community incorporate the learning of successful efforts into current and future efforts?
- Whether community efforts succeed or not, do participants ask what they have learned to help them with the next stage of work?
- Are neighborhood and community histories documented?
- Does the community see its work as an ongoing endeavor or as a "one-shot" effort?

Source: National Civic League, *The Civic Index*, 1999

KPAs can be broken down in a variety of different ways — by sector (e.g., business), by issue area (e.g., homelessness), or by project (e.g., community center).

## FORMING TASK FORCES

Successful visioning projects have formed task forces either by assigning interested stakeholders or by choosing members at random from the stakeholder group. Either way, additional expertise and perspectives are usually added to help balance the group and develop comprehensive plans. Task forces vary in size from as few as 15 people to as many as 50. Each KPA task force should:

- Assign a convener who is responsible for convening the sessions, keeping the group on task and focused, and reporting the updates back to the large stakeholder group.
- Assign a facilitator to run the meetings (he or she may or may not be the convener) and a recorder to keep minutes and write up the work in a presentable format.
- Plan a number of meeting sessions (how many depends on the timeline) around the large stakeholder meetings.

For each key performance area, the task force and the stakeholder group as a whole will complete the following tasks.

*Recruiting outside expertise.* One of the task force's first assignments is to look at the group's composition and ask, "What interests and expertise are missing from our group?" The task force members should generate a list of people who can fill in the gaps and recruit those individuals to participate. Just as balance was important in filling out the large stakeholder group, the same consideration must be given to the smaller task force groups. Although presentations to the larger stakeholder group will often "safeguard" any domination within the task forces by individuals with special interests, developing the plan with diverse perspectives always enhances the plan's credibility and likelihood of implementation.

*Evaluating the community's current performance within*

*the KPA.* Task forces will assess the community's current performance in each priority area using the work of the research committee, surveys, and past discussions in the stakeholder group. This is also the time to integrate the findings of the Civic Index if utilized in the visioning process. Much of the work from this stage will provide the rationale for proposals to address this key area. It will also help members identify what benefits they want to result from implementation of the action plans. These benefits should be developed into a "mini-vision" that will drive the action planning in this specific KPA.

*Developing goals.* Task forces will develop specific goals to reach the desired future for each KPA. There may be numerous goals and objectives within a specific KPA. For instance, for a KPA of economic development the goals may be:

- Starting an incubation program for small business development.
- Attracting new corporations to headquarter in the community.
- Retaining and enhancing current businesses based in the community.
- Building and retaining the skills of the labor force in the community through mentorships and scholarships.

It will be up to the task force to prioritize the goals and make recommendations on which ones should receive the greatest emphasis.

*Specifying "who will do what by when and how."* Task forces must delineate specific action steps, identifying what resources will be required and the options for acquiring them, where they will come from, what the time frame for action is, and who will be responsible for ensuring that implementation occurs. It is during this step that the specific benchmarks and actions of the Civic Index will be integrated into the appropriate KPAs to build community capacity. By now the vision has been translated into practical and attainable outcomes to be achieved through specific tasks and actions. This step crystallizes the vision into a tangible program.

*Reporting back to the stakeholder group and receiving feedback.* As the KPA task forces proceed through the plan development, they will periodically meet with the larger stakeholder group to share their findings and coordinate overlapping efforts as appropriate. When reporting back to the large group, each task force should hand out written summaries of the work done to date, with highlights transferred to overhead transparencies for viewing by the group. The task forces should incorporate feedback from the stakeholder group into their planning to ensure agreement on the direction being taken. Although many of the action plans developed will require financial and other resources, sometimes in significant amounts, communities can take certain actions to increase cooperation or shift to approaches that require little or no financial resource outlay. As the whole stakeholder group reaches consensus on the work of each KPA task force, the high-priority projects must be identified and a rough consensus reached on their inclusion into the final action plan. If successful to this point, the stakeholders will have reached a general agreement on the individual goals, objectives, action plans, implementers, resource needs, and time frames identified by the task forces. Once the KPA evaluations have been completed, it is necessary to integrate all of the goals and recommendations into a final action agenda with a formalized implementation strategy. Certain goals and action steps will be complementary and will need to be combined in some way to create a coherent overall strategy. The stakeholder group should publish a report on its community visioning process and final action plan, but it is essential that the work should not stop here. Too many visioning projects end with a report that eventually gathers dust on the shelf. The community visioning process is designed to produce action and results. Reports do not, in and of themselves, assure any action.

*Building a final consensus.* Consensus on the final action plan is the final – and occasionally most difficult – phase of the community visioning process, the phase in which previous agreements are tested and a final community consensus is reached. The stakeholders meet in large and small groups to confirm the soundness of their goals and plans and the pro-

jected results of their implementation strategies. Some action plans may require initiation of new projects. Others may involve support for existing efforts. Some may entail the termination of an existing activity. Because the actions will be varied in nature, it is essential that the entire community and its diverse sectors be behind them. Some action plans may embrace policy initiatives or changes; some may involve significant financial investment; and others may simply pose new approaches to current practices. All may involve the development of new cross-sectoral partnerships. As the visioning action plans and implementation strategy are finalized, the stakeholders must specify who will take responsibility for what. Some issues will clearly fall within the purview of a specific government agency or nonprofit service provider. Other action steps might not immediately suggest a "champion," and the group will have to engage some entity to take the lead. Although an accountable organization or group of organizations may not initially be found for every action step, this is an essential part of the process that cannot be left incomplete. It may be necessary to assign a group of entities to locate a champion for a specific action area. The general rule is: there will be no action without an implementer. As the formal planning steps of the process draw to a close, stewardship of implementation becomes the responsibility of the stakeholder group and the community as a whole. If the process has been effective at developing a sense of ownership and true consensus, it will be possible to hold the whole community and all its citizens and organizations accountable to their commitments. This point highlights the importance of the community outreach process. The investment of time and resources made earlier to ensure full community representation and participation comes to fruition here.

*True consensus and rough consensus.* At the end of this phase, the stakeholders should have reached consensus on the content of each KPA. Sometimes a full consensus cannot be reached. If a large number of stakeholders cannot live with the plan, then the group must take the time to discuss the reasoning of the disagreeing viewpoint and look at ways to fine-tune the approach so that all participants can live with the final plan. If one or two people continue to dissent after all discussion and alternatives have been addressed, it is important to move ahead while making sure that the differing viewpoint is noted and placed in the final report.

*The community celebration.* Celebration is an essential part of a community-based visioning project. There should be a celebration to acknowledge the commitment of individuals involved in the planning phase of the initiative and the results they achieved. Such an event brings citizens together around shared values and aspirations, and nurtures the seeds of change in building a better community. The city of Lindsay, California, held a celebration that residents are sure to remember for years to come. At the conclusion of its long-term visioning process, the community held a grand festival at the City Park and community center, attracting approximately 1500 people. Featured events included a games arcade for kids; live entertainment on the main stage led by Mariachi Infantil Alma de Mexico; a winding parade through the park; a canine fashion show; a decorated bicycle contest; a "kiss-the-pig" contest in which nearly two dozen of the city's leading citizens gave Blossom, a pot-bellied pig, a big smack on the lips; drawings for a television and a blimp ride for two; and a food booth serving burritos, corn on the cob, strawberry shortcake, and watermelon. The celebration was a great success, allowing the citizens of Lindsay to take pride in their accomplishments and enjoy the fruits of their labor. The celebration should acknowledge the planning work of the stakeholders and various contributors, announce the action plan, and – most importantly – be seen as the commencement of the implementation phase of the project.

## Step five: implementation of the action plans

In the community visioning effort, a minimum of two years following completion of the planning process is recommended for intensive focus on project implementation. For many communities, this will be a multi-decade effort. Successful implementation processes contain the following ingredients:

- The establishment of implementation structure such as a committee with staff that oversees and ensures that a variety of areas (that follow this bullet) are addressed.
- Clarity of goals/desired result for both the implementation committee and implementers.
- Criteria (established by the stakeholders or implementation committee) that will be used to prioritize projects.
- Prioritized projects based on the applied criteria.
- Implementers/champions for each project.
- Identification of barriers to implementation and steps to overcome them.
- An overall timeline based on the prioritized goals, barriers, and resources.
- Coordination of all efforts being implemented from the action plan.
- Ongoing community outreach of successes and ideas.

Community and outside resources will be needed to implement the action plan. The initiating committee should have laid the groundwork for this resource-development process, but more work will likely remain. The implementers named in the action plan will need to champion these efforts. Resource development will be most effective if it begins immediately, capitalizing on the momentum from the publication of the report and the community celebration.

*Choosing or establishing an implementation entity.* Implementation efforts should follow the plans created during the planning process. Lead implementers must confirm the commitments already agreed upon and begin their work, drawing on the momentum created by the celebration and publication of the final report to facilitate rapid progress. From the kickoff until this point, the coordinating committee has provided process management for the community effort. Some of its members will be ready to leave the committee, and others will be ready to serve in a more active manner. This process should leave current participants with a strong sense of accomplishment and invite the participation of others. The coordinating committee may retain its original form and become an implementation committee, or it may choose to change its structure as well as its membership. Typically, retaining the cross-sector, broad-based citizen form is the most successful approach as it avoids controversy and keeps the focus on community-wide participation. Some communities choose to create a separate nonprofit organization to serve the ongoing effort. The coordinating committee may also be embraced by an existing entity deemed neutral and inclusive, although this can be risky if the organization attempts to hoard the effort or takes actions that dampen community-wide ownership in implementation.

*Monitoring and tracking.* There are three primary areas where active, ongoing monitoring and tracking are required in order to:

- Ensure follow-through on the implementation of action plans and policy recommendations.
- Provide ongoing support for implementers.
- Measure changes in the community quality of life indicators developed earlier in the community scan effort.

During the first two years, the implementation committee or other implementation entity should consider providing updates at least quarterly to the community on project and policy actions. In subsequent years such updates can be made annually.

## The final report

The report on the work of the community visioning process serves many of the same objectives as the celebration (i.e., acknowledging contributions to date, building momentum, and enrolling new implementers). At the same time, it is a flexible tool that may be used to inspire organizations and companies to embrace the community vision and frame parts of their own strategic planning around it. The report also serves to remind implementers and the community of their commitments and provides future efforts with something on which to build. The important thing is that it is used, not simply published, bound, and left to gather dust on the shelf.

## Conclusion

It should be stated that the community visioning and implementation process described in this chapter is an overview of a model. This model has been successfully used and tested in different forms in many communities around the nation in recent years. Each community should work closely with experienced facilitators to adapt the model presented here. Use it as a guide to the design of local process; customize it to match specific needs, priority areas, and available resources.

## Keywords

Strategic planning, community visioning, civic infrastructure.

## Review questions

1 What is civic infrastructure, and what is its role in community visioning?
2 What are the five major steps of the community visioning process?
3 How does community visioning relate to community development both in general terms and specific applications?

## Bibliography and additional resources

Chrislip, D. (2002) *The Collaborative Leadership Fieldbook*, New York: John Wiley & Sons.
Doyle, M. and Straus, D. (1982) *How to Make Meetings Work*, New York: Jove Books.
Larsen, C. and Chrislip, D. (1994) *Collaborative Leadership: How Citizens and Civic Leaders Can Make a Difference*, San Francisco, CA: Jossey-Bass.
Leighninger, M. (2006) *The Next Form of Democracy: How Expert Rule Is Giving Way to Shared Governance – and Why Politics Will Never Be the Same*, Nashville, TN: Vanderbilt University Press.

## CASE STUDY: BROOMFIELD, COLORADO

Broomfield is located just north of Denver in the metropolitan region. After a series of annexations in the 1970s, Broomfield eventually found itself in four different counties. This led to challenging logistical issues for the citizens, service providers, and local government when attempting to build consensus, coordinate services, and address community concerns. In 1994, the citizens, with the participation of government officials, businesses, and community organizations, convened a grassroots visioning effort formally adopted by the City Council. The vision addressed the need to develop "a sense of community" and urged the creation of the City and County of Broomfield as one of the important objectives. Doing so would be a daunting task. No city and county had been formed in Colorado since 1902 when the City and County of Denver was formed. Creating the City and County of Broomfield meant seceding from four different counties, meeting the requirements of being their own county (e.g., creating their own court system, health department, building a county jail), and statewide approval of the voters. Still, the idea excited all levels of the community and they chose to press forward.

Shortly after the adoption of the vision, a strategic planning committee created a specific plan of action, including creating a city and county. One of the action steps included testing the idea with Broomfield residents on the ballot of the 1997 election. In this "go/no go" step, nearly 80 percent of the voters said they would support the idea. As a result of the test ballot, the implementation continued. Paying for the transition would be daunting as well. This is where the economic development portion of the strategic plan came into play. A new state-of-the-art, regional mall was planned. Recruitment of high-tech businesses was also envisioned. These revenue sources would help pay for the sustainability of the new county. The implementers of the city/county initiative comprised city staff, council, businesses, and citizens. They carefully followed the action plan designed by the strategic planning committee. In addition to the test ballot, they negotiated with each of the four counties and worked together to lobby the legislators at the state capital to put the initiative on the state-wide ballot. Citizens, business leaders, and elected officials spoke with editorial boards and other media outlets. As a result, every newspaper endorsed Broomfield becoming their own city and county.

On November 2, 1998, Colorado's voters approved the formal creation of the City and County of Broomfield by a resounding margin. In November 2002, the transition was complete and Broomfield became Colorado and the nation's newest county! Other communities in the Denver metro area have tried to follow Broomfield's example but have failed. Why? Because the county initiative was spawned and led by the local government, not the residents of the community. Again, the credo, "People support what they help create" is true.

Okubo, D. (2000) *The Community Visioning and Strategic Planning Handbook*, Denver, CO: National Civic League Press.

National Civic League (1999) *The Civic Index*, Denver, CO: National Civic League Press.

Peck, M.S. (1993) *A World Waiting To Be Born: Civility Rediscovered*, New York: Bantam Books.

Straus, D. and Layton, T. (2002) *How to Make Collaboration Work: Powerful Ways to Build Consensus, Solve Problems, and Make Decisions*, San Francisco, CA: Berrett-Koehler Publishers.

# 7 Establishing community-based organizations

## Monieca West

By definition, community development is about organizing people and resources to accomplish common goals. Therefore, one of the most fundamental questions facing community developers is, "How should we be organized?" This chapter identifies some important issues that should be addressed before a community-based organization (CBO) is formed and provides an overview of different types of CBOs. For those new to the field, students and even experienced community developers, this chapter can serve as a "how-to" guide for establishing CBOs. Examples of different types of CBOs are provided.

## Introduction

So, you want to start a community-based organization (CBO) or a community development corporation (CDC). You've looked around the community and found something you think should be or could be made better, be it economic development, health care, housing or transportation. You've assembled a few similar thinkers and you are all committed to doing something positive. That usually leads to forming an organizational committee that will lead to a formal community-based organization that will lead to....

Stop! Before the first bylaw is written, before the first officer is elected, there are some basic organizational structuring steps that, if taken, will save time and effort and create a more productive and sustainable organization. A little thought and deliberation at this juncture will pay great dividends over time.

Starting a community organization should be approached with as much care as starting a small business ... because that's exactly what you are contemplating. Both must plan, market, manage staff, create revenue streams and maintain cash flow. The important point for CBOs is to give as much attention to the process of running the organization as it has passion for the project.

## Fundamentals of forming a community-based organization

Whether the CBO is organized as an informal steering committee, a traditional board/committee, or as a complex network of organizations, there are several fundamental questions that should be answered by all start-up organizations. Get these right and almost everything else will fall nicely into place.

### What do we want to do?

To determine this, make sure you're asking the correct question. The initial question should not be, "How do we become a CBO?" It should be, "How can we get better and more affordable housing for all?" Keep focused on the destination, not the vehicle in which you will be traveling. This is not always as easy as it sounds, especially in the formative stages of group process when strangers come together to pursue a passion. Some will be very type A and get bogged down in the mechanics of organization. Others will be so driven by the passion for the project that they will want to skip important foundation-building requirements. The remainder may get lost in the debate and go elsewhere with their time and talent. Articulate what it is that you want to happen in the community and use this as the compass to guide all future decisions.

## Is there anyone already doing this?

The next logical question would then be, "Is there anyone else already doing this?" Determine if there are organizations already addressing your concerns because the best interests of the community are not served when too many organizations compete for the same resources. If there are others, you may elect to simply join their group, or to proceed with establishing your organization and form a collaborative arrangement with the existing organization, or to focus on another issue in community development. Just be sure you are not fragmenting scarce community resources with your admirable zeal to do good.

## Mission and purpose

A clearly defined mission statement is critical, but even more critical is its constant use to guide strategic decision making. It establishes what will be done, who will be involved, and how the community will be affected. The mission statement is what keeps your ship in the correct ocean.

While crafting the mission statement, it is important to recognize that there are four different types of community development functions and people – organizers, developers, planners, and resource providers – and that each has different purposes and expected outcomes.

*Organizers* are about advocacy and empowerment, about influence and being heard, about applying political pressure, or staging protests. Organizers are likely to congregate in neighborhood coalitions organized around social issues. These groups can focus attention on a need previously ignored and provoke resource commitments previously unavailable.

*Developers* are project centered and about doing, creating, and building. Development projects can require substantial resources to deliver the products or services making the organization dependent upon outside support and relationships. Developers require a wide range of technical and administrative skills to accomplish complex and time-consuming, long-term projects. Development organizations vary widely in scope and can be independent or networked, simple or complex.

*Planners* are about visioning and charting a course of action. They compare the past to the present and develop a roadmap for the future. Planners examine possibilities and pull the pieces together into a larger picture. While elements of planning are found in all CBOs, some organizations form purely for the purpose of strategic planning and must be able to bring together diverse groups to create a shared vision. It is not uncommon for planning organizations to complete the planning process and then regroup into development organizations.

*Resource providers* are just what the name implies. They are about giving and assisting. Providers include private and public charities, nonprofit organizations, government agencies, private individuals and businesses, faith-based organizations, and human service agencies.

## Is the organization feasible?

A worthy cause supported by a workable plan with sufficient funding and competent leadership will generally generate sufficient organizational capacity. Capacity includes:

1 Short- and long-term strategic plans.
2 Professional and/or volunteer staffing with effective leadership and management skills.
3 Sufficient available facilities and equipment.
4 Sustainable financial resources and sufficient cash flow to provide for the organization's operations.

Capacity also includes the credibility and influence of those leading the organization, a track record of demonstrated ability to achieve results, and a constituency base that can provide political influence.

## Is there a business plan?

The most important use of a business plan is to ground the organization in reality and provide operational guidance for daily decisions. A business plan should include:

1 A description of the organization (what it will do and how it will help).
2 A marketing plan (who will be served and how they will be reached).
3 A financial plan (start-up financing, expenses, revenues, and cash flow).
4 A management plan (description of the management team's experience).

A business plan increases the chances of securing funding, helps identify strengths and weaknesses of the organization, and provides a way to measure actual results against what was planned. There are some who will resist a formal business plan, seeing it as a waste of time. "We don't have time for all of that; we know what needs to be done. Let's just get busy." Reach agreement among the governing body that the business plan will be adhered to. Stand firm for formal planning; it is time well spent.

## Can we pay the bills?

"It takes money to make money" is as true for non-profits as it is for private business. The only difference is that in the nonprofit world, "money" takes many forms. It may be cash, in-kind, volunteers, or institutional support. The bottom line is that the organization must have sufficient start-up support to get organized and secure initial and sustained funding. Most small businesses fail because of inadequate cash flow and CBOs are no different. According to an unknown sage, "The not-for-profit that doesn't make a profit is a bankrupt not-for-profit."

## Partner and grant organizations for local community development

Community-based organizations do not exist in a vacuum. They must establish local relationships as well as partnerships with state and federal agencies and organizations throughout the public, private, and nonprofit sectors.

## Public sector

### Public sector federal programs

A review of federal agencies indicates that resources from federal agencies most commonly flow to their own regional branches, to state and local government entities, to regional economic development authorities, to educational institutions, or to nonprofits. These entities may then distribute the resources further until they are received by a local development organization, an intermediary linking government funding with a private sector entity, a nonprofit or an individual consumer.

The Catalog of Federal Domestic Assistance (see General Services Administration 2004) is an exhaustive listing of federal programs searchable by function, agency, program title, eligibility and beneficiary. Federal agencies with a significant community development mission and some of their most recognizable programs are shown in Table 7.1.

### Public sector state programs

Every state has an agency that delivers economic and community development programs using a variety of organizational models. The superagency structure combines all functions of primary economic and community development activities – such as economic development, workforce development, and tourism – out of a single centralized administrative unit. The regional structure is also a single department but services are delivered through a clearly defined network of regional offices and regional advi-

■ **Table 7.1** Some federal programs for community development

| Federal department | Programs and focus areas |
| --- | --- |
| Department of Agriculture www.usda.gov | Office of Rural and Community Development |
| | National and State Rural Development Councils |
| | National Rural Development Partnership |
| | Rural EZ/EC/CC* |
| Department of Commerce www.doc.gov | Economic Development Administration |
| | Minority Business Development Agency |
| | Economic Development Districts |
| | University Centers |
| | Small Business Innovation Research Program |
| Department of Housing and Urban Development www.hud.gov | Community Renewal Initiatives |
| | Rural and urban EZ/EC/RC* |
| | Rural Housing and Economic Development |
| | Community Development Block Grants |
| Department of Health and Human Services www.hhs.gov | Food and Drug Administration |
| | National Institutes of Health |
| | Administration for Children and Families |
| Department of Labor www.dol.gov | Office of 21st Century Workforce |
| | Center for Faith-Based Community Initiatives |
| National Science Foundation www.nsf.gov | Division of Grants and Agreements |
| | R&D in science and engineering fields |
| Small Business Administration www.sba.gov | Small Business Development Centers and sub-centers |
| | Women's Business Centers |
| | Office of Business and Community Initiatives |

Sources: Individual agency web.

Note
*Enterprise Zones, Enterprise Communities, Champion Communities, Renewal Communities.

sory councils. The umbrella-structured agency centralizes policy and administration with daily operations executed by a network of private nonprofit corporations and related state agencies (e.g., tourism, workforce development). The private sector-structured agency basically outsources its programs through a state-level public–private partnership.

State development agencies also provide community betterment programs with awards and recognition when a community meets certain levels of achievement in community development. These programs may be run entirely by the state agency or in partnership with other organizations. It is also common for an independent organization to take the lead in these programs with support coming from the state agency.

*Public sector university-based programs*

University centers are partnerships between federal government and academia that mobilize the vast resources of universities for development purposes. Partially funded by the Economic Development Administration, these centers perform research to support development of public policy and economic programs. They may also initiate special development projects and consult with government agencies, businesses, media, and the general public.

Local universities play important roles in business development through business incubators for technology or biotechnology and are the funnel through which flow extremely important federal research and development dollars. Fully accessing the resources of colleges and universities should be a priority when organizing the community for development.

The Small Business Development Centers (SBDCs) are partnerships between the Small Business Administration and local universities. The SBDC network has a central state-wide office and also includes regional sub-centers which provide assistance direct to the business owner.

The Cooperative Extension Service (CES) was founded in 1914 and is connected to land-grant universities. CES programs have grown far beyond the original county extension agents and home economists, with programs including Master Gardeners, 4-H, leadership development, business development, and building healthy communities for the twenty-first century.

## Private sector

Private sector companies and organizations participate in community development through grants, technical assistance, and staff involvement in community leadership roles. Utilities, banks, and private developers are the most common private enterprises to undertake community development activities and they are generally involved in activities that relate to their specific business mission or that will create revenue-generating opportunities for them. Electric utilities in particular offer extensive support in prospect leads and responses, community preparation, and leadership development. Corporate grants and contributions are valuable resources and it is important to establish relationships with the local manager or economic development representative.

## Types of community-based organizations

Community-based organizations are too numerous to describe fully but they can generally be categorized as independent organizations, networks, partnerships or regional initiatives.

## Independent organizations

The local Chamber of Commerce is perhaps the most pervasive community-based organization and represents the fundamental organizational structure for CBOs – a volunteer board of directors with committees. Chambers pursue a variety of interests including economic development, leadership programs, community promotions, governmental affairs, and are as diverse as the communities they serve. Because Chambers are such familiar organizations, there is little need to describe them in detail but they should be recognized for the direct and indirect impact they have on a community. A community with a strong, progressive Chamber is likely to be a very successful and competitive one.

Other local civic organizations include Business Improvement Districts, Downtown Partnerships, and Main Street programs. The Rotary Club, PTA, and the Boys and Girls Club are other examples of local CBOs that need to be brought into the mainstream of community development at the local level.

## Community development corporations

A community development corporation (CDC) is a nonprofit organization that serves a particular geographic area and is normally controlled by its residents. A board of directors, usually elected from the membership, governs CDCs and may also have board positions reserved for representatives of key local institutions such as banks, city government, or the hospital. A paid staff and volunteers execute the programs of work.

Some CDCs focus on one issue, such as housing, while others pursue a wide range of activities from grass-roots advocacy to job creation. CDCs may vary in focus but they have one thing in common – they are entirely focused on the issues that are unique to the local community. In many ways, CDCs are the purest example of a grass-roots development organization.

CDCs depend upon collaborative efforts and must be very astute in developing relationships and building partnerships. Typical partners include faith-based institutions, nonprofits, government departments and agencies, private developers and businesses, banks, national and state intermediaries, social service agencies, schools and colleges, and others related to the specific mission of the CDC. These organizations also provide the volunteer base that is critical to CDC success (see Box 7.1).

## BOX 7.1 CDCs IN MASSACHUSETTS

CDCs defy a single definition because they are custom-made to fit each community. A review of CDCs in Massachusetts provides useful examples of successful organizations.

### Community Economic Development Corporation of Southeastern Massachusetts (CEDCSM)

Structured with a board of directors, paid staff, consultants, and volunteers, CEDCSM is representative of a CDC formed primarily for organizing and advocacy. Originally founded by a community action agency, CEDCSM has created partnerships between private sector employers, immigrant workers, organized labor, grass-roots activists, entrepreneurs, banks, local and state government, and many others. Programs include a micro-enterprise network, computer centers, a workers' support network, training and support for all-volunteer grass-roots groups, and affordable housing.

### Allston Brighton Community Development Corporation (ABCDC)

ABCDC is a very sophisticated CDC. Its board of directors is made up of neighborhood residents, a paid staff of 17, plus a network of volunteers to carry out its program of work. The staff includes a director for each of the primary focus areas of affordable housing, community organizing, economic development, home ownership, and open spaces development. Affordable housing utilizes both public and private resources and is made available to the public using a lottery system. Community organizing focuses on conferences, task forces, ethnic festivals, tenant counseling and services, voter registration, and legislative issues support. Economic development activities include small business, access to technology, employment training, and placement. Home ownership involves intervention services that reduce barriers to home ownership as well as facilitating access to financial resources. Open spaces projects include transforming school yards of asphalt into vibrant playing and learning spaces, and developing urban wild spaces, underpasses, and turnpike murals. Primary partners include local banks, city and state agencies, the United Way, and community action agencies.

### Franklin County Community Development Corporation (FCCDC)

FCCDC is the oldest rural CDC in Massachusetts, serving the 26 towns of Franklin County. The board of directors is elected from the membership and employs a staff of nine. The FCCDC mission includes strengthening the local economy through technical assistance to business owners and entrepreneurs, operating a lending program that provides gap financing for business start-ups and expansions, and organizing community groups and neighborhood associations. Partners include local utilities, state agencies, private individuals, local business and industry, representatives of the farming industry, volunteers, and VISTA workers.

## Massachusetts Association of Community Development Corporations (MACDC)

MACDC is the state-wide policy- and capacity-building arm of the CDC movement in Massachusetts. Its board of directors includes representatives from member CDCs and its extensive staff provides support in the areas of economic development, affordable housing, community organizing, advocacy, and building CDC capacity. One relatively new tool that MACDC has deployed is the Individual Development Account (IDA). IDAs are income-eligible savings accounts that are matched anywhere from a 1:1 to 4:1 ratio from a combination of private and public sources. The funds are then used for one of three purposes – education for themselves or their children, business capital, or the purchase or repair of a home.

The importance of a strong support system for local CDCs cannot be overstated. MACDC and the services it provides are largely responsible for the successful CDC movement in Massachusetts.

## Networks

A network is formed when two or more organizations collaborate to achieve common goals; to solve problems or issues too large to face independently; to leverage the power of numbers in exercising influence or flexing political muscle; to maximize limited financial and human resources of a community by reducing duplication of organizations; or to operate more efficiently in concert with others.

### Information exchange networks

Networks may be as simple as an arrangement to exchange information among organizations or businesses that share common interests such as trade associations or Chamber-based Leads Groups. Leads Groups are structured so that there is only one company per business type represented in the group – one bank, one utility, one lawyer. Group members

share business opportunities, and doing business with one another is encouraged.

### Service delivery networks

Other networks, such as the International Council of AIDS Services Organizations (ICASO), are very sophisticated and can be very high maintenance. ICASO is a massive network of numerous CBOs providing prevention and/or treatment services for AIDS patients and is organized from the local to international levels. Networks can be very effective in addressing social issues, especially those that are multi-dimensional.

### Flexible manufacturing networks

Flexible manufacturing networks (FMN) gained popularity in the U.S. in the 1980s and are still an important economic development tool. FMNs may limit their partnerships to simple exchanges of information or may go so far as to jointly own production facilities or share IT infrastructure. They can be structured on the strength of a handshake or through legal, contractual arrangements. The catalyst for the FMN may be an existing organization such as a trade association, a government agency that organizes new private or nonprofit organizations, or a group of firms may organize themselves in response to market conditions. In most cases, an FMN is built with the services of a broker, someone who helps companies form strategic partnerships, organize network activities, and identify business opportunities. The broker may be an employee, a consultant, a government agency, or a nonprofit.

The wood products industry was among the first to embrace the FMN concept. In Arkansas, it was a nonprofit – Winrock International – that was the catalyst for formation of the Arkansas Wood Manufacturers Association (AWMA), which now offers both youth and adult apprenticeship programs and provides collaborative marketing and support for member companies. In Kentucky it was the state legislature that enacted legislation to develop the Kentucky Wood Products Competitiveness Corporation to develop secondary wood industry business networks.

## Business cooperatives

In 1985, a small group of community members in southeastern Ohio established a number of worker cooperatives to help low-income people start worker-owned businesses. This evolved into ACEnet, the Appalachian Center for Economic Networks, and was based upon models found in Spain and Italy. In 1991, ACEnet started a small business incubator to serve the food sector market niche. The network clustered food-processing companies, trucking firms, and restaurants, and provided support services, access to capital, and cooperatively owned equipment.

Over time, ACEnet spun off free-standing organizations to further enhance network services. The Food Ventures Center provides technical and start-up assistance to local organizations and entrepreneurs, and access to shared warehouse and equipment and an automated distribution hub. TechVentures helps business owners integrate computers into their business operations and also opens up the computer training center to the general public. The Computer Opportunities Program trains teachers in local schools to offer entrepreneurship classes and sets up student-owned businesses. A separate nonprofit subsidiary, ACEnet Ventures, provides venture capital to companies in the network.

## Public–private partnerships

Public–private partnerships (PPPs) are collaborative arrangements between government and the private sector that involve the public partner paying, reimbursing, or transferring a public asset to a private partner in return for goods or services. Government today struggles to deliver public services, often forced to choose between harmful reductions or significant tax increases. PPPs can provide a welcome alternative. Table 7.2 depicts some different levels of PPPs.

Outsourcing and privatization are fairly straightforward, but the blending of the two is much more complex. PPPs are not always well received, with strong opinions on both sides. Supporters cite cost savings of up to 40 percent while realizing more innovation and improved quality of service. Detractors fear violations of constitutional or statutory law, private sector greed, lack of accountability, and an increase in unemployment. The debate can be very polarizing, so it is important to focus on actual case studies when exploring the potential benefits of implementing a local public–private partnership.

PPPs often pair up competitors to create a win-win partnership. For example, the U.S. Postal Service recently awarded a contract to Federal Express for the transportation and delivery of its international Global Express Guaranteed mail service, with about 7400 USPS locations offering the

**▥ Table 7.2** Public–private partnerships

| Level of Partnership | Description |
| --- | --- |
| Outsourcing | Contracting by a public agency for the completion of government functions by a private sector organization. For example, contracting janitorial service for a city hall or contracting the design and maintenance of a city's web presence. |
| Public–private partnership | Means of utilizing private sector resources in a way that is a blend of outsourcing and privatization – an interactive, working partnership. |
| Privatization | The sale of government-owned assets to the private sector. For example, when government turns over prison functions to private providers. |

Source: *National Council for Public–Private Partnerships.*

co-branded service. It may not be as extreme as a Coke machine in the Pepsi plant, but it does seem a bit unusual at first thought.

Union Station in Washington, DC, was condemned property when Congress passed the Union Station Redevelopment Act to restore the building and create a functional transportation center. The $160 million price tag did not include one taxpayer dollar. How was this done? Through a public–private partnership, the facility is owned by the U.S. Department of Transportation and is managed by a private development firm that leases Union Station space to 100-plus retail shops and restaurants with annual sales exceeding $70 million. Retail rents pay for the operation of the public facility and for its debt service.

Given the dire conditions of most K-12 public education facilities, PPPs may offer the best opportunity for major relief. Through a PPP between the District of Columbia public school system and a national real estate development company, a new state-of-the-art school and a new apartment building were constructed on an existing school property. The District approved a tax-exempt bond package to be repaid entirely with revenues generated by the apartment building. The children attend a brand new school built at no cost to the taxpayer.

Similar success stories may be found in other public service areas. The first transcontinental railroad was the product of a PPP when Congress chartered private companies which issued stock to finance construction of the railways. Federal lands along the route were granted to the railroads for private development, helping the private companies further recoup their investments. More recently in New Mexico, a private firm supported by issuance of state bonds is expanding a major highway. The private firm also holds a 20-year contract to maintain the road instead of using the state highway department (National Council for Public–private Partnerships).

One of the most visible community-based organizations is usually the economic development agency. Perhaps the most successful model for economic development is the public–private partnership, which comes in various forms. Some communities have separate nonprofit economic development organizations funded by local government monies and private sector contributions. In other communities, the local government sector may provide funding to the economic development agency housed in the Chamber of Commerce. Nationally, the average funding share for local economic development is approximately 50 percent public and 50 percent private.

The litmus test for assessing effectiveness of PPPs is this: Has the partnership enabled government to act more efficiently and better utilize its limited resources to meet critical societal needs? Has the public been well served by the public–private partnership? While there is clear evidence supporting the use of PPPs, there are important cautions to take when delivering essential public services:

- Provide for public involvement in the process through community meetings or public hearings to educate and gain approval from affected constituencies.
- Allow considerable time for the planning process.
- Provide for public disclosure of the PPP agreement including financial arrangements and guarantees.
- Clearly define performance guarantees and associated penalties and/or incentives in contracts to minimize the risk of disreputable contractors in the marketplace.
- Finally, don't eliminate the potential of PPPs because of their complexity or newness. Consider them as one possibility among many that may be utilized for community development.

## Regional initiatives

Regional initiatives are important because they provide a framework for addressing society's most complex problems. Communities are finding that issues of air quality, transportation, infrastructure investments, and economic development are well beyond the ability of one municipality or one organization to handle alone.

Regionalism involves formal institutional arrangements, shared decision making and participation of governing institutions throughout the region, and varies structurally according to objective, project scope, who is involved, and time requirements. Regional collaboration among government units can take many forms such as consolidated government functions (combined police forces with equal authority across multiple jurisdictions), metropolitan planning councils, special service taxing districts, and joint service agreements. Regional collaboration is also possible among citizens' groups, area coalitions, and alternative planning organizations (National Association of Regional Councils).

Regionalism can be especially effective in economic development. More and more cities and counties are banding together to create regional marketing organizations in order to "get on the radar screen" for investment projects. Regionalism can be one of the best tools for rural communities that otherwise would not have the resources to market themselves, or even to develop infrastructure by themselves.

## Regional leadership programs

In order for regional initiatives to succeed, leaders must adopt a regional attitude. Similar to local leadership development programs, regional programs focus on increasing understanding about issues of the region, developing collaborative skills, and building personal and trusted relationships throughout the region. Programs come about due to a variety of reasons.

The Kansas City Metropolitan Leadership Program was created by a number of civic leaders who were upset that progressive public issues, such as bond issues and school assessments, never seemed to be aired. In Atlanta, Georgia, the Council of Governments was about to launch an extensive regional visioning process and it created the Regional Leadership Institute to prepare a cadre of citizens to provide leadership for the project. The IDEAL Program was developed in California's Central Valley to provide skills to a growing immigrant population that was assuming leadership positions. In Denver, a group was butting heads over a controversial issue and discovered that the process was much easier once they got to know and trust one another. From that emerged the Denver Community Leadership Program.

Regional programs are just as varied in their organizational structure. Some are run by regional government groups, by regional nonprofits or citizens' groups, some by a local university, some through a consortium of other civic groups. The underlying foundation, however, is an understanding of the importance of regionalism and building the capacity to achieve positive results at that level (Alliance for Regional Stewardship).

## Metropolitan planning organizations

Metropolitan planning organizations (MPOs) act as development intermediaries for federally funded programs and are governed by city, county, and state governments as well as representatives from various community stakeholder groups. Federal highway and transit statutes require, as a condition for spending federal highway or transit funds in urbanized areas, the designation of MPOs, which have responsibility for planning, programming, and coordination of federal highway and transit investments. These MPOs are composed of local elected officials and state agency representatives who review and approve transportation investments in metropolitan areas.

## Regional government initiatives

Portland Metro is a directly elected regional government, serving 1.3 million residents in 24 cities in the Portland metropolitan area. The Metro Council has a president elected region-wide and six councilors elected by district. The Council is responsible for growth management, environmental conservation, transportation planning, public spaces, solid waste and recycling, and owns and operates the Oregon Zoo (Portland Metro).

## Planning and development districts

With the passage of the Economic Development Act of 1965, a state-wide system of planning and development districts was put into place in many states to provide a single system of planning, development, and programming from a regional approach. Because of its multi-functional capability, the PDD is used by many federal and state agencies as a delivery organization for those programs. Cooperative relationships are maintained through formal and informal contacts, partnerships, and memoranda of understanding which focus on areas such as technical assistance, education, industrial development, recreation, social services, environment, tourism, zoning, housing, agriculture, communications, consulting, and workforce development.

## Special service taxing districts

Special service taxing districts are regional efforts such as regional water and sewer districts, fire districts, postsecondary vocational and technical education, library systems, and transportation systems. Districts are legally constituted and can levy taxes for a specific community improvement. These taxes are collected and then redistributed across the district rather than to the jurisdiction where the tax originated. Districts are administered by a board that may be either elected or appointed.

## Foundations

Community foundations are nonprofit, tax-exempt public charities primarily funded by contributions from individuals, corporations, government units, and private foundations. Community foundations focus more on grant making than on providing direct charitable services. Most importantly, through an organized and deliberate effort, they increase awareness of philanthropy and provide the community with a systematic approach for charitable giving. Community foundations are endowed by individuals or groups and may be for general or specific purposes. It is common for there to be a state-wide community foundation that also makes grants, in addition to providing support infrastructure for the network of local foundations.

The Arkansas Community Foundation (ACF) is an example. ACF coordinates the work of 26 affiliated foundations across the state. In addition, ACF provides staff support and manages the investment portfolio for many of these organizations.

Regional foundations are also established such as

■ **Table 7.3** Charitable foundations

| | |
|---|---|
| *Private foundations* | |
| Family | Donor's relatives plan a significant governing role. |
| Independent | Makes grants based on charitable endowments. |
| Operational | Uses the bulk of its resources to provide a service or run a charitable program of its own. |
| Company | Corporate giving programs. |
| | |
| *Public foundations* | |
| Community foundations | Funded by public giving and focuses more on grant making than on providing direct charitable services. |

Source: *Council on Foundations.*

■ **Table 7.4** Faith-based organizations

| Type | Description |
|---|---|
| Congregational | Individual congregations and their denominational organizations, such as the Roman Catholic archdiocese, and their service organizations, such as the Jewish Federation or Notre Dame University. |
| National networks | National networks of special purpose providers formed to mobilize energies of individuals and congregations around specific projects such as Habitat for Humanity or the Christian Coalition. |
| Freestanding | Independent organizations that are not part of a congregation but have a religious connection organized to pursue special development objectives. |

Source: *U.S. Department of Housing and Urban Development.*

the Foundation for the Mid South (FMS). These regional foundations create partnerships between state-wide associations of nonprofits, community development corporations, faith-based organizations, and other CBOs.

In addition to public community foundations, there are several types of private foundations that play an integral role in community development. The Council on Foundations (COF) is a national organization that supports foundations development. Table 7.3 contains a brief description of foundations.

## Faith-based community organizations

The attention given to the role of the faith-based organization (FBO) in community development has increased in recent years. Should you doubt this, consider that FBOs represent the third largest group in the nonprofit sector, surpassed only by health and education. There are some 350,000 congregations in the United States with estimated yearly expenditures exceeding $47 billion. Places of worship have long been noted for their prominence in providing food, clothing, and shelter to people in need. More recently, however, FBOs have expanded their areas of involvement and are now quite active in workforce training and housing initiatives and, with federal funding of FBOs gaining wider acceptance, the trend is likely to continue.

The three types of FBOs – congregationally based, national networks, and free-standing religious organizations – are described in Table 7.4.

Community development corporations are common examples of congregational sponsorship of a free-standing organization. The Abyssinian Development Corporation was established by the Abyssinian Baptist Church in Harlem and received major in-kind support from the church including office space and infrastructure, donated management services from skilled congregational members, and extensive volunteerism from among church members. The Allen Methodist Episcopal Church in Jamaica, New York, was originally structured as a coalition of urban and suburban churches following the 1968 Newark riots, and has grown to offer its own affiliate organizations that provide economic development and social services initiatives.

Other examples of faith-based projects also include economic development projects such as commercial real estate developments, full service credit unions, micro-loan funds, and workforce training programs. The Jobs Partnership began in Raleigh, North Carolina, when the owner of a construction company had to turn down business because he didn't have enough qualified employees and the pastor of his church had congregants desperate for work. Beginning very informally, the Partnership has now been replicated in at least 27 cities and has placed over 1500 individuals in good paying positions.

Faith-based programs have several advantages including access to volunteers, access to financial and other types of resources resident within the members of the congregation, and a reputation as trustworthy and working for the public good. FBOs also have the disadvantage of being narrowly viewed as "church" and have the potential for conflict between religious views and the secular marketplace.

The best-case scenario for probable success is when faith-based organizations secure the services of seasoned community developers or non-faith-based partners. In 1989, the Lilly Endowment launched a national program that encouraged congregations to form partnerships with experienced community development organizations. Lilly funded 28 programs and evaluated progress in 1991. Within this time period, 1300 housing units had been built, rehabilitated, or were under construction; 11 revolving loan funds with almost $6 million in assets had been established; eight new businesses had been created; and seven faith-based credit unions had increased their assets by $500,000. Perhaps as important, and certainly consistent with the principles of good community development practice, evidence was found that the bringing together of religious institutions and community organizations opened up increased possibilities for bridging barriers across racial and class lines (U.S. Department of Housing and Urban Development 2001).

## Conclusion

So, do you still want to start a community-based organization? If you choose to proceed, there is one final point to keep in mind. Community development organizations should always remember that they are only part of a broader community. The strength and diversity of these groups, their relationship to one another, and their ability to form alliances with organizations outside their community greatly influence the level of success experienced locally. Cities with comparable development organizations will experience different results based upon how development is organized and the working relationships among the CBOs.

The concept of comprehensive community development recognizes the varied sectors of a community – social, economic, physical, governmental, cultural, educational, and environmental – and the need to address them in holistic rather than piecemeal fashion.

So, elect a chair, build a budget, and get organized for a most rewarding experience – effecting positive community change.

## Keywords

Community-based organization, community development corporation, business plan, nonprofit, public–private partnerships, community foundations, faith-based programs.

## Review questions

1 What are some fundamental questions that should be addressed before a community-based organization (CBO) is established?

2 What are some of the public and private sector programs that can support the establishment and operation of CBOs? Where do these programs reside?

3 How do independent community-based organizations differ from community development corporations (CDCs)? What are some examples of each?

4 What are the different kinds of public–private partnerships and how can they be used in community development?

5 Why are regional initiatives useful in community? What are some examples?

6 What are the different types of foundations and how can they help with community development?

7 How can faith-based organizations help with community development?

## Bibliography and additional resources

Alliance for Regional Stewardship (2002) *Best Practices Scan: Regional Leadership Development Initiatives.* Available online at http://www.regionalstewardship.org/Documents/LeadershipDev_BestPract.pdf (accessed June 2, 2004).

Allston Brighton Community Development Corporation. Available online at http://www.allstonbrightoncdc.org/ (accessed April 14, 2004).

Appalachian Center for Economic Networks. Available online at http://www.acenetworks.org/ (accessed June 1, 2004).

Arkansas Community Foundation. Available online at http://www.arcf.org/ (accessed June, 2004).

Arkansas Wood Manufacturers Association. Available online at http://www.arkwood.org/ (accessed June 1, 2004).

Association of Metropolitan Planning Organizations. Available online at www.ampo.org.

California Center for Regional Leadership. Available online at www.calregions.org.

Center for Community Change (1985) *Organizing for Neighborhood Development.* Available online at http://tenant.net/organize/orgdev/html (accessed May 11, 2004).

CEOs for Cities. Available online at www.ceosforcities.org/index.htm.

Community Building Resource Exchange. Available online at www.commbuild.org.

Community Development Society. Available online at www.comm-dev.org.

Community Economic Development Corporation of Southeastern Massachusetts. Available online at http://members.bellatlantic.net/vze3h2jm/ (accessed April 14, 2004).

Community Foundations of America. Available online at www.cfamerica.org.

Council on Foundations. Available online at http://www.cof.org/ (accessed June 2, 2004).

Foundation Center. Available online at www.fdncenter.org.

Foundation for the Mid South. Available online at http://www.fndmidsouth.org/ (accessed June 11, 2004).

Franklin County Community Development Corporation. Available online at http://fcdc.org/ (accessed April 14, 2004).

General Services Administration. *The Catalog of Federal Domestic Assistance.* Available online at http://www.cfda.gov/ (accessed April 26, 2004).

Holley, J. *Creating Flexible Manufacturing Networks in North America: The Co-evolution of Technology and Industrial Organizations.* Available online at http://www.acenet-works.org/juneholley/docs.html/concept.htm (accessed June 1, 2004).

International Community Development Council. Available online at www.cdcouncil.com.

International Council of AIDS Services Network (1997) *HIV/AIDS Networking Guide.* Available online at http://www.icasco.org/ (accessed May 5, 2004).

James Irvine Foundation. (1999) *Getting Results and Facing New Challenges: California's Civic Entrepreneur Movement.* Available online at http://www.calregions.org/publications.html (accessed May 29, 2004).

Lane, B. and Dorfman, D. (1997) Northwest Regional Educational Laboratory. *Strengthening Community Networks: the Basis for Sustainable Community Renewal.* Available online at http://www.nwrel.org/ruraled/stengthening.html (accessed June 1, 2004).

Massachusetts Association of Community Development Corporations. Available online at http://www.macdc.org/ (accessed April 14, 2004).

Missouri Department of Economic Development. *Starting a Nonprofit Organization.* Available online at http://www.ded.mo.gov/business/startabusiness/ (accessed June 5, 2004).

National Association of Development Organizations. Available online at www.nado.org.

National Association of Manufacturers. *Manufacturing Networks: Partnerships for Success.* Available online at http://www.nam.org/ (accessed April 14, 2004).

National Association of Regional Councils. Available online at http://www.narc.org/ (accessed June 4, 2004).

National Council for Public–private Partnerships. *For the Good of the People: Using Public–private Partnerships to Meet America's Essential Needs.* Available online at http://ncppp.org/presskit/ncpppwhitepaper.pdf (accessed June 4, 2004).

New Jersey Turnpike Authority. *History of Metropolitan Planning Organizations.* Available online at www.njtpa.org/Pub/Report/hist_mpo/default.aspx (accessed June 10, 2004).

Parzen, J. (1997) Center for Neighborhood Technology. *Innovations in Metropolitan Cooperation.* Available online at http://info.cnt.org/mi/inovate.htm (accessed May 13, 2004.

Portland Metro. Available online at http://www.metro-region.org/ (accessed June 10, 2004).

U.S. Chamber of Commerce. Available online at www.uschamber.com.

U.S. Department of Housing and Urban Development (2001) *Faith-Based Organizations in Community Development.* Available online at http://www.huduser.org/publications/commdevl/faithbased.html (accessed May 15, 2004).

U.S. Government's Official Web Portal. Available online at www.usa.gov.

# 8 Developing community leadership skills

## David R. Kolzow

To be successful, community development programs must have input and support from the broad spectrum of stakeholders. Because these stakeholders often have different agendas, good leadership is necessary to bring all parties to the table, reconcile differences, achieve consensus on a vision and path forward, and then lead the charge. Leadership skills are not a birthright – they must be learned and honed continuously. Successful communities make leadership development programs an integral part of their community development plan. Good leadership in both the public and private sectors is necessary to maximize success in community and economic development.

## Introduction

In all rural areas, community after community is struggling to figure out what needs to be done to create a healthy economy and an attractive quality of life. The problem is complicated by the accelerating rate of change in the nature of economic activity, globalization, growing concerns for local educational capacity, increasing demand for a broader array of municipal services, the impact of the Internet, and so on.

In particular, the dynamics of the new economy are placing additional pressure on communities to develop leaders who understand what it will take to survive as a community in this new environment. This new economy is knowledge-based and requires knowledgable and involved local leadership to take advantage of it. In addition, the shift from a manufacturing or resource-based local economy to a service economy has left many communities, and particularly rural ones, struggling to determine their destiny.

Anyone traveling around the U.S. or other countries will come across a number of communities that are responding more effectively than others to these challenging times. Why are some communities able to rise above their problems and achieve a more sustainable living and working environment, while others are bogged down in poverty, high unemployment, and a deteriorating quality of life?

The answer is most likely to be found in the quality of local leadership, often the critical ingredient that determines whether a community overcomes its limitations or remains mired in them. Evidence strongly supports the conclusion that it is the effective involvement of local leaders that leads to successful communities. In these rapidly changing times, it is critical that local leaders understand how to manage change to the benefit of their community. Furthermore, if these leaders lay out careful plans of action and pursue them with some measure of unity, it is amazing what can be accomplished to change the destiny of the community.

The premise of this chapter is that the positive impact of local leaders in community and economic development can be dramatically improved if that leadership is "enlightened" and committed. This premise seems reinforced by the growing interest in community leadership, as seen by the increasing number of books and articles on the topic. This chapter provides a brief overview of how to create a stronger and more effective set of leaders in the community.

## What does it mean to be a leader?

It is difficult to discuss the development of a stronger leadership base without common agreement of what "leadership" means. Unfortunately, the definition of "leader" or "leadership" is frequently misused or misunderstood. For that reason, it is important to begin by attempting to get a better grasp of what these terms mean.

## Leadership defined

An extensive review of the literature reveals that no one has satisfactorily defined what leadership is, especially in the context of a community. The term often refers to anyone in the community who has relatively high visibility. However, a leader should be identified as someone who is more than a widely recognized individual or a local official. Recognition alone does not constitute leadership.

Most articles and books on the topic of leadership conclude that it involves influencing the actions of others. According to Vance Packard, "leadership appears to be the art of getting others to want to do something you are convinced should be done." Leadership implies "followership." A community leader emerges when he or she is able to get a number of community residents and/or businessmen to strive together willingly for leaders' goals.

How does one become a leader in a community? This doesn't occur by simply declaring oneself a leader; others need to acknowledge that leadership. The followers actually determine whatever "real" power the leader may have. No simple formulas or models exist to guarantee that one can achieve leadership. So, how does one attract followers? It would appear that the level of credibility of an individual is the single most significant determinant of whether he or she will be followed over time. Leaders create followers because they are able to bring about positive change in others' understanding. This is in contrast to leaders who emerge from a group of citizens who react to an adverse situation, often by

doing nothing more than getting in front of the parade.

A leader's ability to get people to do something comes from a strong desire to see a particular outcome and a commitment to pursue that conviction. It has been said that "leaders are just ordinary people with extraordinary determination." Participants in a state-wide "town hall" in Arizona agreed that *a leader is someone with the inherent ability to visualize the goals that must be accomplished, the courage to accept the risks associated with the struggle to attain these goals, and the skills to develop a consensus.*

Another important characteristic of effective leaders is their ability to inspire people. Inspire them to what, however? Community leaders have to believe in and communicate something worthy of inspiration. People will only be inspired if they strongly want and believe in what the leader stands for. They have to perceive that the leader is going somewhere worthwhile if they are to be persuaded to go along. The ability to communicate and to stimulate action is probably more important than any other specific personal leadership style or characteristic. Leaders must be able to make a compelling case for the current need for change, or their followers will stay satisfied with the existing situation, no matter how bad it is.

Leadership is not innate. Many seem to believe that leaders come into this world with that capability but, in reality, they are nurtured. Although the *basic desire and motivation* to be a leader cannot be taught, if a child develops self-worth, self-confidence, and a concern for others, that individual begins the process of becoming a leader. Furthermore, basic leadership skills – such as the ability to listen actively, to clearly state one's position, to work with others collaboratively, and to negotiate solutions – can be developed through a combination of education, training, and experience.

Few communities are totally lacking local leaders. This leadership shows itself in the activities of local organizations, government, school districts, civic clubs, and so on. Typically, a wide range of talent exists that can be put to good use in moving the community forward. Unfortunately, far too often, local leaders don't work together toward common goals and visions for the community, and little improvement is accomplished. The good news is that, if appropriately directed, leaders can develop a shared vision and goals, and can work together to determine priorities and strategies to achieve the desired future. Additionally, when leaders of both public and private sectors work together, a community can accomplish more than it may ever have dreamed possible.

## Power and influence

As noted, leadership involves influencing others, particularly with respect to embracing the leader's ideas. Since "power" may be defined as the ability to influence the actions of others, leadership may be viewed as the effective use of power.

Power, as defined by Stephen Covey (1991), is having the strength and courage to accomplish something. It is the ability and the desire to make choices and decisions. As a leader, the appropriate use of this power should be influencing others to embrace those choices and decisions. Actually, it is the perception others hold about a leader's power that enables that person to influence behavior.

A number of ways exist to describe leadership power. For the purposes of this chapter, only two types of this power will be examined: position power and personal power. *Position power* belongs to people who are able to get other individuals to do something because of their position in the community or organization. Frequently, a mayor of a city is a good example of this type of power. Often, individuals holding a position use the power of that position to try to exert the influence they want.

*Personal power* is awarded to individuals who derive it primarily from their followers. Personal power is the extent to which followers respect, feel good about, and are committed to their leader, and see their goals as being shared and satisfied by the goals of their leader. In other words, personal power is the extent to which people are willing to follow a particular leader. As Irwin Federman, the CEO of

"Monolothic Memories," wisely said, "Your job gives you authority. Your behavior gives you respect."

Some individuals have both position power and personal power. Regardless of the type, the amount of power or influence a leader has is related to a number of factors, including:

- Personal qualities (appearance, age, family background, reputation, and communicativeness).
- Control of or access to such resources as jobs, wealth, property, businesses, and prestige.
- Their professional reputation and their technical expertise.
- Their ability to get things done.
- Their confidence and positive outlook.

## The visionary leader

Recently, much research has been conducted which bears out the previous statement that leaders need vision. In the context of community development, leaders should have a strong idea about a new and desirable direction for the community. Generally, local residents receive their vision of the community from their leadership since most have not developed their own. However, unless local leaders are able to clearly articulate their visions, those will remain important only to the individual leader.

The leader's own belief in and enthusiasm for the vision is what inspires others. People will be inspired as they see how the vision can directly benefit them and how their specific needs can be satisfied. By clearly articulating a desirable vision, leaders help the community to see its potential. However, if people are to trust in the vision, they must trust those who promote it. They need to see the commitment and follow-through of their leaders. In other words, these visionary leaders must "walk the talk."

Ultimately, the many visions of the various key community leaders need to be merged into a compelling and "shared" vision held by the majority of them and, therefore, by most of the citizenry. Typically, local leaders have different ideas about what they want to see happen in their community.

However, if there is little agreement among these leaders as to what the future should hold, it is unlikely that much will happen to ensure that future. It is similar to being on a ship with no one in agreement as to its course.

In this author's experience, the most effective means for obtaining a shared vision is to move the key local leaders and stakeholders through a consensus building process in a workshop or retreat environment. Using small group techniques, the basic elements of a common vision can be quickly and efficiently developed. The drafting of the vision can be done by a select committee or the development organization using the input received.

Despite the best efforts of visionary leaders, most of the residents of a community tend to resist getting excited over a vision that takes them into new places. They are most comfortable with what they are familiar with, and are likely to resist change. Unfortunately, it sometimes takes a crisis in the community to shake people and officials out of their complacency. In his book, *Thriving on Chaos* (1987), Tom Peters states that "the most obvious benefit of unsettled times is the unique opportunity they afford to create rapid change. For those of vision, chaos can facilitate innovation." Unsettled or crisis conditions can open doors to new thinking from new leaders. The community needs individuals to step forward who are willing to embrace change and take some risks, thereby helping move the community into a more sustainable and healthy future.

## Community leadership

### Leadership in transition

Good leadership has always been important. Leaders have been critical to communities since people first started living communally. However, the nature of the local "community" has been changing, which requires new thinking about effective community leadership. In the last half of the twentieth century and into the present, the mobility of the nation's population has often resulted in a loss of a "sense of community" in the traditional sense. This has often made it easier for residents to identify with the more *highly visible political* leaders than with the *less visible business* leaders who so frequently played a key role in the past.

It is difficult for many in the community to be aware of potential private sector leadership because many of these leaders may be more involved with their businesses rather than with their local community. Their place of business has often become their "community of interest." Furthermore, as a result of globalization and competition, the business leader today typically experiences a long work day. A segment of potential leadership is thus less available during "regular business hours" when many traditional community development activities take place.

Many potential leaders are also being lost to the community due to the loss of local businesses that they had developed and managed there. By the same token, the influx of branch manufacturing plants, mergers and acquisitions of businesses, retail chains, and so on create a more fluid and mobile set of managers and executives. As these potential leaders are transplanted into the community from some other location, their lack of familiarity with their new surroundings reduces their leadership impact. They are not likely to have as much desire or opportunity to build a following among community residents, nor feel much loyalty to what may be viewed as a temporary assignment.

The shortage of effective leadership is evident in business, government, education, churches, volunteer groups, and every other form of local organization. Complaints regarding the scarcity of leadership talent in communities do not relate to the lack of people to fill organizational positions but to the lack of those who will and can be significant leaders. This presents a major challenge for the community development professionals as they try to create a higher level of leadership involvement.

## Leadership effectiveness

Leaders can have the most significant impact if they represent the "culture" of the community. If the goals and values that the leader pursues are not in sync with those who will be affected by his or her decisions and actions, it is likely that not much attention will be paid to that so-called leader. This means that effective leaders need to know what the local culture really is. Useful feedback on local thinking may be obtained from constituents through such sources as the media, the mail, e-mail, faxes, and face-to-face contact. In addition, polls and surveys may be used. However, a leader must remember that polls tend to reflect rather superficial opinions that are subject to change. In addition, results from polls may not necessarily be the same as the desires and expectations of most local residents.

It would also appear that a fundamental requirement of good leadership is a set of ethical standards that reflect the moral standards of the community. Stephen Covey has stated that "moral authority is the key to leadership." In essence, an effective leader must be able to tell the difference between right and wrong and to know when self-interest is impeding the greater good of the community. It is critical that leaders develop trust among their constituents, a quality which results from their perception of his or her ability, level of commitment, and integrity. The trust and respect given to leaders reflects not only what they do but how they do it.

An effective leader is *transformational*. When the leader has a desired impact on followers, positive change occurs in the community. Effective leadership, then, should be measured by the extent to which a leader has helped to produce actual social or economic change – a change in the way people live and work. According to the "Move the Mountain Leadership Center," transformational leaders are able to articulate a clear and powerful agenda of change that answers the question, "Where, exactly, are you leading us over the next five to ten years?"

As was stated earlier, this transformation is more likely if the goals and desires of the leaders are compatible if not identical with those of their followers.

The concept of "movers and shakers" – a leadership elite who make the major decisions that others blindly follow – is becoming outdated, particularly in the nation's larger communities.

## Collaboration for progress

The most successful community development programs are usually effective public–private partnerships that include a high level of participation by key leaders across all sectors. In addition, these leaders/stakeholders collaborate with each other, which means they work together to creatively solve local problems. The idea is to bring people to the table who have different experiences, knowledge, and perspectives so that new and innovative ways may be found to bring about change. Instead of majority rule, a collaborative process seeks consensus on critical issues. This process of collaboration among the leadership in a community must be nurtured, strengthened, and managed; it doesn't happen by chance.

Unfortunately, the most powerful leaders or stakeholders in a community are rarely open to working cooperatively in solving local problems. Instead, their self-interest drives them to control local initiatives. Those who perceive that they have the most to lose will be the least likely to support significant changes, making collaboration difficult. For example, local manufacturing firms in a rural county might oppose attracting new companies because that might raise wage levels and require new expenditures on infrastructure. Or local leaders might resist expanding the level of tourism, preferring to limit the intrusion of "outsiders" into the community.

Collaborative leaders are the leaders of the present and especially the future. This new type of leadership involves setting aside personal egos and working with others to come to a better decision than any individual leader would have been able to achieve him or herself. It is not telling others what to do, but facilitating a productive "team" process.

## Identifying leaders

### Who are the leaders?

As a community grows in size and economic complexity, it becomes less likely that it will be dominated by just a few leaders. Instead, a multidimensional leadership structure tends to emerge, comprised of a large number of individuals with a variety of specialized interests and leadership skills. As a result, it becomes more difficult to identify many of these leaders so they can be effectively integrated into the local community development process.

Generally, successful long-term community development requires identifying and engaging the most important and influential leaders including those individuals who are highly visible as well as those less obvious individual. It also involves assessing who among these leaders may resist collaborative efforts, how strongly, and why. Finally, it entails developing working relationships with these individuals to facilitate effective action toward community improvement.

Identifying and involving leaders in the community development planning process has two purposes:

* First, to enlist their support in the initiatives needed to stimulate the local economy, improve the quality of life, or manage local growth.
* Second, to educate them so that their vision is based on a realistic set of assumptions that may be shared among the key leadership.

It is particularly important that local business leaders are recruited. Generally, the business power structure is the only local interest group that has a continuing and consistent stake in local development policies and programs. While local policy initiatives generally come from city or county government, these agencies feel more secure about local development policy if they have the support of the major business organizations. It is usually well recognized that these people make or influence private investment and stimulate a healthy local economy.

### Developing local leadership

The problem of developing adequate local leadership may be considered an issue of "civic capacity" (often referred to as social capital). Extensive research on this topic has demonstrated a clear contrast between strong and weak communities. The strong communities have the following in common: a leadership that works collaboratively, active citizen participation, a high level of volunteerism and philanthropy, and frequent public dialogue. It takes a lot of work to make this happen on a sustained basis.

Civic leadership clearly requires strong and competent individuals. However, the overall intent of these leaders should be to empower others and increase participation in community life in order to create a better future for all (transformational leaders). The strategies used to enhance civic capacity include education and training, strategic thinking and planning, creating opportunities for effective volunteer involvement, and a good flow of information about the needs and assets of the community.

Smaller communities in particular tend to have a unique leadership need. Their organizations are often run by volunteer leaders with few if any paid professional staff. This reliance on volunteers for such areas as fiscal management, economic development, education, and program development and implementation, makes the need for leadership training in rural areas even more imperative.

Why have a program to improve the competence of local leadership in the community? The rationale is as follows:

* Consensus decision making by a large group of participants is more desirable than centralized decision making by a small group of officials or professionals.

- Planned change, involving widespread public participation, is possible and more desirable than unplanned or *ad hoc* change.
- The process that produces planned change can be learned and used by a wide variety of leaders who are interested in improving the quality of the community.
- Potential leadership competence exists throughout the population of most communities and can be developed through education, skills training, and leadership experience.

There is a strong relationship between leadership and civic capacity. Fredericksen (2004) points out that effective leaders can help develop communications, networks, trust, and shared values among different groups in a community, thereby facilitating community and economic development.

## Leadership training programs

There are now literally thousands of local, regional, and state-level leadership training programs in the U.S. and other countries. Funding for these programs comes from a variety of sources including membership dues, state and local tax dollars, and foundation support. Many of them are initiated and run by local Chambers of Commerce or nonprofit organizations set up to facilitate local leadership development.

Participants for many of the community programs are chosen either based on their established history of local participation, from within the community, or by an employer. These programs are as diverse in length and scope as they are in number. Typically, the leadership training program lasts one year and involves day-long seminars once a month on specific high-priority topics.

The ultimate purpose of the leadership training program is to bring new individuals into community leadership and to assist existing leaders in being more effective. It is also a means for facilitating communication among leaders by providing an opportunity to work and learn together. Frequently, the leadership program involves team projects and information sharing that provide hands-on experience and a chance to develop awareness of local problems and possible solutions. Most programs have associations or networks for their graduates so that they can continue their learning and their contributions to build stronger communities.

The leadership training programs have proven to be a good method for involving leaders and training potential ones; however, the programs in themselves do not necessarily resolve the local leadership problem. Issues of local fragmentation and lack of unity of purpose can only be effectively dealt with through a consensus building and collaborative process that involves all the relevant leadership. The community leaders need to see themselves as in the same boat, rowing together to get where they want to go.

In addition, much of agenda of the leadership training programs appears to be focused on acquiring knowledge. Participants learn how the community functions, how to plan, what community development is, and so forth. Certainly, knowledge is a necessary first step to changing leadership behavior but, by itself, it is not enough to develop new leaders. The new knowledge must be put into action. That usually requires the acquisition of new *skills*. To fully learn a new skill, people must first understand it, then practice it and get feedback on their performance. Much more attention needs to be given to providing ongoing training in leadership skills at the community level.

Certain leadership skills that can be developed and maintained are:

- problem solving
- planning and goal setting
- critical thinking
- effective decision making
- conflict management tactics
- negotiation techniques
- team building and team management
- group process techniques
- effective listening.

Simply having these skills, however, does not guarantee effective leadership. The successful leader is one who is inspired and motivated to use these skills in the appropriate ways at the appropriate times.

## Getting leadership involved

As important as training local leaders is, it is not enough if the community is to benefit from their skills and knowledge. Frequently, the comment is heard: "I spent a year in this leadership training program but I don't know what to do with what I learned." The follow-up to all of this education and training is as important as the learning process itself. The key is to determine how to most effectively involve these emerging leaders in the local community development process. A number of community leadership programs have set up projects that are important to their development, and involve the new leaders in them. These projects may be an important contribution to their respective communities but should be weighed against the priority development issues that each community faces. Consideration should be given to channeling the effort of these new leaders into activities that will address key development problems.

## Motivating local leaders

Leaders can be energized and motivated if they come to realize that their personal lives and their business activities are dependent on how well the community is doing. Their enthusiasm for getting involved will be greatly increased if they see that they can play an important and decisive role in creating a more positive living and business environment.

In a more practical sense, the appeal to the volunteer leader can be explained in the description that he or she will "give up a dollar today with the hope of making two dollars tomorrow." Hopefully, leaders will look at their investment of time, energy, and money as a means of gaining more for themselves as the community prospers. The "selfish" aspects of this are offset by those of hope. Although there is no assurance of gain for the individual, it may be hoped that if effort is made the community is likely to come out ahead.

It is important not to understate the need for personal sacrifice in soliciting the effort of the volunteer leader. On the other hand, for the community and the leader alike, the possibilities of both financial and other gain should be demonstrated clearly. In the typical community, the number of leaders willing to sacrifice their time and talent is likely to be small. However, these few can be the catalyst for making major improvements in the community.

### Workshops/retreats

One approach to motivate and involve local leaders is to conduct a workshop or retreat to teach them why their involvement is needed and to solicit their input regarding the development needs of the community. The key issues identified by the leaders can also be prioritized through a structured group process. The results from this process can be incorporated into the community development plan. This process is also useful for the board members of the development organization so that they can establish organizational priorities consistent with those of the community at large.

It is usually desirable to hold a retreat at some distance from the community in an unusual surrounding away from ordinary distractions. This fosters a sense of togetherness that facilitates dialogue and problem solving. Over the course of a day or more, this relative isolation also permits a continuity of discussion and input among the participants that is often impossible to achieve in a series of meetings, or within a localized workshop that allows participants to come and go as their own perceived schedules dictate.

### Advisory councils

Another way to involve local leaders is to enlist their participation in various organizational advisory councils. These are frequently formed by various agencies in the community to encourage input and support of their programs. Local government agencies may establish private sector advisory boards. Community development organizations may use the councils or boards for obtaining advice on a variety of key program areas.

Although advisory councils can be helpful in providing insight and a forum for ideas, it is difficult to sustain interest in such group participation. Busy leaders may tire of being part of something that is merely advisory. Furthermore, the advice given may not be heeded by the organization to which it is given, which is frustrating to the giver. As a caution, without a mechanism to address these issues systematically, advisory councils can be reduced to public forums for airing grievances.

### Leadership trips

A number of communities around the nation have initiated what have become known as "leadership trips." Typically, elected leaders or volunteer participants focus on a particular problem in their community and then find a city that has tried and ostensibly succeeded in tackling the same problem. Visits to these selected cities can be very productive, as these visiting leaders are able to see what the community has done and hear from local leaders how they were able to accomplish it. New ideas can then be brought back and, through a process of dialogue and planning, possibly initiated in their community.

### Task forces

A higher and more sustained level of leadership involvement can occur through establishing task forces to deal with specific aspects of community development. They are called "task forces" because their assignments are usually focused on "tasks" or areas of need and members are generally appointed for a specific and reasonable length of time. Task forces can present opportunities to incorporate the expertise and interest of leaders and volunteers into productive channels of activity through direct membership in or consultation to the task force.

Enthusiasm toward their task force appointments can be much higher if the members know they have been carefully selected and are made to feel that their participation is an honor as well as an important contribution. A task force approach provides the advantages of: (1) concentrated activity within a designated time frame; (2) working in a collaborative group; and (3) being part of a network of influence.

Typical concerns of task forces established for community development programs are:

- population growth and housing availability
- workforce development
- stability of the local economic base (e.g., business retention and expansion)
- land planning and environmental considerations (e.g., "sustainable development," "smart growth")
- quality of life trends
- infrastructural demands and capabilities
- local education improvement
- marketing strategies
- program funding
- technology development.

The precise size of such a group will be determined to a large extent by its purpose and the representation needed in its membership. Unless wider representation is needed, most task forces or committees should comprise fewer than 20 people. A smaller task force or committee increases the opportunities for each member to take part and business can be handled in a more informal fashion. Smaller committees not only assemble more efficiently for meetings but tend to be more flexible and able to handle sensitive or controversial topics.

It should be kept in mind that task forces or committees are usually not meant to be action-oriented organizations. These leadership groups are policy makers, thinkers, and planners; they are not doers. Their satisfaction comes from seeing things happen in the community as a result of actions in which they took part. Their role is to help the community development organization: (1) think through problems,

(2) encourage education when needed, (3) determine priorities and timing, (4) mobilize financial support as appropriate, and (5) support and promote a plan of action. The job of implementing the programs then falls on the governmental agencies or private sector organizations responsible for these programs.

### Board positions

One of the most important roles that community leaders can play is membership on the board of directors of the local development organization. However, the board should be an integral part of the organization rather than only a group to whom the executive director reports periodically. The skills and contacts of these leaders usually offer a great deal toward implementing the community development program.

On the other hand, the board should understand that its primary responsibilities are to set policy and evaluate results. They should not take an active role in the day-to-day implementation of that policy ("micromanagement"), which is strictly a staff function.

A more detailed discussion on the role of the board, how it is selected, and how it functions would require a document of its own to do this topic justice.

## Leadership in community and economic development

Leadership and community and economic development share many common elements. As discussed in this chapter, leaders study their community situation, solicit widespread input from all concerned parties, build consensus, create and communicate a shared vision, and motivate people to implement the community plan. Rather than immediately dictating a plan of action, leaders learn about the issues facing their community through an assessment process, as described in Chapter 9. As defined in Chapter 1, building consensus, creating a shared vision, and communicating effectively are all cornerstones of the community development process, as well as being critical to successful economic development. Such development flows from creating a "development-ready" community that not only has a strong physical infrastructure for businesses (prepared sites, good transportation, utilities), but offers good education, health care, and an overall good quality of life.

All too often, citizens do not fully understand the importance of making their community development-ready and quick decisions are made without adequate thought and planning. In some communities, the attitude is "let's find a piece of land and develop an industrial park immediately," which frequently leads to a rash decision to buy a less marketable (and often cheaper) piece of property. Worse yet, the community's attitude may be "this is a great community and business should want to join us; we'll help them find a site when they decide to locate here." The fallacy of this attitude is that most businesses would chose instead to find a good prepared site, or building in another community, rather than run the inherent risk of delay in new site development and, in turn, incur a financial penalty from a later start-up.

With effective leadership, such a scenario would be less likely to occur. Exercising the skills discussed in this chapter, a good leader would make the community development-ready by encouraging it to develop a vision and plan, conducting an assessment of its strengths and weaknesses, and developing both a step-by-step strategic plan and a marketing plan to attract industry to that community.

## Conclusion

It is essential that local leaders become more effectively involved in local community development activity if their communities are to grow and prosper. However, the competitive pressures on companies and government are creating a shortfall in the availability of key leadership. The development profession cannot afford to ignore or gloss over this growing crisis in community leadership. It cannot simply be assumed that the problem will take care of itself.

Therefore, it is important that development organizations give serious consideration to the way they might facilitate the leadership in community development or serve as catalysts in a more positive direction. It is hoped that the information in this chapter will aid in creating more effective local leadership.

---

## CASE STUDY: ATLANTA AND "THE PHOENIX"

Atlanta, Georgia, a 28-county metropolitan area (see Figure 8.1) with a population of five million, is often referred to as the "Capital of the New South." It is one of the fastest growing metro areas in the U.S. From 2000 to 2006, the Atlanta Metro Area gained over 800,000 people, more than any other metropolitan area during that time.

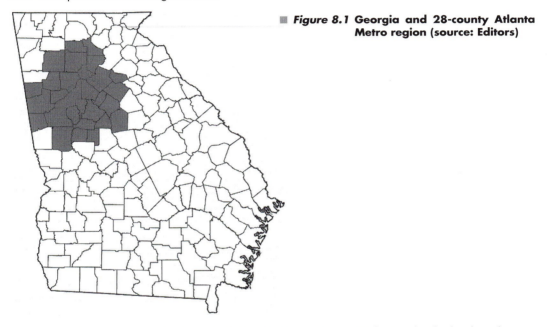

■ *Figure 8.1* **Georgia and 28-county Atlanta Metro region (source: Editors)**

Atlanta is known worldwide for its Civil War history. It was made famous by the book and movie *Gone With the Wind*, which tells the story of the burning of Atlanta by General William Sherman of the Union Army. Following the Civil War, Atlanta was quickly rebuilt and prospered again, due in part to its geographic location and railroad service. The seal of the city features a Phoenix – the mythical bird that rose from its own ashes (see Figure 8.2).

As is the case with many other urban areas, the City of Atlanta itself did not experience the same growth and prosperity as did suburban cities and metro areas as a whole over the past three decades. From 1980 to 1990, the City's population actually declined by 7.3 percent as residents moved to the suburbs. Atlanta was beset by many urban problems such as a high crime rate, a growing homeless population, and a shrinking tax base. As the rate of population growth and the tax base declined, Atlanta failed to maintain and upgrade much of its infrastructure.

Atlanta made international headlines in 1996 by hosting the Summer Olympics. However, Mayor Bill Campbell and his administration were not given high marks for the planning and operational logistics of the games. Allegations of political and financial improprieties dogged the Campbell administration throughout the late 1990s. Race relations and cooperation between the public and private sectors deteriorated as the City became more polarized. After leaving office, Mayor Campbell was indicted on numerous counts, convicted of tax evasion, and given a prison sentence.

Following the failed administration of Mayor Campbell, Shirley Franklin, an African-American and Democrat, became Atlanta's first woman mayor in 2002. Many of the "power players" in the business community did not support her candidacy, believing she was an inexperienced political crony of Maynard Jackson, Atlanta's first black mayor. As Mayor Franklin's supporters celebrated her inauguration at a downtown convention center, they were blissfully unaware of the problems they would have to face. Within weeks of taking office, Mayor Franklin announced that the City faced a budget gap of as much as $90 million, or almost 20 percent of the main operating budget. In addition to budget woes, the City faced the problem of crumbling infrastructure due to years of neglect. The estimate to repair the aging water and sewer system, which in places still relied on wooden pipes, came in at $3 billion.

Mayor Franklin proved equal to the challenge. Using strong leadership skills, she turned a $90 million deficit into a $47 million surplus during her first year in office. The broad consensus in Atlanta is that Mayor Franklin was instrumental in the financial turnaround (Mollenkamp and Rose 2002). The first thing she did was communicate the magnitude of the problem to all constituents, making countless telephone calls to key stakeholders. The consulting firm of J. Bain & Co. volunteered to assist the City with a comprehensive audit and budget analysis. This public–private partnership provided credibility to the findings and recommendations – the highly respected Bain & Co. was regarded as a neutral party that would provide accurate information.

Mayor Franklin developed a broad consensus on her plan to solve Atlanta's problems. She had no choice but to raise property taxes and cut City government jobs and services. To gain the support of business leaders for the tax increase, she demonstrated her willingness to share in the sacrifice by cutting nearly 16 percent of the City's jobs and cutting her own salary by $40,000. She reached out to Republicans and Democrats alike and convinced them of what had to be done. She met personally with labor leaders from public sector unions and showed them the Bain & Co. conclusions that government employment and costs would have to be cut and taxes raised.

Leaders must be familiar with their situation and Mayor Franklin came into office with much experience in Atlanta government. Not only had she worked for the Jackson Administration, she had also served as chief administrative assistant to Mayor Andrew Young. Mayor Franklin successfully

■ *Figure 8.2* **The Phoenix has risen over Atlanta**

created and communicated a vision for the future of Atlanta as a growing, fiscally sound city. Through her credibility as a leader, she generated enthusiasm and widespread support for the vision.

After solving the immediate budget problem, the credibility and consensus support she built helped get a major bond issue passed to repair and upgrade the water and sewer infrastructure. For her role in turning the City of Atlanta around and to acknowledge her leadership skills, *Time* magazine named her one of the year's top five mayors in 2002. In 2005, she received the "John F. Kennedy Profile in Courage Award." Once again, the Phoenix has risen over Atlanta.

The Editors

## Keywords

Power, influence, visionary leader, collaboration, civic capacity, advisory councils, workshops, task forces, consensus.

## Review questions

1  What are some characteristics of good leaders?
2  How is power different from influence in the leadership context?
3  What is transformational leadership?
4  How are leadership and civic capacity related?
5  What are some leadership skills that can be taught in leadership training programs?
6  What are some ways to get leaders involved in community affairs?
7  What can leaders do to facilitate community and economic development?

## Bibliography and additional resources

Bennis, W. (1989) *Why Leaders Can't Lead*, San Francisco, CA: Jossey-Bass.

—— and Goldsmith, J. (1997) *Learning to Lead: A Workbook on Becoming a Leader*, Reading, MA: Perseus Books.

Bryson, J.M. and Crosby, B.C. (1992) *Leadership for the Common Good: Tackling Public Problems in a Shared-Power World*, San Francisco, CA: Jossey-Bass.

Carnegie, D. et al. (1993) *The Leader in You: How to Win Friends, Influence People and Succeed in a Changing World*, New York: Simon & Schuster.

Chrislip, D.D. (2002) *The Collaborative Leadership Fieldbook*, San Francisco, CA: Jossey-Bass.

—— and Larson, C.E. (1994) *Collaborative Leadership: How Citizens and Civic Leaders Can Make a Difference*, San Francisco, CA: Jossey-Bass.

Conger, J.A. (1989) *The Charismatic Leader: Behind the Mystique of Exceptional Leadership*, San Francisco, CA: Jossey-Bass.

Covey, S. (1991) *Principle-centered Leadership*, New York: Simon & Schuster.

Fredericksen, P.J. (2004) "Building Sustainable Communities: Leadership Development Along the U.S.–Mexico Border," *Public Administration Quarterly*, 28: 148–181.

Galloway, R.F. (1997) "Community Leadership Programs: New Implications for Local Leadership Enhancement, Economic Development, and Benefits for Regional Industries," *Economic Development Review*, 15(2): 6–9.

Greely, Paul J. (1995) "Energizing Boards, Commissions, Task Forces, and Volunteer Groups," *Economic Development Review*, 13(3): 24–27.

Herman, R.D. and Heimovics, R.D. (1991) *Executive Leadership in Nonprofit Organizations: New Strategies for Shaping Executive-Board Dynamics*, San Francisco, CA: Jossey-Bass.

Kolzow, D.R. (1999) "A Perspective on Strategic Planning: What's Your Vision?," *Economic Development Review*, 16(2): 5–11.

—— (2002) *Leadership: An Essential for Dynamic Community Development*, Atlanta, GA: Southern Economic Development Council.

Kouzes, J.M. and Posner, B.Z. (1987) *The Leadership Challenge: How to Get Extraordinary Things Done in Organizations*, San Francisco, CA: Jossey-Bass.

Luther, V. and Wall, M. (1994; 2nd edn 2001) *Building Local Leadership: How to Start a Program for Your Town or County*, Lincoln, NE: Heartland Center for Leadership Development.

Lynch, R. (1993) *Lead! How Public and Nonprofit Managers Can Bring Out the Best in Themselves and Their Organizations*, San Francisco, CA: Jossey-Bass.

McCauley, C.D. et al. (eds) (1998) *The Center for Creative Leadership Handbook of Leadership Development*, San Francisco, CA: Jossey-Bass.

Mollenkamp, C. and Rose, R. (2002) "Municipal Bond: New Atlanta Mayor Points the Way Out of the City Fiscal Hole," *Wall Street Journal*, February 14, p. A1.

Morgan, Glen M. (1996) "Creating a Local Leadership Development Program," *Economic Development Review*, 14(1): 36–38.

Mumford, M.D. et al. (2000) "Leadership Skills for a Changing World: Solving Complex Social Problems," *Leadership Quarterly*, 11(1): 11–25.

Nanus, B. (1992) *Visionary Leadership*, San Francisco, CA: Jossey-Bass.

Northouse, P.G. (1997) *Leadership: Theory and Practice.* Thousand Oaks, CA: Sage.

Peters, T. (1987) *Thriving on Chaos: Handbook for a Management Revolution*, New York: Alfred A. Knopf.

Senge, P. (1990) "The Leader's New Work: Building Learning Organizations," *Sloan Management Review*, 32(1): 7–23.

Sogunro, O.A. (1997) "Impact of Training on Leadership Development," *Evaluation Review*, 21(6): 713–737.

Sorensen, A. and Epps, R. (1996) "Community Leadership and Local Development: Dimensions of Leadership in Four Central Queensland Towns," *Journal of Rural Studies*, 12(2): 113–125.

Wheatley, M.J. (1994) *Leadership and the New Science: Learning About Organization from an Orderly Universe*, San Francisco, CA: Berrett-Koehler Publishers.

Yukl, G.A. (1981) *Leadership in Organizations.* Englewood Cliffs, NJ: Prentice-Hall.

# 9 Community development assessments

## John W. (Jack) Vincent II

Before a community develops a strategic plan, before it develops a marketing plan to attract new jobs, before it develops action steps to address community problems – in short, before it does anything – it should complete a community development assessment. A good assessment forms the foundation for a successful community and economic development effort. Without a good assessment, communities are usually "driving blind" with regard to the feasibility of their strategic planning and marketing programs.

## Introduction

Community leaders often approach the community development process with a "let's get started" mentality. "We know what's wrong. Why delay getting started? We have talked about this for years; that's all we do is talk." They believe that an assessment is not needed to decide when to begin implementing changes in their community. They are anxious to get started and usually want to begin by creating a strategic plan or implementing some very specific initiatives. These often address the most visible of problems: to improve the community. While they may be correct, sometimes they may not be. Whether or not their actions will result in positive change is another matter.

## Why conduct an assessment?

Before beginning a community development effort or creating a strategic plan, an assessment should be performed to determine what assets are present for development and what liabilities exist that need to be addressed in order for desired improvements to occur. This assessment will identify the strengths on which planned development can be built and identify the weaknesses that need to be eliminated or mitigated as much as possible to give the community the best probabilities for success. The community development assessment process is directed toward supporting the creation of a strategic plan that will guide a comprehensive development effort and involve other citizens in the process. This assessment also provides specific information that helps leaders identify opportunities to be exploited and threats that need to be considered when creating and implementing a strategic plan.

Beginning a community development process without conducting an assessment is like a doctor prescribing prescription medicines without first giving a patient a thorough examination in order to get a correct diagnosis. He or she could be treating symptoms rather than root causes of problems, and the treatments may interact with other conditions to cause additional problems. In the same way, community leaders who act without conducting an assessment can spend valuable time and scarce resources treating only the symptoms. They may not see the long-term improvements they desire if they do not identify and work on root causes of problems. Their efforts could simply become another failed community effort, added to those that preceded it.

A common goal in many communities, for example, is the creation of new, high-quality jobs that offer advancement and career opportunities. Citizens want job opportunities for themselves and their children. In support of that goal, a community development group may identify and market available buildings and greenfield sites, and tout their transportation and services infrastructure. They may also offer development incentives. Yet the desired development may still not come because the leaders may not have addressed the need for workforce development, the poor appearance of their town, mismanagement of public revenues and government operations or problems in the local housing stock. There could be problems in the local school system that make the community a less desirable location for workers and management who would transfer in to operate a new facility. There may also be a poor quality of life due to a lack of recreational and cultural activities. Considering and working on only a few of the factors related to successful development will probably not result in bringing in the new jobs that are sought after.

## Other benefits of assessment

In addition to the data produced for decision making, a formal and comprehensive community development assessment can provide major secondary benefits in support of the community development process. It can help initiate a community dialogue in which citizens discuss problems and issues and agree on the future direction of their community. Many citizens are often surprised to find that they share similar values with others who they had perceived to be quite different from them. This realization helps citizens focus on problem solving versus focusing blame within the community. This initial dialogue feeds directly into the production of a vision statement toward which all citizens can identify and work.

Many citizens may not even be aware of their community's problems or how poorly their community compares to others. Feelings of frustration and dissatisfaction are often focused inward, and this disunity can dissuade outside investors and new residents from investing in or moving into the community. Community forums held in support of the assessment help identify problems and help citizens understand what is wrong with the status quo. This awareness creates additional momentum for planned change when it is properly focused on problem solving rather than on personal attacks or hidden agendas. Further, the final assessment report also serves as a data source for creating a marketing plan or a community profile for responding to inquiries for information.

## Project planning vs. strategic planning

Assessments are performed in communities for many reasons. Some are performed by internal groups in support of a particular community project, program, or initiative such as a Main Street program. Others are performed by individuals from outside the community such as business representatives seeking sites for relocation or expansion. These external assessments, usually performed by site selection specialists, are likely to be very specific, examining the economics of a specific building or greenfield location and the area labor force. It may involve a general overview of all community factors related to a company's operations such as workforce availability, prevailing wage rates, transportation infrastructure, utility capacities and costs, business development incentives, and tax rates. The assessor attempts to answer specific questions related to that one project or initiative.

Community development assessments, however, are comprehensive reviews of the community aimed at supporting a host of initiatives, programs, and projects. They are normally performed for (and possibly by) community leaders to help guide the creation of a comprehensive community development plan. That plan usually includes several goals and a multitude of objectives (the programs, projects, and initiatives that support the plan). This chapter will provide an overview of a community development assessment and provide some guidance as to how one might be performed.

## Quantitative and qualitative data

Traditional economic development assessments focus primarily on quantitative data. They include population demographics (e.g., education and income levels), tax rates, wage rates, and other objective data on which business decisions can be made. These focus on such business-related topics as cost–benefit ratios, return on investment, cost of operations, and profitability of operations.

However, many business location decisions toward a community's successful development are also influenced by more subjective or qualitative factors. Because they are subjective, however, does not necessarily mean that they cannot be measured. Citizen opinions, for example, can be measured and tracked over time. Community spirit and a progressive "can-do" attitude can also be observed and measured. Qualitative factors include the underlying attitudes in a community, the way citizens feel about themselves and their community, and how those feelings are interpreted into visual expressions of pride, cleanliness, friendliness, pro-business attitudes, and can-do spirit.

Many times qualitative factors are what drive a business decision about locations with similar quantitative benefits. The location selected may be chosen because it is perceived to be a quaint, historic, and safe community. It could be that the community has wonderful curb appeal due to its cleanliness, well-landscaped public places, and private property. It may offer a variety of recreational and cultural activities that make it a great place for leisure activities. Often, a community's qualitative factors can tell an assessor far more than the objective data alone.

## Defining the community

So what is a community? It is important to define the community before beginning the assessment. At first, many communities performing assessments or conducting strategic planning define themselves by legal boundaries. However, that is usually not an accurate description of their community. It defines the community too narrowly and results in artificially excluding resources and allies that have a vital stake in the community's success. A community is as large as the area it impacts or from which it draws its existence.

Many sister cities, such as St. Paul and Minneapolis, Minnesota, are really one large community. Similarly, West Memphis, Arkansas and Memphis, Tennessee, are part of the same community even though they are in different states. Legal boundaries only play a small role in determining what needs to be considered when conducting a community development assessment. It should be remembered that the word "unity" is literally a part of the word "community." Community development is not about "him and me" or "us and them." As John F. Kennedy said in his inauguration speech, "A rising tide raises all boats." What affects me also affects my neighbors.

An individual who lives outside a town's boundaries yet owns and operates a business in town is a stakeholder in that community. A business located just outside the town's boundary, but using the town's water, wastewater, and utility system, is also a stakeholder in the community. Rural residents who rely on the town for shopping and services are also stakeholders. Being a resident of a town is different from being a citizen of a community. As a community goes, so go the fortunes of all its citizens.

The assessment and planning process needs to be broadened to include all stakeholders – those within the legal boundaries as well as those within the community's "impact" boundaries. Limiting the community to the legal boundaries of a town, city, or county also limits its ability to claim nearby resources as part of its assets. Is the large plant just outside the city limits in the community? What about the interstate highway that is five miles away?

A more appropriate definition for a community would be the geographic area with which people identify themselves as well as the area served by a town's retail and service sectors. The actual community might be a portion of one or more counties. It could include farmland and unincorporated villages and may span a state line.

## Comprehensive assessment

Assessment is basically a process of asking and answering questions about key factors that influence the community's potential for planned growth and development. A community development assessment is a broad assessment, since many factors within a community are interrelated and influence each other.

If a community seeks new residents, for example, it must consider such development factors as housing availability, construction time for new housing, price, and quality. If it is seeking families with school-age children, it must consider the quality of local schools, since housing location decisions are often driven by school districts and their perceived performance. When retirees are sought as new residents, other factors such as health care facilities and recreational alternatives become more important. The existence of such resources as retirement homes, assisted living facilities, organized leisure activities, physical therapists, waterways and lakes, and golf courses can also influence retirees' location decisions.

If a community is seeking new business or expansion of existing businesses, community leaders must consider the availability of labor and their skills, since these skills relate to the type of work required by the industries being sought. In addition to the current skills possessed by the workforce, communities planning to attract business must consider workforce training capabilities. Are there educational institutions that can provide specialized training to support the industry types being sought? Will local educational institutions work with business to develop and deliver new training courses?

## Data-collection methods

Community development assessments use a variety of methods for collecting and analyzing data. Any data-collection technique can result in misinformation and errors. It is not the author's intent to provide details of all scientific data-collection methods but to provide an overview of some of the more common ones. For most community development activities, qualitative data collection does not need to meet strict scientific research standards. What is usually being sought is information that provides general direction and that identifies broad categories of community advantages or problem areas. To minimize the chances of collecting incorrect information, it is suggested that multiple methods of qualitative data collection be used and the information collected by one method be verified through another method.

Whenever data are collected from individuals, respondents should be told that their identities will be protected. They can be told that while some of their comments or suggestions may appear in the final assessment report, their identity will not be associated with the comments. In that way, the assessor can provide a safe environment for the respondent to share candid opinions. Specific quotes and comments that represent the general feelings and opinions found in the community may also be included in the report on an anonymous basis. Some data collection methods are listed and described below.

*Research and read*

The Internet provides a huge amount of information that supports the community assessment process. It puts the power of professional research in the hands of volunteer community leaders. A professional community developer can use the Internet to glean a great deal of information about a community before he or she even visits it. The federal census site (www.census.gov) provides a vast array of demographic data on housing, income, race, education, and a variety of other areas. When compared with past census data, the information can provide trends that also help identify strengths and weaknesses. Similar data may be found on the sites of other federal agencies such as the Department of Energy and the Department of Labor.

A search engine can also provide links to other sources of data including a state's department of education and department of economic development. For example, under the "no child left behind" initiatives, many state education departments publish online data about school systems and individual school performance. Another good source of data is the state's labor department or job service office which also provides current information on unemployment statistics and workforce demographics. If there is a topic of interest, usually an Internet search can result in the identification of a reliable and respected source of data, and much of it is free.

Other sources of data include a community's public utilities, enforcement agencies, and local employers and merchants. If a community assessor calls and introduces him- or herself over the phone, many utility representatives will provide data in support of the community development initiative. For example, utilities can provide data on number of customers as well as growth and decline in various indices such as power, water, residential and commercial telephone subscribers, and wastewater use.

## Observe and listen

This author often drives into and all around a community on a "windshield tour" the afternoon before the day he is expected. He visits local coffee shops and stores. By politely eavesdropping on conversations and asking a few questions during the visit, a great deal of information is collected. These methods often produce many qualitative insights about a community's strengths and weaknesses.

If "a picture is worth a thousand words," then a drive around all areas of the town can be an eye-opening experience. Be certain to travel not only the main streets but back streets as well. Later, you will likely find that even some locals do not know all that may be found in their own town's back streets.

## Use a camera

Digital and 35mm cameras are valuable tools to record support for visual observations. Later, when making public presentations or when compiling reports, it is possible to use pictures to reinforce

major findings. Photographic images give residents the ability to visually visit all areas of the community including areas that most citizens do not visit or see. The images provide them with a more comprehensive understanding of their community's strengths and weaknesses.

## One-on-one interviews

The benefit of one-on-one interviews is that interviewees often feel free to be open and honest, especially if they do not have to be concerned about their identity being compromised. Usually there are a number of key informants in a community who are extremely knowledgeable. This usually includes the school superintendent, school board members, the mayor, council members, the police chief, the fire chief, the city engineer, major property owners, Chamber of Commerce executives, economic development executives, ministers and priests, public housing officials, neighborhood and civic organization leaders, major business owners/managers, long-term residents, bankers, and leaders of minority groups.

Before beginning the interviews, it is best to develop a structured interview form. This provides the interviewer with a consistent set of questions for all respondents. Asking these same questions often produces different answers and perceptions that can be explored further. Usually, one or two responses to questions will lead the interviewer to ask follow-on questions. On the structured interview form, list some open-ended questions that allow respondents to tell you what they want you to know. Leave some blank spaces to record comments and notes. Comparing responses when tabulating the results also helps corroborate qualitative data.

Some interviewees don't like to just respond to questions from your interview form; they have several key issues they want to "vent" about. You can usually tell from body language and clipped responses when they are chomping at the bit to talk about their issues. In these situations, it is usually better to let the interviewee lead the discussion. Let him address his major issues (he will appreciate the fact that someone is listening) and try to get answers from him on some other areas of interest from your survey form if you have time. Often these one- or two-issue interviews will give you a wealth of in-depth knowledge about the community.

### Community meetings

Community meetings open to all interested citizens can be held to gather information, opinions, and ideas. These meetings need to be planned and managed very carefully so as to avoid their becoming divisive and disruptive. At the very beginning, an agenda and topics of discussion need to be laid out as well as ground rules that govern the meeting. An announcement can be made that a major community revitalization effort is being launched and that the citizens' ideas and opinions are being solicited so that the most important issues may be addressed.

First, participants might be asked to identify the things that they believe are community strengths. Then they may be asked to identify things they believe are weaknesses. These topics can be catego-rized, and the group can be asked for ways to build on or take advantage of strengths and what needs to be done to eliminate or lessen the impact of weak-nesses. Participants might also be asked to identify factors outside the community that present opportunities that should be pursued or threats that should be considered.

The focus of these meetings should be positive, not on "How did we get where we are?" but on "What do we want our community to become?" and "What do we need to do to make it that way?" If a professional facilitator is not used to manage the meeting, it is recommended that a well-respected community leader lead the meeting to prevent it from becoming a target for any dissatisfaction or frustration that may relate to the topic. This indi-vidual should be very tactful, focused, outcome-oriented, and experienced at managing meetings. Above all, he or she should not be perceived as strongly on one side or another of a critical community issue.

### Focus group meetings

Focus group meetings are similar to community meetings but are directed toward one topic or a few related topics such as job creation and economic development, tourism and recreation, or education and workforce development. Citizens who attend these meetings usually have sincere and deep feelings about the importance of the topics being discussed. Many may be experts in these fields. They also have ideas and suggestions for what needs to be done to solve the problems related to the topic.

The strategy for these meetings should be similar to that of community meetings. The leader needs to have a specific agenda and objectives for the meet-ings. This should be shared with the group along with any ground rules for managing the meeting, such as "no personalizing any comments or attacking any individuals." Focus group meetings need to be problem identification- and solution-oriented. The agenda should involve a set of questions or topic areas in which participants can give their opinions about problems, and offer observations about and solutions to problems.

These meetings also help identify individuals who may later work on strategic planning commit-tees, provide information or services to strategic planning committees, or volunteer to help work on projects, programs, and initiatives developed during a strategic planning process that follows the assessment.

### Questionnaires and opinion surveys

Questionnaires and opinion surveys are another data-collection method that may be used to collect qualitative information from citizens. While there are data-collection problems associated with these types of methods, they still provide another way to collect data and involve citizens. These methods may also appeal to those who would not normally attend a meeting or participate in an interview.

Distribution and collection of questionnaires and opinion surveys should be planned carefully so as not to exclude any particular group of citizens and to ensure a good cross-section of community representation. It is sometimes possible to distribute them through prominent citizens who bring them to civic organization and club meetings where they are completed, collected, and returned. The local newspaper may even publish the survey so that it can be mailed in by respondents for tabulation. Pastors may distribute them at churches.

Thanks to modern technology, these can also be posted on the Internet at very low cost. As of this writing, for example, a short questionnaire can be set up at www.branenet.com for under US$5 a month. As each vote is submitted, the results are tabulated and reported in a graphic format. Caution should be used with online surveys due to the fact that those without computer access cannot participate, and anonymous "ballot-box stuffing" may occur.

Traditional questionnaires and surveys can also produce a large volume of paper that must be processed. Care must be taken in tabulating the results to be sure that all responses are accurately captured and reported. However, surveys are useful, since they offer confidentiality to respondents and may reach a whole group of individuals who, for many reasons, will or cannot participate in the assessment process through other means.

Citizens groups should not be overly concerned about collecting data through written surveys, meetings, and interviews. Care and thought should be given to designing the questions to be asked or in planning and managing the meetings to be held. There are many books available at libraries and bookstores that deal with tests and measurements (in education) and surveys and research methods (in business and science). Many of these provide chapters on question design and ample guidance on response options. All things considered, it is always best to use a number of data-collection sources and methods so that information collected during the assessment can be corroborated through multiple sources.

## Community assessment topics

As previously mentioned, assessment is a comprehensive process that involves a review of all major sectors of a community. The process attempts to involve a broad cross-section of stakeholders in identifying the factors to be considered in planned growth and development. It examines four broad areas: physical infrastructure, social infrastructure, economic development infrastructure, and human infrastructure. Within each of those major categories, many factors are considered. Each of these factors can be reported as chapters in the assessment report, and include both the quantitative and qualitative data collected from the many methods described above.

## Physical infrastructure

When considering the physical infrastructure, community leaders need to examine the factors that will influence a business's operations in their area. Shown below are the areas that are evaluated and a sample of the types of questions that might be asked about each factor.

*Transportation system for moving goods*

*Highways* – Are the highways to and from town in good condition and well maintained on a regular basis? Are they two lanes or four lanes? Are they primary or secondary roads? Do they have shoulders on which disabled trucks may pull over? Are any future improvements planned? How far is the community from the nearest interstate highway?

*Rail* – Is there rail? Is it a main or secondary line? Is there a siding on which cars may be stored? Are the rail rates competitive?

*Airfreight* – Is there a local general aviation facility or international airport? Is local airfreight available at that facility (other than from FedEx, DHL, and UPS)? Are their rates competitive? What are the heaviest items that service providers can take?

*Transportation network for business travel*

Is there an international or hub airport nearby? Is there a general aviation facility (GAF) nearby? What is the driving distance and time? What is the length of the runway at the GAF? What flight and landing services are provided at the GAF? Does the GAF have instrument flight rules (IFR) support equipment? Is it lit at night? Is the GAF tower manned 24 hours? Does the GAF have aircraft maintenance services? Does the GAF have hangar space available? What leasing arrangements are possible at the GAF?

*Weather and geography*

Geography and weather often drive location decisions. Many northern companies have chosen southern locations over the past 30 years. The Sunbelt, particularly the southeast, is the leading growth region in the United States. When moving to new locations, residents and businesses alike often look for areas with sunny weather and weather that lacks extreme snow and rain events. Businesses in particular often seek sites with land that is well suited for development and that does not require large sums of site preparation money. They also usually seek locations in or near major metropolitan centers so that they can access the amenities and workforce skill found in these areas. Therefore, geographic and weather conditions can play a major role in relocations and expansions. Community leaders would do well to ask the following types of questions about their areas:

> Is the community isolated or near other towns? Is the area near a Standard Metropolitan Statistical Area (SMSA)? Is the area prone to flooding? Are there other hazards to structures in the area such as earthquakes, forest and grass fires? Does the area have relatively flat land that can be easily developed? What is the average rainfall and snowfall for the area? What are the average high and low temperatures? How many days a year, on average, does the temperature fall

below freezing? How many days, on average, does the temperature exceed 100 degrees Fahrenheit? Will the weather lead to transportation and construction delays or work stoppages? What are the soil and subterranean conditions in the area (e.g., will there be higher construction costs related to site preparation due to the soil conditions)? Are there any EPA emission restrictions that govern the area (e.g., is the community in an ozone attainment zone)? Are there sources of fresh water? Are they available for distribution or accessible by private well? Do water sources need to be heavily treated or processed with simple filtration and chlorination?

## Social infrastructure

Listed below are some of the key social infrastructure factors influencing the development of communities.

*Availability of quality health care*

With declines in rural population and migration to major metropolitan areas, fewer doctors are opening practices in rural areas. This can make it difficult for residents in rural areas to get quality health care. As the world's baby boomers age, gerontological health care is becoming increasingly more important. Many now have all their children leaving home, but find themselves taking care of elderly parents, many of whom have health problems that need regular management. Community leaders need to explore their community's health care services by asking the following type of questions:

> Are there local clinics and doctors? How many? Does the community have ambulance service? Are emergency medical technicians present on the ambulances? Does the community have dentists? Are there medical specialists practicing in the community? Are there physical therapist and rehabilitation therapists in the community? Is there a hospice? How far is it to the nearest

hospital with an emergency room/trauma center?

### Safety of investment

Business location as well as home-buying decisions are driven by economics. While it may be possible to get an excellent low-cost building or piece of land, the operating costs of that location need to be examined. High insurance rates and losses due to vandalism or theft can quickly erode the initial cost advantage. How safe is our investment at this location? Are there hidden costs of doing business related to threats in the community? Community leaders need to know the answers to these questions and others below when planning a development initiative or marketing the community to investors.

What is the public's perception of safety in their community? Is it still an environment where many people do not lock things up or is there an obvious abundance of burglar alarm company signs, home window bars, and other indicators of criminal activity or fear? What is the fire insurance rating for the area? What is the crime rate? Are there signs of vandalism or gang activity? Do individuals gather on street corners for illicit activities? Is their an obvious drug problem in the community? Is the community well lit? How many police officers are on duty? How many firemen serve on the local fire department? Is the fire department a volunteer unit or does it have some full-time personnel?

### Quality of school system

The quality of local schools not only supports workforce capabilities but often drives residential growth. Given a choice, parents will usually seek a residential location in a school district with a good reputation for academics and administration. Many state education departments have data available on school performance by district, and sometimes on individual schools. The Census Bureau (www.census.gov) can also provide data on citizens' educational levels as well. Listed below are some of the questions that community leaders should ask about their local schools:

How well do students in the local system perform on standardized tests? What is the dropout rate compared to state and regional rates? What is the condition of the schools' physical plant and facilities? What is the pupil–teacher ratio? Do the high schools offer a full range of academic subjects? In addition to academic courses, do the high schools offer vocational training? Where do most major employers' high school graduate employees come from? Is the community a residential demand area because of its school system?

### Parks, recreational, and cultural opportunities

Community leaders need to determine what their community offers compared to other communities of its size. This may be done by asking the following type of questions:

Where do most residents go for recreation? Are there periodic cultural events in the community for locals and visitors alike? Is there a community center? Are planned recreational and sporting activities held for youth and adults? Are the playgrounds well maintained, neat and clean? Are facilities well lit? Is the play equipment safe, attractive and functioning? Are boundary fences in good repair? Are neighboring buildings and structures in good repair and attractive? Are nearby and adjacent properties residential or compatible with commercial use? When making a visual inspection of the location, are people present? What are they doing? Are recreational areas defaced and vandalized? Are there recreational activities for people of all ages?

### Availability, affordability, and quality of housing

A drive through most communities will tell you a lot about their housing stock. Since many residents travel the same routes to and from their home and work locations, there may be conditions on other streets or in other areas of the community of which they are not aware. Realtors can provide very good insight to the condition of and demand for area housing. Another good source of somewhat dated housing information is the Census Bureau (www.census.gov). Listed below are some questions that community leaders should ask about their area's housing:

> What percentage of the county's housing consists of mobile homes? What is the value of housing compared to the adjacent counties? Is there zoning in the community? Are local ordinances consistently and strictly enforced? What percentage of residents rent compared to adjacent counties? What is the occupancy rate of local housing? What is the cost of a three-bedroom home with two baths? What is the condition of vacant buildings and housing? What is the time needed for construction of a new home? What is the cost of construction per square foot for residences and commercial structures? What percentage of local housing is in need of renovation or demolition? Are there abandoned commercial buildings in the community? Are they potential brownfield sites? Are there abandoned commercial buildings that could be quickly rehabilitated and put back into use? Are there housing subdivisions with infrastructure in place in which new homes could be constructed? Are there quality apartments or rental properties in the community that could serve as temporary housing until individuals buy or build homes? Are there existing commercial buildings that could be converted into multi-family housing? Is there public housing? In what condition and how well-managed are the public housing units?

*Quality college/university nearby*

Institutions of higher learning provide the skilled workforce needed for most of today's higher paying jobs. Being located in close proximity to a college or university can be an advantage to a community because it provides a ready source of candidates for skilled jobs and gives locals an opportunity to upgrade their skills for advancement. Listed below are some of the questions that community leaders should ask about nearby institutions of higher learning:

> Is there a community college, four-year college or university nearby that serves the community? What types of technical and vocational degrees or training do they offer? Are there technical and scientific courses offered that can support high-tech manufacturing and businesses? How many graduates are produced in each of the programs annually? Do the institutions have an active placement office serving new and past graduates? Where are these graduates employed? Are the institutions willing to and capable of adding more programs in support of new businesses coming to the area? How far away are the institutions? From where do most of the students come? Where do most of the students live? Are there knowledgable and expert professors who can provide research support for business operations?

## Economic development infrastructure

*Low cost of living*

The cost of living in the United States varies greatly from region to region. Housing, taxes, food, insurance, taxes, and other expenses are significantly higher in places like San Francisco, Los Angeles, and New York. Employees relocating to these areas often receive salary bonuses or higher salaries to help compensate for the cost associated with moving there. Communities with a lower cost of living have an advantage when recruiting new residents and businesses from areas with higher living costs.

Community leaders working for planned development need to ask and answer questions similar to those that follow:

What is the cost of living in the community? Will those moving into the region experience much higher living expenses than the area from which they are moving? How does the cost of living here compare to similarly situated communities in the regions from which we hope to draw businesses and residents?

*Quality and competitiveness of public utilities*

Energy is becoming a major concern for most businesses. It is often less expensive to operate in foreign countries where energy costs can be lower, particularly when oil, coal, or natural gas are needed for production processes. In spite of recent major outrages, the United States still has an advantage over the rest of the world with regard to reliability of electric and gas distribution systems. In time, however, that reliability advantage may also decline. When assessing their community's utilities, leaders should ask questions similar to the following:

> Who are the current utility providers serving the community? How well maintained are the current distribution and transmission systems? Is the community on a transmission and distribution network? What are the reliability percentages for the electrical and gas service? Has there been a history or prolonged interruptions? Are the costs competitive? Are there any factors that will result in significant increases in the near future? Are long-term contracts available? What is the potential for long-term rate stability? What are the sources of generation (nuclear, oil, gas, or coal)?

*Availability of water and waste systems*

Many fast-growing communities have outpaced their community's public water and wastewater system's ability to expand. As in many older communities, their systems may not be in compliance with recent EPA standards and may soon require major upgrades that will result in rate increases. Further, new restrictions by the Environmental Protection Agency may have limited the capacity of existing systems so that the community has no growth capacity. Additional residents and businesses coming to a community may result in the inability of these systems to meet demand. Some areas of the U.S. are experiencing water shortages and rationing (particularly arid areas of the west southwest and, recently, even the southeast U.S.).

Similarly, solid waste processors and facilities are raising their rates, and some landfills are closing because they have either reached the end of their useful life or can no longer operate due to environmental regulations. It is not uncommon for many communities to export their solid waste many miles for disposal.

Listed below are some of the questions leaders should ask about their community's water and waste systems:

> What are the current capacities of the water and wastewater system and, if they are not operating at capacity, how much surplus capacity do they have? Are the current water, sewer, and solid waste disposal rates competitive? If the community owns its systems, are they subsidized by taxes or operating on their own at a "profit"? If they are operating at a "profit," are the funds being wisely managed? Are the rates being charged reflective of fair market value or are they artificially low? What are the long-term operating outlooks for the systems (replacement, major renovations, or closure)?

*Telecommunications*

From hard-line (fiber-optic and copper) telephone and cable systems to Internet service providers and wireless broadband, telecommunications are becoming increasingly important for both business operations and personal convenience. Businesses use these systems to transfer electronic data and to maintain contact with other offices and customers. Communities with a variety of telecommunications services and providers are more competitive as sites for business relocation and expansion than those with limited services. Listed below are the types of questions that business leaders should ask about their community's telecommunications capabilities:

> Is cell phone coverage and service good in the community? Does the community have a cable provider? Is broadband (high-speed) access to the Internet available in most areas of the community through cable, DSL, or perhaps satellite? Is the community served by fiber-optic cable service? Are rates for these services competitive? Is there a loop network supporting these systems?

### Available commercial buildings

It has long been recognized that communities with available commercial buildings are much more likely to locate a new or expanding business than those with greenfield sites. Many businesses seeking relocation or expansion want to make the move quickly and find greater value in existing buildings that can be quickly refitted to their needs. Community leaders would do well to work with their area's commercial realtors to identify and maintain an inventory on available commercial buildings. Here are some of the questions that leaders should ask about the available buildings in their community:

> How many commercial buildings are available in the community? What size are they? What are the features of each (e.g., office space, warehouse space, truck bays, ceiling height, floor load, utility capacities)? Who is the current owner? What is the asking price? Is the price negotiable? Will the owner enter into a right of first refusal? Will the owner sell or lease? Is owner financing available? (Note: the International Economic Development Council provides a template for collecting building data at www.iedconline.org.)

### Lack of governmental "red tape"

Complicated building permits, compliance with local zoning and other ordinances can make attracting new businesses and residents difficult. Local leaders should carefully examine the process for setting up new businesses, building new commercial buildings, and building new residences. The following are some of the useful questions that will provide insight about the process:

> Who needs to be contacted to apply for permits, approvals, and licenses? Is there a one-stop shop that can guide an applicant through the process? Is there one document that clearly walks an applicant through all requirements and provides contact names and numbers? What is the cost of each? Are the forms easy to understand? What support needs to be provided in addition to the forms? How long do the forms take to process? Once approved, can they be revoked or amended? In the event of a mistake on the part of a governmental agency, who will pay the cost of bringing the project into compliance? Are kickbacks and payoffs a hidden part of the system? Are government clerks and officials "customer-focused" or "process-focused?" (Do they seek to help the applicant comply with the law or seek themselves to comply with the law?)

### Tax rates on business

When considering a location for an expansion or relocation site, business managers look at the total cost of doing business at the prospective location. While the location may have all that is needed with regard to location, utilities, building, and workforce, the tax structure may be such that it makes a particular business unfeasible. Some taxes originate at the state level. However, others, such as local sales and franchise taxes, are under local control. Local leaders need to examine the impact of their tax structure on the types of businesses that are the best fit for their location by asking questions such as the following:

Do businesses pay a grossly inequitable share of taxes compared to other taxpayers? Do the public services and infrastructure supported by business taxes run efficiently, using the tax income specifically for services delivered? Does the local tax structure have a chilling effect on the location of companies for whom the community would otherwise be an excellent location? Is the current business tax structure competitive compared to other locations?

### First-rate scientific community in area

As the United States transitions from an industrial to an information service and high-technology environment, more emphasis will fall on emerging technology and scientific research that supports business competitiveness and opportunities. An assessment should consider what resources are available to a community to support its businesses research and scientific needs. Questions similar to those which follow should be asked during the assessment:

Are high-tech businesses or scientific research facilities nearby? Are there any high-tech or scientific companies that are seeking partners for testing or marketing the work in which they are involved? Is there a technology transfer center that can assist businesses in adopting new technologies that will lower operating costs and increase competitiveness? What fields of excellence are available at local universities and major hospitals? Are these facilities willing to partner with businesses to bring new technology to the marketplace?

### Availability of fully developed sites

Communities with a fully developed industrial or business park have a distinct advantage over those without. That advantage can increase when the park is owned or managed by a government or quasi-governmental organization responsible for economic development. Where necessary, rather than relying on private property owners to market and develop their commercial property, local leaders can take the lead in developing a site ready for development that can be marketed to existing and new businesses. If a community has an industrial or business park, all utility, lot, and building information should be included in the assessment. Shown below are the types of additional questions that should be answered about the park:

Is the property available for lease or sale? Can speculative buildings be constructed? Is there a revolving loan fund resulting from park revenues that is available for new businesses? What incentives can the park offer for businesses locating there? What is the cost of lots and buildings? Are those prices competitive with other buildings and sites? Is the property governed by a restricted covenant? What types of businesses are being sought for the park? What types of businesses are not of interest to the community?

### Strong existing businesses

A healthy local economy and tax base are heavily dependent on existing local businesses. While existing business owners and operators may be concerned about new businesses coming in and competing for their employees, they can also provide support for and benefit from the building of new or expansion of existing businesses in the community. Local retailers, for example, often fear the competition from major retailers such as Wal-Mart. However, the savvy retailers realize that they can benefit from the increased traffic to the community if they can realign their operations to take advantage of the change. Having knowledge of a community's businesses can also help identify the types of businesses that should be recruited.

A community assessor should examine local businesses, make judgments about their strengths, and include his or her findings in the assessment report.

Are they competitive and successful? Has there been a lot of turnover in local businesses (i.e., a lot of openings and closings) or have many of the local businesses been in business for a long time? Is there a large inventory of commercial buildings due to the many closed businesses? What is affecting the health of local businesses (e.g., population decline)? What types of businesses can use the services and products delivered or made by local businesses?

## Human infrastructure

### Broad-based "can-do spirit"

Many times the prevailing local attitudes can poison the development well. Communities that have a history of failed development activities or ill-conceived projects often see themselves as helpless. One of the most important things that a professional community developer can do is help instill the belief that citizens are not helpless and that by conducting an assessment and carefully planning and working together, they can see positive change in their community.

Positive attitudes are a powerful force for planned change. Community leaders and citizens who believe that they can make things happen, and see opportunities and possibilities rather than problems and threats, often help separate their community from those that fail.

A community assessor should ask questions about the community's past projects, programs, and initiatives to determine what citizens and leaders believe. When citizens offer ideas about what can be done or discuss the history of past events, what are their responses? Is there a history of failed events? Has there been a lack of action to address community problems? Are local leaders hesitant to try new things or do they lack the political will to make hard and sometimes controversial decisions? If comments such as the following are heard, in all likelihood the community has a poor "can-do spirit":

"That will never happen here."
"We discontinued the town festival because of lack of support."

"We have tried to get organized several times in the past."
"We had an assessment and developed a plan once but I don't really know what happened to it."
"We don't have the local leadership to get that done."
"We tried that once and..."
"We just don't have the resources like other small towns."

Conversely, if comments such as the following are heard, in all likelihood citizens and their leaders have the confidence needed to bring about positive change:

"We have a very successful local festival every year."
"We were able to leverage public and private money to improve the local community center and playgrounds."
"Our bankers and government officials can work together to put a financial and incentive package together for a major new business."
"We know that grants and funding are available for developing our business park."
"Our local government is well run and efficient. Not only do we have a surplus, but our utility rates and fees are among the lowest in the area."
"Our elected officials, business and civic leaders have a history of working together on projects. We have..."

### Desire for development

It is important for a professional community developer to remember that his or her mission is not to direct the community development process or tell citizens what their community should become. The professional developer's role is an advisory one. Professional developers listen to and reflect back the values and desires of the community; they provide options, ideas, alternatives. It is also appropriate to describe strategies and projects that have worked well for other communities. However, in the end, citizens have to decide for themselves what they want their community to become and the actions they will take.

There is often an interesting paradox found in quaint and attractive rural communities. Many citizens want the amenities and conveniences associated with larger communities, but they do not want to do what is needed in order to have them or deal with the side-effects that come as a result. They want more retail variety, for example, but local business leaders do not want the competition from chains and franchise operations, and many citizens do not want new residents who will contribute to traffic congestion and overcrowding.

Desire for development sounds like a very "soft" subjective measure. However, a community's collective desire for development can and must be determined as well as the nature of the development that citizens would like. This determination can be made quite quickly and easily by asking some very simple questions such as, "What would you like to see take place here?" or "What do you think needs to be done to make the community a better place for citizens?" If the majority of comments are similar to those below, then low-impact activities and strategies such as residential and retirement community development, recreational and tourism development, Main Street specialty and theme retail, or a "bedroom" community strategy in support of a larger nearby community are the likely development goals.

"I moved here because it is clean and green and that's the way I want it to stay."

"We don't want this to become like a city."

"We don't want more people moving in."

"We like it just the way it is, there is nothing that we should change."

"I moved from the city because this was a quiet and safe community. I don't want the growth and problems associated with the city."

"We don't want any smoky smelly businesses here and all the truck traffic related to their operations."

On the other hand, if the majority of the comments are similar to those that follow, then a community's citizens are very supportive of change and desire positive change and growth.

"We need more highly paid technical jobs in our community, jobs that help keep our children here and that bring in new residents."

"We need to increase the tax base with more business development that will help support the public infrastructure."

"We need to develop a business park and actively market our community to business investors as a location."

"It would be great if we could get a Wal-Mart and some chain restaurants in town."

"We really need to improve our streets and public utilities to support new growth. Our systems are old and do not have the capacity for new businesses and homes."

"We really need to improve our town's appearance and be stricter in enforcing abatement regulations."

### Competitive wage rate/salary rates

Wages and personnel costs usually represent a significant portion of most manufactured goods and services. Regardless of all the political rhetoric, it is a simple economic fact that, in order for businesses to stay competitive in a world marketplace, they must seek locations with the lowest possible operating costs for a given labor force availability. This economic principle not only applies to businesses but to consumers. Given a choice between buying the same quality goods in their community or in another location, consumers will make their purchases where they get the most value for their dollar.

Not only do communities compete against those in other states or regions of a country for business investments, but they compete against the labor force in other nations. The wage rates paid in a local market should not be so great that they discourage new businesses from relocating or opening a new venture there.

As an example, the petrochemical industry tends to be clustered in locations such as Louisiana, New

Jersey, and Texas, where there is oil and/or gas production or importation or where the refineries are located. The wages paid to workers in most of these unionized facilities are quite high. These operations are able to attract the most qualified and highly trainable employees from the regions in which they are located. Those attempting to locate other businesses within these high wage rate areas will face tough competition for the best employees and may be forced to pay a premium above what they would in other locations. This can have a chilling effect on the attraction of new businesses for which labor costs are a significant percentage of their operating costs.

Usually, a state's department of labor and its job services office can provide demographics on wages and salaries in an area and offer insights into the competitiveness of local wage rates.

### Labor availability

Not only must labor be affordable, but it must be available in order to attract new businesses. Not only must the unemployment rates be identified, but an estimate of those in the workforce who are underemployed (having skills greater than those required for their current positions) should be made. When researching the labor force available, a community assessor should consider the community's own citizens, and those in nearby communities and counties from which workers might realistically be drawn in support of business operations. As discussed above, the state's labor department and local job service offices can provide a great deal of data. Similarly, the U.S. Census Bureau can provide additional workforce data.

### Quality of workforce

"Cheap labor" is not always the answer to successful business expansion and attraction. Workforces in developed countries cannot compete with the wage rates offered in developing countries such as those in South and Central America, and Asia. Over the past several decades, the developed countries have lost many low and semi-skilled jobs (such as cut-and-sew operations and unskilled manufacturing assembly) to

developing nations whose wage rates are less than US$1 per hour.

Much information about the quality of labor can be provided or found in the sources previously listed. However, local employers are often an excellent source of information about the capabilities of a community's workers. Asking local employers about their workers can provide anecdotal evidence regarding the ability of employees to learn new skills; their dependability; honesty; drug abuse; and overall performance. About five years ago, for example, the author was told by the branch manager of a national police equipment manufacturer that when the company went through a cost-cutting effort, another plant was closed and the work transferred to his location, which then expanded. The reasons workers were more productive (as measured by the cost of production including raw materials consumed) were that their absentee rate and turnover were low, and the quality of their work (as measured by their rejection rate) was very high.

### Labor climate

Most enlightened business managers and labor leaders realize the need for a team environment rather than a combative one that adds to the cost of operations. For labor, management, and shareholders to benefit, they must work together. This not only makes them competitive with other domestic and international operations, but makes them likely candidates for reinvestment, expansion, and continued operations.

When conducting the assessment, local leaders should be asked questions about work stoppages, sabotage, strikes, and other labor unrest. Is this a location where production and services can be reliably provided? Or is this a location that will consume valuable management time and corporate resources, continually distracting attention from the primary function of business operations? Communities with a reputation for high employer/worker teamwork will have an advantage to retain and expand business operations over those locations that are more disruptive and combative.

## SWOT analysis

For many years, business leaders have used the "SWOT analysis" as a method of identifying the *strengths*, *weaknesses*, *opportunities*, and *threats* impacting their commercial ventures. The SWOT process provides them with a systematic approach for analyzing options and making decisions. By researching and laying out the SWOT factors, they can prioritize their actions and focus their efforts for the greatest impact.

The same process can be used in support of the community development assessment. Strengths and weaknesses are the direct factors impacting a community, those that can be directly controlled or influenced by the local leadership. The condition of local streets, water, and wastewater systems, fire protection, and emergency services are all within the control of local leaders. They are examples of areas in which strengths and weaknesses may be found (Table 9.1).

Strengths describe the assets that are already present and upon which development may be built.

**Table 9.1** Subjective rating of location factors for Anytown: possible ratings – Good (2) Average (1) Poor (0)

| Location factors | Ratings | Point value |
|---|---|---|
| *Physical infrastructure* | | |
| Transportation system for moving goods | Good | 2 |
| Transportation network for business travel | Average | 1 |
| Weather and geography | Good | 2 |
| *Social infrastructure* | | |
| Availability of quality health care | Average | 1 |
| Safety of investment (police, fire, emergency services) | Average | 1 |
| Quality of school system | Average | 1 |
| Parks, recreational, and cultural opportunities | Average | 1 |
| Availability, affordability, and quality of housing | Poor | 0 |
| Quality college/university nearby | Good | 2 |
| *Economic development infrastructure* | | |
| Low cost of living | Poor | 0 |
| Quality of public utilities (electricity and gas) | Good | 2 |
| Availability of water and waste systems | Good | 2 |
| Telecommunications | Good | 2 |
| Available commercial buildings | Poor | 0 |
| Lack of governmental "red tape" | Average | 1 |
| Tax rates on business | Good | 2 |
| First rate scientific community in area | Good | 2 |
| Availability of fully developed, publicly owned sites | Poor | 0 |
| Strong existing businesses | Good | 2 |
| *Human infrastructure* | | |
| Broad-based "can-do" spirit | poor | 0 |
| Desire for development | Good | 2 |
| Relatively low wage/salary rates | Good | 2 |
| Labor availability | Average | 1 |
| Quality of workforce | Average | 1 |
| Labor climate | Good | 2 |
| *Overall rating* | Average | 1.28 |

Note
Overall rating scale: 0–0.69 = Poor 0.7–1.39 = Average 1.4 or higher = Good.

What do we have here in abundance? Who needs it? For example, a community may have a bountiful supply of fresh water that requires very little processing before distribution. It may also have excess capacity in its municipal wastewater system. With regard to solid waste disposal, it may have a large, well-managed sanitary landfill with many years of capacity. Such communities are an attractive location for businesses in need of fresh water supplies and whose operations result in considerable solid waste, such as those of food and beverage processors.

Weaknesses are also factors that influence business location decisions. Will the assets mentioned above be countered by major community weaknesses such as poor housing; a declining and aging population; high tax rates; unreliable utilities; badly deteriorated streets; a lack of land-use planning and zoning; and poor local elected leadership?

Strengths and weaknesses are primarily under the control of local leaders. Once identified, strategies can be implemented to build on existing strengths, to turn weaknesses into strengths, and to minimize if not eliminate weaknesses.

Opportunities and threats tend to be factors outside the control of local leaders but that can impact the community's development efforts. Geopolitical events and economic trends, national political events, environmental factors, and corporate realignments tend to present opportunities and threats outside of the control and possibly the influence of communities. However, once such factors are identified, communities can create new strategies, or adapt their existing strategies, to manage threats to the extent that the negative impacts either do not occur or are minimized when they do.

If, for example, a town is a "company town" with one major employer providing most of the local jobs and supporting most of the local service businesses, this presents a threat. The community does not have control over the corporate management's decisions. If the plant becomes less cost-effective to operate, management may decide to close the facility. Prior to that threat becoming a reality, local leadership might have regular meetings with plant and corporate managers to identify potential problems so that

they could be mitigated. For example, they might seek state incentives and tax abatements to help the facility expand or modernize. Further, they might initiate an economic diversity strategy to attract new, unrelated businesses to the local economy so that the community is no longer a "company town."

Other threats are environmental. In the early 2000s, West Nile Virus entered and began spreading throughout the United States. This event created a threat to summer tourist areas that had large populations of mosquitoes. While action should be taken at the local level to reduce the threat, the problem was still largely out of local leaders' control. It is a threat that would have to be managed by increased mosquito-control efforts, reminiscent of those against malaria and yellow fever. Some communities also increased their advertising budgets and changed their advertising messages.

Sometimes threats can also produce opportunities. On September 11, 2001, terrorists attacked and destroyed major offices in the World Trade Center in New York. Some of the firms there, such as Kantor-Fitzgerald, had their entire operations disrupted and lost a significant percentage of their total employees through the attack because all their corporate functions were at one location. As a result, other companies began to consider dispersing their vital business functions to different and, in some instances, less threatening locations.

The 2001 terrorist attacks, as tragic as they were, created opportunities for communities that might be destinations for dispersed corporate operations. The realignment of the federal government and creation of the Homeland Security Department (HSD) will result in the creation of regional HSD offices, another opportunity for some communities.

## The assessment report

The product of a community assessment is usually a written report. This report generally consists of four parts. First, it provides an analysis of each of the development factors listed and described above and possibly others that are specific to the assessment location. This section is a blend of objective data, observations and subjective findings, and may include pictures supporting the findings. Second, the report should include a SWOT analysis that describes the strengths and weaknesses identified during the assessment and the opportunities and threats identified by local leaders. Third, the report provides a section on "possibilities." This is a compilation of wants and desires as expressed by citizens (what they would like to see happen in the future) and possible courses of action (programs, projects, and initiatives) to help the community reach its full potential. In addition to describing possibilities, it may also include pictures of projects from other communities, so that citizens and leaders can see what similarly situated communities have accomplished. Finally, the report would include appendices of statistical data that support the findings including opinion surveys, demographic data, and other evidence on which the findings are based.

It has been this author's experience that community leaders and citizens also want a summary of the findings. They seek a quick reference and a numerical rating that they can use to make comparisons between themselves and other communities. Often, statistical data are used to compare a community with other similar or sometimes "competing" communities in the SWOT analysis. Communities also want a benchmark from which to measure improvements. Table 9.1 is an overview of the development categories for a small community as it would appear in the report. This table would typically appear in the first part of the assessment report, possibly at the end of the findings. The ratings would be supported by the descriptive statistics and observations of the assessment team. There are potentially hundreds of community factors that could be included in an assessment. Each situation is unique.

The SWOT analysis categories in this chapter and Table 9.1 are illustrative of many of the potential categories but are not exhaustive. The categories chosen in an actual SWOT analysis (see the Case study below) will depend in part on the exact nature and purpose of the assessment.

## Conclusion

Starting economic development and community development initiatives without an assessment can often lead to addressing problems' symptoms rather than root problems, misidentifying strengths and weaknesses or failing to identify opportunities and threats. Assessments are an important precursor to strategic planning and can help start the community dialogue that fosters planned change. They may be performed by community leaders themselves, allies such as state and federal agencies, and professional developers.

The objectives of community development assessments are:

- To support a community dialogue that involves citizens in determining how they would like to see their community develop.
- To identify the factors that influence the potential development of the community.
- To identify the strengths, weaknesses, opportunities, and threats that are or can influence the community's development.
- To provide information for those leading a strategic planning process so that a workable and focused set of goals and objectives may be created to help the community achieve its potential as well as help citizens realize their shared vision of the community.

### CASE STUDY: BAYSHORE COUNTY COMPETITIVE ASSESSMENT

Bayshore County, located in a coastal area, has a strong tourist industry as well as a variety of man-ufacturing and service industries. The county realizes that tourism is highly competitive and jobs in this industry are not high-paying. The Bayshore County Economic Development Agency hired a con-sultant to assess the county strengths and weaknesses for recruiting manufacturing and service indus-tries. The consultant gathered and analyzed demographic, economic, and social data for the county and compared it to similar counties that might compete for the same types of industries. In addition, the consultant conducted over 30 interviews with local company executives, elected officials, and other community stakeholders. In these confidential interviews, the consultant asked questions and probed to determine what the interviewees considered to be the county's strengths and weaknesses from the economic development standpoint.

Based on the data analysis, interviews, and a tour of the county, the consultant rated the county in terms of its strengths, weaknesses, or neutral on several key factors relevant to attracting new manu-facturing and service industries. Table 9.2 shows the summary factor rating matrix. The consultant's report was comprehensive and included a narrative on each factor and explanation of how the rating was determined. Excerpts from the consultant's report are also included below.

### K-12 education: strength

- K-12 education in Bayshore County is a strength in comparison to the rest of the state.
- The biggest issue facing K-12 education in Bayshore County is the perception of the local community. Individuals who come from school systems outside of the state seem to have the per-ception that the school system is not very good.
- The dual enrollment program offers high school students the ability to earn a two-year degree prior to graduating from high school. This program provides a mechanism for post-secondary training for students entering the workforce upon graduation from high school.
- The International Baccalaureate program at Bayshore High School is a nationally recognized program that provides a competitive college preparatory curriculum.
- Offering choice of school in the county is a positive attribute and has led to a strongly competit-ive environment within the high school system. Each school must be able to recruit students in order to maintain levels of funding. This provides for a strong education system, since the best and brightest are being recruited to choose a high school rather than simply being districted to one school or another.
- There is a countywide school system. This is a strength, since there are no competing school dis-tricts within the county.

### Highways and roads – weakness

- That there is no access to an interstate or interstate-quality roadway is a weakness for Bayshore County. This limits the types of industries that might be recruited to the area.
- There are only three main highways into and out of the county: U.S. 79 and U.S. 262. The east–west route of U.S. 24 is bottlenecked at the Harrison Bridge crossing in West Bayshore. The bridge is currently under construction for expansion.
- Traffic congestion could pose a problem in recruiting industry to the area. It is not currently a

critical factor in the county, since a majority of trips take, at the most, 20 minutes. Adequate planning for traffic growth could cause additional problems in the future.

- Along with the new airport it is understood that a four-lane, potentially limited access highway is planned to connect the new airport to the interstate highway. This would greatly improve Bayshore County's position in regard to highways and roads.

■ **Table 9.2** Factor rating: Bayshore County – summary strengths and weaknesses matrix

| Category | Strength | Neutral | Weakness |
|---|---|---|---|
| *Available land and buildings* | | | |
| Available land | | X | |
| Available buildings | | X | |
| *Labor* | | | |
| Labor COST | | X | |
| Labor AVAILABILITY | X | | |
| Labor–management relations | | X | |
| Labor productivity and work ethic | | X | |
| *Utilities* | | | |
| Electricity | X | | |
| Water | | X | |
| Sewer | | X | |
| Natural gas | | X | |
| Telecommunications | X | | |
| *Transportation and market access* | | | |
| Air | | X | |
| Roads and highways | | | X |
| Rail | | X | |
| Water | X | | |
| Market access | | | X |
| *Business, political and economic climate* | | | |
| Local taxes and incentives | | X | |
| State taxes and incentives | | X | |
| Local government support | | X | |
| Economic base | | | X |
| Permitting | | | X |
| Zoning and land-use planning | | | X |
| Image and appearance | | | X |
| Economic development vs. tourism | | | X |
| Regional cooperation | X | | |
| *Education* | | | |
| K-12 | X | | |
| Vocational and technical training | | X | |
| Higher education | X | | |
| *Quality of Life* | | | |
| Cost of living | | X | |
| Housing availability | X | | |
| Health care | | X | |
| Cultural activities | | X | |
| Recreational activities | X | | |
| *Other* | | | |
| Economic development program | X | | |

The Editors

## Keywords

Community development, asset identification, economic development, data collection, assessment report, SWOT analysis, human infrastructure, community profile.

## Review questions

1  What is the purpose of a community assessment? Why conduct one?
2  How do you collect information for an assessment?
3  What are some of the characteristics of a community an assessment should cover?
4  What is a SWOT analysis and why is it useful?
5  What should be included in the assessment report?

## Bibliography and additional resources

Clark, M.J. (2007) *Community Assessment Reference Guide*, Upper Saddle River, NJ: Prentice Hall.

Cornell-Ohl, S., McMahon, P.M. and Peck, J.E. (1991) "Local Assessment of the Industrial Development Process: A Case Study," *Economic Development Review*, 9: 53–57.

Luther, V. (1999) *A Practical Guide to Community Assessment*, Lincoln, NE: Heartland Center for Leadership Development.

Stans, M.H., Siciliano, R.C. and Podesta, R.A. (1991) "How to Make an Industrial Site Survey," *Economic Development Review*, 4: 65–69.

U.S. Census Bureau. Available online at www.census.gov.

U.S. Department of Energy. Available online at www.energy.gov.

U.S. Department of Labor. Available online at www.dol.gov.

Williams, R.L. and Yanoshik, K. (2001) "Can You Do a Community Assessment Without Talking to the Community?," *Journal of Community Health*, 26: 233–248.

# 10 Community asset mapping and surveys

## Gary P. Green

Although a variety of approaches may be used to begin the community development process, those used most often are either a needs assessment or an asset-mapping study. This chapter focuses on several techniques for mapping community resources/assets such as asset inventories, identifying potential partners and collaborators, and various survey instruments and data-collection methods. Much of the discussion is also appropriate for conducting needs assessment studies and covers some issues organizers may want to consider in conducting surveys of individuals, organizations, and institutions. This chapter provides details on how to design and conduct mapping and surveys and is an instrumental component of the community development process described in Chapters 3 and 6.

## Introduction

The community development process is often initiated with some form of needs assessment or asset mapping. Needs assessment is a method for identifying local problems or issues. These "needs" in turn become the basis for a strategic action plan (Johnson et al. 1987). Community organizers mobilize residents around the specific issue in order to seek new resources, obtain information and expertise, or to pressure local officials to solve the problem. It is assumed that residents will act to address the perceived deficiencies in their neighborhood or community. Organizers hope that residents will gain a sense of efficacy by ameliorating problems, which in turn will help them to build the confidence to address other issues as well.

Kretzmann and McKnight's (1993) asset-based development model, however, turns this approach on its head and starts with a map of local resources which provide the basis for a community vision and action plan. As discussed in Chapter 3, an asset-based development approach focuses on the strengths rather than the weaknesses of the community. It also

relies heavily on developing partnerships and organizing across the community in order to identify ways of building on local assets to improve the quality of life. The goal is to develop trust among various partners and to operate on the basis of consensus.

Both models of community development rely heavily on survey data. Needs assessment uses surveys to obtain individuals' evaluations of local services, social problems, and other local issues. Asset-based development relies on surveys of local residents to identify skills and talents that may be underutilized or not often recognized. Surveys may also be used to conduct assessments of organizational and institutional resources within the community. With both models, the survey is more than just a means of obtaining information – it is also considered as a way of securing public participation.

This chapter focuses on several techniques for mapping community assets. Much of the discussion is also appropriate for conducting needs assessment and covers some of the issues organizers may want to consider in conducting surveys of individuals, organizations, and institutions.

## Asset mobilization

There are several steps in mobilizing community assets (Kretzmann and McKnight 1993). First, it is important to map the capacities of individuals, organizations, and institutions within the community. This process helps identify the resources that are available for development. Not only are these assets frequently overlooked, but residents have a tendency to focus more on how external resources can address local deficiencies. Second, organizers build relationships across the community in order to generate support. Most often community organizers work through existing organizations and associations (Chambers 2003). The goal is to identify common values and concerns that can form the basis of strategic action. Third, the community develops a vision and an action plan for achieving its goals. The vision should be based on the values of local residents and the resources available to them. Finally, communities can leverage their resources to gain outside support. Although the asset-building approach relies on mobilizing local resources, it does not ignore the importance of tapping into external resources and sources of information. However, the focus is on how community action can build on the expertise, experiences, and resources that are already available.

A core premise of asset-based development is that local resources are often overlooked in community development. One of the best examples of asset-based development is the Dudley Street Initiative in Boston (Medoff and Sklar 1994). Like many inner city areas, this neighborhood had numerous vacant lots that had become illegal dumping sites. Rather than viewing these vacant lots as a problem, residents worked with city officials to give the neighborhood association the right of eminent domain to claim this property for affordable housing projects. They formed a land trust for the property that essentially gave the community control over the development of the land. Rather than letting developers decide on the best use for the land, residents found ways of developing affordable housing that was sorely needed.

Organizations and associations play a critical role in the community development process. In addition to their resources and membership, they can provide community organizers with legitimization in the community. Rather than working to start new organizations, it is possible to build on existing ones. This strategy is typically used by the Industrial Areas Foundation (IAF), which relies heavily on mobilizing churches, unions, and community-based organizations. Mark Warren's (2001) book *Dry Bones Rattling* does an excellent job of demonstrating how schools and churches worked together to build a powerful network of organizations in Texas. The network not only crossed various denominations and organizations, but also formed a multiracial constituency.

Finally, local institutions are important actors in communities. They facilitate regular interaction among residents and link the locality to the broader society. Local institutions can also affect the neighborhood or community through its practices, and many institutions have resources that remain untapped by local residents. For example, schools may purchase goods and services outside the area, and banks may be investing their resources in non-local investments as well. With greater understanding of these resources, community residents can potentially shape local institutions so that they serve local residents better.

## Methods for mapping assets

Communities can use several different methods for mapping assets. The purpose of the project should guide the decisions about which method to use and what specific information should be collected. The goals may range from promoting local economic development or community health to supporting youth programs. The community and/or organizations need to clearly state the purpose of the project.

Next, it is important to define the territory of the neighborhood or community. This decision will affect almost everything else that is done and should specify which individuals, organizations, or institutions should be included in the project as well as what issues face the community. In most cases there will not be a consensus regarding the boundaries of the community or neighborhood. Natural barriers, such as rivers or lakes, often serve as a boundary. Major streets or highways can help define a neighborhood. Many people will define neighborhoods by key institutions such as schools and churches. School districts provide a useful way to define a community because the population is often fairly homogeneous with regard to socioeconomic status and home ownership. Schools generate interaction among residents which can facilitate the community development process. They also have the advantage of having clearly demarcated boundaries. However boundaries are chosen, they should reflect residents' perception of the community and promote interaction on issues of common interest.

After identifying the purpose and the geographic boundaries of the asset-building community development project, it is important to consider the appropriate method(s) for conducting the project. There are several issues to consider including the available resources, timing, geographic area, and so on. Weighing different factors can be difficult, and there is no easy way to balance the various considerations. One of the most difficult tradeoffs is between cost and quality. For example, how much will the increased cost of a second wave of surveys improve the quality of the data? Each community needs to decide how it wants to balance this tradeoff.

Most communities rely on surveys to document local assets. However, it may be more appropriate to use other methods such as focus groups. Focus groups have the advantage of being relatively inexpensive and can be conducted more quickly than surveys. They can also provide more in-depth information on why people feel the way they do on various issues. On the other hand, focus groups do not give most residents an opportunity to participate in the process, and the findings may not be very representative of the larger population.

Regardless, it is not necessary to choose between surveys and focus groups. In fact, it might be useful to use both focus groups and surveys. Some communities may initially use focus groups to identify what types of experiences and skills may be available locally. Alternatively, conducting focus groups after a survey may permit organizers to ask follow-up questions about issues and questions raised through the findings of the survey.

Surveys can be administered face-to-face, in group settings, over the phone, or through the mail. Conducting a survey requires time and a financial commitment from community members. They need to ask themselves several questions before embarking on such a project: Do we want to conduct a survey or use some other technique for obtaining public participation? What is the best way to obtain the information that is needed? What do we want to know? How will this information be used? Is there sufficient time and financial commitment on the part of residents to conduct a survey? Does the information already exist through bureaucratic records, census data, or some other survey that has recently been collected?

## When is a survey appropriate?

Most communities use surveys to map community assets. If the goal is strictly to obtain public participation on a policy issue, there may be a variety of other techniques that may be more appropriate or cost-efficient. For example, it may be quicker and easier to hold public meetings or to conduct focus groups. Focus groups may be more appropriate in a situation where it is necessary to understand why people feel they way they do about particular issues. Public meetings provide an opportunity for residents to voice their opinions about issues and listen to the perspectives of their neighbors. A survey instrument may not provide the type of information obtained from these two other techniques.

Communities also need to consider whether they have sufficient resources for conducting a survey. There is always a tradeoff between the cost and quality of conducting surveys. By conducting a survey as cheaply as possible, communities may end up with a low response rate, results that are nonrepresentative, or poor data. As discussed below, there are several low-cost strategies that can significantly improve the quality of data that are collected.

## What is the best technique for conducting a survey?

There is no single "best" technique for conducting surveys. The appropriate technique depends on the resources available, the type of information that is desired, and the sampling strategies. Probably the best resource for conducting surveys and constructing questionnaires is Don Dillman's (1978) *Mail and Telephone Surveys: The Total Design Method*. In this book, Dillman provides a step-by-step procedure for conducting mail and telephone surveys. His procedure has been used extensively by survey researchers, and he provides several excellent ideas for obtaining the highest response rate possible. However, most researchers do not completely follow his procedure for mail surveys, which calls for several waves of mailings with a certified letter after several unsuccessful attempts to obtain a response, because most communities cannot afford the complete procedure he recommends. Another good resource for conducting surveys is Priscilla Salant and Don Dillman's (1994) *How to Conduct Your Own Survey*. This book is more accessible to residents who have little experience with survey research.

The advantages and disadvantages of four commonly used survey techniques – face-to-face interviews, mail surveys, telephone surveys, and group-administered surveys – are discussed below. While most communities tend to rely on a single method of conducting surveys, it may be possible to combine several methods.

*Face-to-face interviews* generally provide the best response rate (usually more than 70 percent) among the four survey techniques considered, and permit the interviewer to use visual aids and/or complex questions. This technique is often used with very long or complex questionnaires. Face-to-face interviews are also used to obtain information from groups that would not likely respond to other methods. For example, it may be easier to contact low-income residents through this method than a mail survey or some other technique. In addition, when interviewing employers, it may be preferable

to use face-to-face interviews rather than mail surveys. In all cases, interviewers can follow up on responses to get a better understanding of why a given response is provided.

However, face-to-face interviews are the most expensive of the four techniques, and there may be problems with "interviewer bias." Interviewers will need training, and there is often more coordination of those involved in face-to-face interviews than with other techniques.

*Mail surveys* are probably the most frequently used and the cheapest method for conducting community surveys. Mail surveys are usually shorter in length than face-to-face surveys and may include maps and other visuals aids, but the instructions need to be concise and understandable. The response rate for mail surveys will vary depending on several factors such as how many follow-up letters are sent, the extent to which the material is personalized, the length of the survey, and whether or not incentives are provided. Many communities mail only one wave of questionnaires which generally produces a response rate of between 30 to 50 percent. A follow-up postcard can yield another 10 percent, and a replacement questionnaire will generate another 10 to 20 percent. However, there are several disadvantages to using mail surveys. They are more limited in regards to the length of the survey than in other techniques. It is also very difficult to ask complex questions, and there are no opportunities for follow-up questions or clarifications.

One effective approach in limited situations is to drop off the questionnaire and ask respondents either mail them back or have volunteers pick them up at a later date. This approach requires coordination of volunteers as well as local knowledge of the neighborhoods involved. With this technique it is be possible to explain to respondents the purpose of the survey and how it will be used, and to clarify any issues or questions. In addition, when respondents know that someone will be dropping by to pick up the survey, it improves the likelihood that they will complete it. The response rate is typically higher than that of the standard mail survey.

*Telephone surveys* can be completed quickly and generally have a higher response rate than mail surveys. The cost may vary depending upon whether or not respondents are randomly sampled. The response rate among telephone surveys is not as good as face-to-face interviews, but they have the advantage of possible follow-up by interviewers. One of the chief disadvantages is that the interviewer cannot use any visual materials or complex questions. Phone surveys can be difficult to organize when using volunteers to conduct the surveys, and it is more difficult to manage interviewer bias with this survey technique.

*Group-administered surveys* may be used in situations where the targeted population is likely to attend a meeting where the survey could be administered. For example, a survey could be administered at a neighborhood meeting of residents. The chief advantage is that a large number of respondents can be reached quickly with very little cost. This approach to administering the survey can introduce several problems in terms of the representativeness of the results. For example, the people attending the meeting may not be representative of all members of the association or organization. Similarly, if the goal is to provide information on residents, members of a group or association in the area may not be representative of all residents. However, group-administered surveys can be a cost-effective way to conduct a survey under certain conditions. It is also possible to provide complex instructions and to use visual aids, such as maps, with this method.

These methods of conducting surveys should not be considered mutually exclusive. In many cases it may make sense to mix methods. For example, it may be possible to conduct a mail survey of neighborhood residents and then supplement it with either a phone survey or face-to-face interview with people who have not responded to the mail survey. This strategy of combining survey techniques usually improves the response rate and enables communities to collect information from various groups that may not respond to one particular survey approach.

## What is the best way to draw a sample for a survey?

Often communities struggle with developing a random sample of residents for their survey. The problem is that there is no easy way of identifying the population in a neighborhood or community. Using telephone books or even random digit dialing is becoming a major problem for survey research. Many low-income residents do have telephones. A growing number of households use cell phones and/or have unlisted numbers. Caller ID makes it more difficult to complete telephone interviews since many residents can screen their calls. These issues make it increasingly problematic to obtain random samples from telephone surveys. And for the purposes of neighborhood organizations, it is somewhat difficult to use this method to interview residents.

Property tax records are inadequate in settings where there are a large number of renters. It often takes a lot of work to get the lists in a useable form because business and absentee owners will be included in the list. There may also be multiple entries with the same names (i.e., individuals who own several properties).

Since almost everyone has a utility hookup, utility company records are probably one of the best sources for drawing a sample of households. However, these records can be difficult to obtain, and, in some multi-family units there will only be one name for the entire housing unit. But almost everyone has electricity, and this source can be supplemented with others to provide a good list of the population.

So, what is the best strategy for developing a sample of households? One approach is to combine lists or methods to draw the sample. For example, many communities rely on property tax records to identify property owners, supplement these lists by locating rental units in the neighborhood/community, and then conduct face-to-face or drop-off surveys among these households. It is possible to purchase a random list of residents from firms that compile these lists, but this will be too expensive for many community groups.

Another possibility is to generate a list by identifying all housing units in a neighborhood or community. From this list of housing units/structures, a sample can be drawn. This is a very time-consuming process, and, obviously, this strategy cannot be used on a large-scale area if there are limited resources available.

If at all possible, the community may want to simply conduct a survey of all residents. One of the questions I am always asked when presenting the results of a survey is: How could this sample be representative if I did not receive one? Some people do not believe it is possible to develop a scientific sample of a population that does not have a bias. If one of the goals is to obtain public participation in the process, it is probably better to conduct the survey among all households. Surveying the complete population will put to rest the concerns about the representativeness of the sample and promote interest in the community development project.

Finally, in some circumstances it is appropriate to draw what is referred to as a purposive sample. For example, it may be possible to conduct interviews with residents at a neighborhood event. It must be recognized that this is not a random sample and may not be representative of the community at large, but it may be sufficient in many cases.

How large should a sample be? The main goal should be to develop a sample that is sufficiently large to provide an adequate number of responses for each group in the community that need to be considered in the analysis. For example, in order to compare the responses of youth and adults, it is necessary to have a sufficient number of responses from both categories of respondents. Depending on the types of comparisons that will be made, the sample should be larger. The larger the sample size, the smaller the margin of error in the results.

Sometimes it may be advantageous not to use a random sample at all. If, for example, one of the goals is to look at the assets of the working poor, a random sample may not pick up enough residents in this category. If this is the case, it may be useful to develop a stratified sample that has a disproportionate number of residents in the groups under consideration. This approach can work if it is known where the working-class residents are most likely to live in the neighborhood and then target that area. It may be more difficult to use a stratified sample using other characteristics.

One important issue that is often neglected by communities is the *unit of analysis* of the survey. Is the focus of the study on individuals, families, or households? If the goal is to obtain an accurate random sample of individuals, it may be necessary to conduct a random sample of adults in the household/family. One method of obtaining a random sample is to conduct the interview with the person in the household/family who has had the most recent birthday. Although this strategy reduces the problem of gender bias in responses, when conducting a phone survey, it will increase survey cost because it may be necessary to make several calls before reaching the person who is to be interviewed.

Decisions about the best way to develop a sample of households in a neighborhood or a community are intimately tied to the resources available for the project. Clearly, the best method for sampling households and conducting a survey would be to identify each household and conduct face-to-face interviews with a random sample of individuals. This strategy would be very expensive and impractical for most community organizations. In most cases, there is a need to balance competing demands of cost, data quality, and resident participation.

The quality of community surveys can be significantly improved with some preparatory work. One of the most important things to do in advance is to set up an advisory committee to help construct the survey and build support for it in the community. An advisory board can help raise funds as well as possibly recruit volunteers for the survey. Another role for the advisory committee is working with the media to publicize the survey. The advisory committee can also help plan the feedback sessions to residents.

Pre-tests are essential to a successful survey. Typically, volunteers can administer this survey face-to-face. It is important to do this face-to-face in order to assess whether the respondents are confused about the meaning of any of the questions. Pre-tests can help communities avoid the embarrassing situation of collecting data that has limited usefulness.

Marketing the survey improves the response rate and helps residents understand how the survey information will be used. One strategy is to place an advertisement in local newspapers that explains the purpose of the survey. A cover letter should accompany the survey and explain the objectives of the survey and identify supporting organizations and/or institutions. This letter should also identify a contact person if residents have any questions about the survey. Contacting local organizations – such as churches, schools, and civic organizations – may be another way of explaining the purpose of the survey and gaining support for the effort in the community.

Providing feedback to the neighborhood or community can be a useful way to gain some additional insights into the results of the survey and to reward residents for participating in the survey. It is preferable to provide residents with a written report of the results. Some discussion of feedback in the cover letter may improve the response to the survey.

## Mapping capacities

### Individuals

A central premise of the asset-based community development approach is that all individuals have a capacity to contribute to community well-being. However, assets of youth, seniors, and people with disabilities are frequently ignored. The most obvious assets are formal labor market skills such as work experience, leadership, and organizational skills. Other assets include experiences that individuals may have had outside the formal labor market such as care-giving skills, construction skills, or repair skills. Abilities also include "art, story-telling, crafts, gardening, teaching, sports, political interest, organizing, volunteering and more" (Kretzmann et al. 1997: 4). Another component of individual capacity is interest in participating in various community

organizations and working on local issues. All of these capacities need to be documented and analyzed for their potential contribution to the community.

An excellent resource for such an analysis is the workbook on mobilizing community skills of local residents authored by John Kretzmann et al. (1997). This workbook provides several sample surveys that have been used in the past to document the skills and experience of local residents. Although these surveys can be a useful beginning point, it is important to consider the characteristics and dimensions of any unique population in the community. In other words, do not assume that these sample surveys will necessarily work in all communities or that they adequately tap all the individual assets among various groups in the population.

The economic capacities of individuals in the community may also be considered. Residents often purchase goods and services outside of their community. These purchases could contribute to the local economy. Part of the mapping process is to evaluate where consumers purchase goods and services and how much they spend in- and outside of the community. When aggregated, these figures should provide community organizations with a sense of what goods or services could be provided locally. For example, if most neighborhood residents purchase groceries outside of the area, this may be a signal that there is a potential for establishing a grocery store in the neighborhood. It should be understood that there are many factors that affect the potential for retail establishments and this type of information is just one of those considerations. Aside from proximity, consumers may choose to purchase goods and services based on quality, service, and/or convenience (they may work close to the establishment). Some sample surveys for documenting these assets may be found in Kretzmann et al. (1996b).

How is this information used? First, the data may be used to help existing retail establishments identify goods and services that local residents consume but do not purchased locally. Second, this information is useful to potential entrepreneurs interested in starting businesses in the neighborhood. Finally, survey data collected on expenditures may be used as an educational tool to help residents understand the power their local expenditures have on the neighborhood's economy.

## Associations

Associations and organizations can facilitate community mobilization. Many efforts to mobilize communities begin with existing organizations because they have established relationships, trust, and resources that can be used in the asset-based community development effort. Although formal organizations are often visible and well established, there are many more organizations without paid staff that are not as easily identified. Some examples of these informal organizations are block clubs, neighborhood watches, garden clubs, baby-sitting cooperatives, youth peer groups, recreation clubs, and building tenant associations (Ferguson and Stoutland 1999).

How are these informal associations mapped? Most local organizations and associations do not show up on any official lists of nonprofit organizations because they are not incorporated or have any paid staff. Probably the best way to identify these associations is to conduct a survey among residents to identify any associations/groups they have heard of or belong to in the community. This method not only enables communities to identify nonprofit organizations but also the interorganizational networks that exist. For example, it may be useful to understand how organizations are linked through overlapping membership as a means of creating potential partnerships and collaborations.

Another reason why it is useful to map organizational resources is to help identify who should/could be mobilized in the asset-based community development project. One of the keys to the success of community development projects is inclusion of a wide range of residents. It may be possible to use a list of organizations and associations as a means of checking which community groups and interests are represented in the project. Obviously, it would be impossible to invite representatives from all organizations and associations to participate, but it may be useful to at least ensure that individuals from various areas, such as environmental, health care, economic development, or other areas, are selected.

What type of information should be collected about associations and organizations? The most important issue is identifying individual participation. Individuals should be asked what associations and organizations they belong to inside and outside of the community. This exercise should produce a relatively comprehensive list of organizations and associations. It is also important to identify an individual's leadership role (e.g., served as an officer) and level of participation.

Based on the list that is generated from the surveys of individuals, it is also possible to collect information from these organizations and associations. A list of board members and officers is useful to identify potential leadership in the community. It is helpful to collect information on the issues and concerns the organization has and what types of programs they have developed or implemented. Finally, some basic information on the resources of the organization will be helpful in order to identify assets that can be mobilized.

## Community institutions

Community institutions hold important resources that could potentially contribute to asset-based development projects. Institutions that are typically most important include parks, libraries, schools, community colleges, police, and hospitals. For example, each of these institutions purchases goods and services that could be directed at local businesses to improve the economy. They may have facilities and equipment that could be used by residents for community events. These institutions could adjust their employment practices so as to benefit local residents. Or, they could offer programs that could be redirected toward local residents. The main goal of mapping local institutions is to identify their resources and mobilize them in a way to benefit the community.

The first step in mapping community institutions is to develop an inventory of the institutions in the community. In most cases this is fairly straightforward, but it needs to be done systematically so that nothing is overlooked. In a small neighborhood this can be done quickly; but it may take more work and time in a larger region.

The next step is to identify institutional assets. Depending on the goals, this may involve identifying the spending patterns of the institution (i.e., where goods and services are purchased). This typically involves conducting surveys of these institutions and identifying key underutilized resources. One example of this would be the school-to-farm programs that have developed across the country. Community groups are assessing where schools are purchasing food and whether this food could be purchased locally through farms in the region. School-to-farm programs improve the markets for local farms and the quality of food in the schools. However, there can be institutional obstacles such as cost, scale, and price to use locally produced goods in the schools.

Another example of institutional mapping is the community reinvestment activities that many neighborhoods have been involved in over the past 20 years. These neighborhoods often begin with a thorough analysis of the lending patterns of local financial institutions to understand the capital flows in their area. Analyses of capital markets permit local residents to understand how well financial institutions are meeting the needs of local residents and whether their savings are being invested elsewhere. With these analyses, residents can work with local institutions to invest more in the neighborhood, use the data to challenge bank mergers, and identify the potential for other community-based lending institutions in the neighborhood.

Mapping institutions could require documentation of the talents and skills of the personnel who may or may not be community residents. It is also useful to assess other resources, such as meeting space, services, or equipment, that could be used for community purposes. One area that is often overlooked is the hiring practices of institutions. In many cases local institutions have little control over hiring because it is done through some central location such as a city or state office. However, there may be opportunities at these institutions to hire local residents for jobs that are available.

Finally, organizations can build strategies based on the identified resources in the community. These strategies need to be consistent with the broad set of goals and vision established at the beginning of the project. Kretzmann and McKnight's (1993) workbook, *Building Communities from the Inside Out*, is an excellent resource for identifying examples and methods of capturing local institutions for community building.

## Conclusion

Asset-based development differs from traditional community development strategies in several key ways. Local assets drive the community's plan for development and mobilization rather than problems or needs. It relies heavily on building relationships and local leadership. The beginning point for the asset-based approach is a map of the community's assets. This mapping effort requires residents to go beyond their preconceived notions of what they believe exists in the community and asks them to identify resources that could be used to achieve their vision for the future of their community. Developing an accurate assessment of resources is critical to the success of asset building. This also shifts the focus from the problems to the opportunities that face the community.

It should be stressed that mobilizing communities around assets can be a difficult process. Most residents want to move quickly to identifying solutions without adequately assessing the issues, understanding the resources that are available to them, or developing a vision of what the community should be in the future. While organizing communities around issues and problems often works in the short term, it is difficult to maintain in the long run. Mobilizing communities around partnerships and developing new leadership should provide a basis for long-term community action.

A final note regarding the quality of community surveys. One of the surest ways to halt a community development project is to start with questionable data. Because everything that follows is based on an accurate assessment of community resources, it is essential to conduct a survey that residents can trust. Many communities outsource the survey to professional or support staff. This has the advantage of improving the perception that the survey will be neutral. Yet the survey is never actually value-free, and residents may still feel that the process is biased. By contracting outside of the community for the survey, residents are missing out on an opportunity to build interest and support for their neighborhood. Residents may gain more insights into how they might promote asset-based development by creating their own questionnaire and conducting some of the interviews and/or pre-tests.

---

**CASE STUDY: HAZELWOOD COMMUNITY ASSET MAP: ASSESSING THE SERVICES, NEEDS, AND STRENGTHS OF HAZELWOOD'S COMMUNITY SERVICE PROVIDERS**

Communities frequently struggle to identify the available services and service providers in their area. Several factors contribute to this problem. Residents are often bewildered by the complex bureaucracy involved in social services. There is frequently a great deal of overlap in responsibilities across organizations and institutions. At the same time, many residents are not aware of the array of local resources available to them. In 2005, the Hazelwood neighborhood of Pittsburgh, Pennsylvania decided to map their assets to better understand the existing organizational resources in their area as well as to document the concerns among residents. In collaboration with the University of Pittsburg Community Outreach Partnership Center (COPC), the neighborhood conducted face-to-face and phone interviews with the contact person from every organization, church, and agency responsible for providing social services. The neighborhood was able to successfully complete 28 out of 40 interviews.

One of the most important findings was that there was very little overlap in social services provided to the neighborhood. While on the survey it appeared that several programs distributed emergency food to residents, they offered them at different times and places. Similarly, after-school programs were fairly well distributed across the neighborhood. Most residents were not aware of where and when these services were provided.

Thus, there was a need to establish a more formalized method of coordinating services. One of the conclusions from the asset- mapping exercise was that most of the service providers were located outside of the community and some effort needed to be made to provide a common space for these providers in the neighborhood. Another "need" that emerged from this study was that residents discovered that public transportation was a major hindrance in accessing the services that were identified. The asset-mapping exercise also identified some gaps in services, especially homebound services, transportation inside and outside of the community, and health needs of the residents. For more information about the project see: http://www.pitt.edu/~copc/Hazelwood_Asset_Map.doc.

## Keywords

Needs assessment, data collection, surveys, data analysis, asset-based development, asset mapping, institutional mapping, asset mobilization.

## Review questions

1  Why is it important to map assets?
2  What is the process for mapping individual capabilities?
3  What are some of the techniques for surveying communities?

## Bibliography and additional resources

Beaulieu, L.J. (2002) *Mapping the Assets of Your Community: A Key Component for Building Local Capacity*, Mississippi State, MS: Southern Rural Development Center. Available online at http://srdc.msstate.edu/publications/227/227_asset_mapping.pdf (accessed June 8, 2004).

Chambers, E.T. (2003) *Roots for Radicals: Organizing for Power, Action, and Justice*, New York: Continuum.

Dillman, D.A. (1978) *Mail and Telephone Surveys: The Total Design Method*, New York: Wiley.

Ferguson, R.F. and Stoutland, S.E. (1999) "Reconceiving the Community Development Field," in R.F. Ferguson and W. Dickens (eds) *Urban Problems and Community Development*, Washington, DC: Brooking Institution Press, pp. 33–76.

Green, G.P. and Haines, A. (2007) *Asset Building and Community Development*, 2nd edn, Thousand Oaks, CA: Sage.

Johnson, D.E., Meiller, L.R., Miller, L.C. and Summers, G.F. (eds) (1987) *Needs Assessment: Theory and Methods*, Ames: Iowa State University Press.

Kretzmann, J.P. and McKnight, J.L. (1993) *Building Communities from the Inside Out: A Path Toward Finding and Mobilizing a Community's Assets*, Chicago, IL: ACTA Publications.

Kretzmann, J.P., McKnight, J.L. and Puntenney, D. (1996a) *A Guide to Mapping and Mobilizing the Economic Capacities of Local Residents*, Chicago, IL: ACTA Publications.

Kretzmann, J.P., McKnight, J.L. and Puntenney, D. (1996b) *A Guide to Mapping Consumer Expenditures and Mobilizing Consumer Expenditure Capacities*, Chicago, IL: ACTA Publications.

Kretzmann, J.P., McKnight, J.L. and Puntenney, D. (1996c) *A Guide to Mapping Local Business Assets and Mobilizing Local Business Capacities*, Chicago, IL: ACTA Publications.

Kretzmann, J.P., McKnight, J.L. and Sheehan, G. (1997) *A Guide to Capacity Inventories: Mobilizing the Community Skills of Local Residents*, Chicago, IL: ACTA Publications.

Medoff, P. and Sklar, H. (1994) *Streets of Hope: The Fall and Rise of an Urban Neighborhood*, Boston, MA: South End Press.

Salant, P. and Dillman, D.A. (1994) *How to Conduct Your Own Survey*, New York: Wiley.

Warren, M.R. (2001) *Dry Bones Rattling: Community Building to Revitalize American Democracy*, Princeton, NJ: Princeton University Press.

# 11 Assessing your local economy

## Industry composition and economic impact analysis

## William Hearn and Tom Tanner

One of the most fundamental questions a community must ask about itself is: "How strong is the local economy?" A strong local economy providing good jobs and income is a primary determinant of the quality of life and standard of living for a community's residents. The first step in assessing the local economy is to understand what industries comprise the economic base. Are these industries related or diversified? Are they growing or declining locally, nationally, and internationally? Often the answers to these questions are surprising to local officials and citizens, and serve as a wake-up call to pursue a more vigorous community and economic development policy.

Economic impact analysis is an important component of local economic assessment. It provides a valuable public policy tool for selecting the best economic development alternatives. It also plays a leading role in helping communities decide on the level of economic development incentives to offer new or expanding companies.

## Introduction

This chapter will provide an introduction to two key areas of economic assessment. The first is industry composition, using industry classifications and commonly available measures. Essentially, it includes the kinds of basic data that any person involved in community and economic development might want to know in order to gain an overview of the local situation. A basic understanding of the local economic situation will facilitate informed decisions by board members, elected officials, and other community development stakeholders. With this knowledge, community and economic developers can better represent their areas and gain the necessary grounding to work with colleagues in the private sector, and within allied economic and community development agencies.

The second area covered in this chapter is impact assessment, as a tool to evaluate the regional economy and support economic development decisions, especially the investments communities make to attract new businesses or retain existing businesses. Economic impact models have existed for many years and there are numerous companies that provide this type of service. Increasingly, communities are operating their own economic impact models so they can get immediate feedback on important economic development questions. Community and economic developers should be aware of these tools and understand their major underlying assumptions. Economic impact models can provide critical information when the time comes to vote or voice an opinion on a community infrastructure or business incentive decision.

## Industry composition: learning about the local/regional economy

Economic development agencies and related organizations generally maintain demographic and economic information about regional and local economies for use in assessing local conditions, recruiting companies, retaining and expanding existing businesses, helping start-up new companies, and a variety of other uses. Economic development agencies reside within many types of organizations including state, county, and local government, educational institutions, and the private sector (e.g., utilities, Chambers of Commerce). Given these many resources, it is beneficial to determine what types of existing information can help provide a basic understanding of local economic conditions and opportunities. Often a wealth of information is readily accessible through allied economic development agencies. The following might be good avenues:

- *State economic development.* State departments of commerce or economic development often have research staff who will have prepared reports of interest on economic development. There may be studies prepared by consultants on behalf of the state. These studies may be focused on industry clusters of excellence in a state or provide an analysis of the business operating environment region by region. There may also be competitive analyses that consider the state's assets and liabilities versus those of other states or comparisons of incentive programs. Some states have outstanding staffs who conduct their own internal research and produce high- quality reports, often on specialized industries. These reports may be published on an annual basis and take a critical look at a certain industry such as automotive, precision manufacturing, call centers, or other strategic industries. These reports can serve as good general background information and as marketing material. On the other hand, it may turn out that the state does very little economic development research or relies on other organizations for

this (e.g., universities or Chambers of Commerce), but it is worthwhile to find out what is available. The data will likely cover the entire state, so this may provide an opportunity to learn more about what other communities in the state are doing.

- *Utility companies.* Many utility companies, particularly the energy companies, are heavily engaged in economic development activities. Some of the areas they support include: direct funding of economic development agencies; specialized studies; project management and marketing; pre-planning and planning of industrial parks; community economic development services, and company expansion services. In many areas, utility companies act as secondary state agencies supporting economic development. In other markets, the utility companies are not as active. In general, it is worthwhile to get to know regional representatives of the utilities and tap into the information they have available. In some cases, the local utilities may have access to national databases on labor costs and other specialized resources, such as mapping and demographic capabilities, that will provide new perspectives on the local economy.

- *Universities.* Typically, there are resources within area universities that can be tapped – each state has a designated university that is its data center. Many states have dedicated economic development resources that reside within their university system. These may include industrial extension and engagement programs, programs focused on manufacturing processes and technology, as well as programs focused on small business. Many universities have affiliated research parks that serve a core economic development mission. This can be a tremendous asset to a community, especially when a university has public service and economic development as a core mission. Communities with such a university should work diligently to ensure that the university is appropriately engaged in economic development. Larger universities with public policy, public administration, and/or planning schools may also have specialized centers or institutes focused on areas of interest. It would be worthwhile to tap into these areas and determine what type of information they have available. In some cases, there are specialized reports that have been prepared for communities in the state. There may also be specialized ongoing economic development initiatives that are of interest, and there may be access to specialized resources such as economic impact modeling. Some of the data, especially within state universities, are free. It may also be possible to find faculty members with expertise in a given field who consult nationally on economic development-related issues.

- *The World Wide Web.* The Internet can provide a wide diversity of information on virtually any economic development topic. There are many specialty data services that provide information on topics of interest to economic development. These include: area, cultural and demographic information; specialized data sources; and information pertaining to schools and employee relocation. The Web can also be a tremendous source of information on the companies in a community, including recent business activity and financial disclosures.
- *Commercial services.* There are numerous fee-based or subscription services that one may tap for statistical, financial, and other information on companies. Those such as Dun & Bradstreet, Hoover's, and Harris Infosource have detailed information about even the smallest companies. These sources may be used to find out the industry composition and names of companies in a state, region, county, or even zip code.

Assessing the industry composition of the local economy is not only important for designing an effective economic development program, but for increasing a community's success rate in recruiting new companies and retaining and expanding existing companies. Each company will have different location factors and data needs. For example, the needs of an existing company going through an expansion will be different from those of a company looking to relocate or place a new plant or business unit in a community. In many instances, companies looking to site new operations will have dedicated teams going through a systematic process. These will be either internal cross-functional teams or external consultants. In either case, they may know more about a community's demographic and economic situation than the residents themselves. The more homework the community does, the more likely it is to win the project.

As the site selection process unfolds, a community may be asked to play a role in the location decision. A community will not be selected because it is a great place to be but because it has survived rigorous screening and elimination rounds. What may ultimately help a community most is an understanding of the types of functions and enterprises that work best in the area. This sense of "knowing what works" is developed from the years that citizens have resided in an area, and it can be supported through some simple data exercises that will also enable the community to speak with an informed perspective. Maintaining a comprehensive demographic and economic database, or at the very least knowing where to get this information immediately, is critical to the process of responding to inquiries from companies seeking locations or state or regional economic development agencies working with these companies. If communities do not respond in a timely manner to data requests or if their data are inaccurate, they will usually be eliminated from the search process.

## Evaluating local/regional industry composition – an example

Industry composition assessment will help communities understand the types of industries in the area, their growth characteristics, their degree of concentration, and other important aspects. The inputs for this type of analysis may be sourced from several private data service providers as discussed above, and the data are also available from government sources such as "County Business Patterns" (www.census.gov), an annual series that provides local economic data by industry. County Business Patterns provides detailed industry employment, total payroll, the number of establishments, and a breakdown of establishment size. These data are available going back in time. However, historical analysis does require some basic knowledge of the conversion of the Standard Industrial Classification System (SIC) to the recently adopted North American Industry Classification System (NAICS). Conversion tables are available, but the conversions sometimes occur at the subindustry level and require detailed manipulation of the data.

When using economic information to assess local industry, it is necessary to decide upon the area of study. If an economic development agency represents a county, then it would certainly want to look at county-level data but would also need to look at surrounding counties as well. In selecting sites, companies and consultants would evaluate regions based on 30- or 45-minute commute areas, not just individual counties. It is also useful to compare a region or grouping of nearby counties to state or national averages.

Table 11.1 is a sample regional industry assessment that represents one of eight regions of a state. It is an example of the type of industry composition analysis that will provide a useful starting point.

Table 11.1 is sorted by regional location quotient (a measure of industry concentration in the region as compared to U.S. employment), which is explained below. The following information is shown for the region:

- Industry NAICS Code – the broadest NAICS industry designation.
- Regional employment as well as the percentage of state employment in the region by industry. This provides insight into the portion of total state employment in each industry in the region.
- Earnings and the percentage of state earnings in the region by industry.
- Average earnings by industry (this can be a useful piece of information for economic development planning as well as industry recruitment).

- Employment concentration measures (industries sorted in descending order) or location quotients (LQ) for the two-digit industry in the region and the state. The location quotient is a measure of industry geographic concentration that compares industry employment in the region to a reference region. In this example, both location quotients are provided: one for the region to determine local industry concentration compared to the state (regional LQ), and one to assess state concentration compared to the U.S. (state LQ). The region location quotient is calculated by dividing industry employment in the region by total employment in the region (which gives the industry's percent share of total employment in that region). The result is divided by industry employment in the state by total employment in the state region. This measure is essentially comparing the ratio of employment for a given industry in the region to the ratio of employment in the reference region (state). An LQ of 1.0 would mean that the same concentration of employment exists in the region as in the reference region (state) – a higher LQ would mean a higher industry concentration and vice versa for a lower LQ. This measure provides insight into which industries are over- or under-represented in the region.
- Column totals and, where appropriate, averages are shown. These numbers can be meaningful when comparing to other areas.

This framework is generally useful as a starting point in order to get an idea of which industries are important to the economic base of a region and which industries have local concentrations or clusters. Forming an economic development strategy around this technique requires more in-depth analysis of resources that are available in the community, but this is an important first step. It is also very useful to add the dimension of time to this analysis (not done here) by looking at historical growth rates and, where possible, projections of future growth, especially where local growth outpaces national growth. Measuring growth can be tricky, however, especially where there are changes in the business

**Table 11.1** Sample regional industry composition assessment

Sample regional analysis

| 2 digit NAICS | Industry | 20xx employment | | 20xx output ($000s) | | 20xx earnings ($000s) | | | Employment concentration | |
|---|---|---|---|---|---|---|---|---|---|---|
| | | # | % of state | Total | % of state | Total | Average | % of state | Region LQ | State LQ |
| 31 | Manufacturing (food) | 40,006 | 17.9 | $6,112,057 | 16.3 | $1,020,393 | $25,506 | 15.6 | 2.09 | 2.20 |
| 11 | Agriculture, forestry, fishing and hunting | 13,404 | 17.0 | $2,226,192 | 16.7 | $347,111 | $25,896 | 17.1 | 2.00 | 0.82 |
| 33 | Manufacturing (metals and machinery) | 36,245 | 14.9 | $7,436,218 | 13.6 | $1,512,050 | $41,717 | 13.9 | 1.75 | 0.71 |
| 21 | Mining | 1,234 | 14.7 | $137,195 | 10.8 | $39,901 | $32,335 | 10.8 | 1.72 | 0.50 |
| 32 | Manufacturing (materials) | 24,312 | 12.3 | $4,369,717 | 11.3 | $874,842 | $35,984 | 11.2 | 1.44 | 1.17 |
| 44 | Retail trade | 45,446 | 10.5 | $2,369,900 | 9.1 | $967,749 | $21,294 | 9.1 | 1.23 | 1.09 |
| 23 | Construction | 26,864 | 10.2 | $2,656,385 | 8.5 | $832,756 | $30,999 | 8.1 | 1.20 | 1.03 |
| 72 | Accommodation and food services | 37,105 | 9.5 | $998,950 | 7.7 | $397,705 | $10,718 | 7.6 | 1.11 | 1.00 |
| 22 | Utilities | 2,685 | 9.4 | $960,806 | 8.5 | $179,063 | $66,690 | 8.4 | 1.10 | 1.20 |
| 45 | Retail trade | 15,264 | 8.9 | $643,667 | 7.9 | $261,424 | $17,127 | 7.9 | 1.05 | 0.97 |
| 71 | Arts, entertainment, and recreation | 4,135 | 8.6 | $218,783 | 7.2 | $84,008 | $20,316 | 6.9 | 1.01 | 0.66 |
| 61 | Educational services | 7,043 | 8.6 | $220,777 | 6.9 | $158,891 | $22,560 | 6.9 | 1.01 | 0.81 |
| 62 | Health care and social assistance | 38,303 | 8.2 | $1,718,861 | 7.4 | $1,155,138 | $30,158 | 7.5 | 0.97 | 0.80 |
| 81 | Other services (except public administration) | 15,257 | 8.2 | $752,121 | 6.8 | $267,535 | $17,535 | 6.3 | 0.96 | 0.90 |
| 42 | Wholesale trade | 18,004 | 7.2 | $1,744,229 | 5.4 | $752,020 | $41,770 | 5.4 | 0.85 | 1.06 |
| 92 | Public administration | 43,642 | 7.1 | $1,888,883 | 5.9 | $1,429,054 | $32,745 | 5.9 | 0.83 | 1.06 |
| 56 | Administrative/support/waste management/remediation | 26,397 | 6.2 | $897,869 | 6.0 | $527,703 | $19,991 | 4.8 | 0.73 | 1.12 |
| 52 | Finance and insurance | 11,577 | 5.8 | $1,297,086 | 4.8 | $473,878 | $40,933 | 4.0 | 0.68 | 0.88 |
| 53 | Real estate and rental and leasing | 4,128 | 5.5 | $761,239 | 3.4 | $103,460 | $25,063 | 3.6 | 0.64 | 0.97 |
| 48 | Transportation and warehousing | 5,718 | 4.1 | $597,703 | 3.4 | $186,367 | $32,593 | 3.0 | 0.48 | 1.14 |
| 54 | Professional, scientific, and technical services | 10,955 | 4.0 | $696,479 | 2.4 | $366,356 | $33,442 | 2.2 | 0.47 | 0.94 |
| 51 | Information | 6,531 | 3.7 | $945,710 | 2.6 | $347,879 | $53,266 | 2.9 | 0.43 | 1.22 |
| 49 | Transportation and warehousing | 2,446 | 3.6 | $175,059 | 3.1 | $93,583 | $38,260 | 3.1 | 0.42 | 1.20 |
| 55 | Management of companies and enterprises | 3,156 | 2.7 | $63,113 | 1.7 | $175,759 | $55,690 | 1.7 | 0.31 | 1.06 |
| | Total | 439,857 | | $39,888,999 | | $12,554,625 | | | | |
| | Average | | 8.7 | | 7.4 | | $32,191 | 7.3 | | |

cycle over the selected time period. To evaluate structural changes, it is important to select points in time that are approximately at the same point in the business cycle.

Table 11.2 builds upon Table 11.1 and introduces the more detailed three-digit industry employment NAICS Codes (which go all the way to the eight-digit level) for the most highly concentrated two-digit industries in the region (based on LQ) from Table 11.1. Table 11.2 provides insight into the more specific industries where there are favorable employment concentrations. For many communities, this level of analysis (three-digit NAICS) is sufficient to begin to get a sense and feel for the types of industries that play a prominent role in the economic base.

At this level, one can take the next natural step and ask: "Who are the major employers in our community in these industries?" This type of industry assessment is only one approach in evaluating industry composition. Further assessments might involve keeping track of different company functions in the area (manufacturing, headquarters, back office, research and development, warehouse and distribution, and the like). At a minimum, it is important to be able to understand industry composition (as laid out above) and location quotients, and to know the major employers and their functions.

While even this simple level of data analysis may seem complex at first, it is really quite simple to accomplish. Public and private data sources, such as those described above, can provide the raw data at little or no cost directly from the Internet, compact disks, or data books. The data can be imported into a spreadsheet to perform the calculations and measurements shown in Tables 11.1 and 11.2. Data and research specialists,

■ **Table 11.2** Three-digit NAICS industries for local concentrated industries

| 2-digit NAICS | 3-digit NAICS | Industry | 20xx employment |
|---|---|---|---|
| 11 | 111 | Crop production | 12,622 |
| 11 | 113 | Forestry and logging | 634 |
| 31 | 311 | Food manufacturing | 22,092 |
| 31 | 313 | Textile mills | 12,040 |
| 31 | 315 | Apparel manufacturing | 4,215 |
| 31 | 314 | Textile Product mills | 1,319 |
| 32 | 321 | Wood product manufacturing | 6,126 |
| 32 | 325 | Chemical manufacturing | 5,138 |
| 32 | 327 | Nonmetallic mineral product manufacturing | 4,973 |
| 32 | 326 | Plastics and rubber products manufacturing | 4,665 |
| 32 | 323 | Printing and related support activities | 2,340 |
| 32 | 322 | Paper manufacturing | 1,031 |
| 33 | 336 | Transportation equipment manufacturing | 8,712 |
| 33 | 332 | Fabricated metal product manufacturing | 7,335 |
| 33 | 333 | Machinery manufacturing | 5,530 |
| 33 | 335 | Electrical equipment, appliance, and component manufacturing | 4,684 |
| 33 | 339 | Miscellaneous manufacturing | 4,521 |
| 33 | 337 | Furniture and related product manufacturing | 2,349 |
| 33 | 334 | Computer and electronic product manufacturing | 2,033 |
| 33 | 331 | Primary metal manufacturing | 1,081 |
| 62 | 622 | Hospitals | 15,532 |
| 62 | 621 | Ambulatory health care services | 11,344 |
| 62 | 624 | Social assistance | 5,918 |
| 62 | 623 | Nursing and residential care facilities | 5,509 |

from state and local departments of economic development, utilities, universities, and other organizations mentioned above, can assist in obtaining and using these data.

Consider the following exercise to assess the local economy:

• Obtain the data on industry composition (employment, output, sales, income) for the region and county or city at the aggregate level (two- or three-digit NAICS as shown in Tables 11.1 and 11.2).
• Import these data into a spreadsheet, calculate measures of concentration (location quotient), averages, ranges, and so on.
• Arrange the industries by any measure such as number of employees, percentage of total employment, or concentration (location quotient) using the spreadsheet sort function.
• Insert into the spreadsheet national, state, and/or regional data on historical and forecasted growth by industry to determine if key local industries are growing or declining nationally and locally.

As stated above, this simple economic snapshot of a community will help determine whether it has a strong economic base comprised of growth industries or a weak one comprised of industries that are declining at the state, national or international level. Again, this simple level of industry composition analysis can often act as a wake-up call for a more aggressive economic development program.

## Evaluating industry clusters – refocusing industry assessment

Given the nature of economic development today, and specifically the convergence of industries that has occurred, it is often useful to regroup industries to form more meaningful "industry clusters." More information on industry clusters may be found by searching the Internet or development literature, or talking with an economic development research

specialist. Figure 11.1 is a sample of an industry cluster, aerospace and defense, and shows its employment across a state by region (percentage of state employment in the region). Table 11.3 shows the four-digit NAICS composition of that particular industry cluster. In some cases, it may be more meaningful to evaluate employment by a particular cluster than by an industry, depending on the exercise. Cluster analysis may be more meaningful to deploy in support of an industry-targeting exercise along with a detailed inventory of available assets. Some of the more common industry clusters are:

1 aerospace and defense
2 agribusiness
3 business and financial services
4 energy and environmental
5 health care
6 homeland security
7 life sciences
8 advanced telecommunications
9 multimedia.

Each of these clusters may have assets in the region that can form the core of an economic development strategy. Many communities and regions have developed successful marketing programs based on industry clusters.

## An introduction to economic impact analysis

Economic impact analysis is an increasingly critical tool in guiding economic development and parallel efforts in smart growth, land-use planning and the like. The tools of economic impact analysis help identify the potential economic outcomes communities can expect from civic projects or new companies, and helps them explore the alternative futures they might realize. The specifics of impact analysis may not be widely understood and appreciated, but it can be very useful in addressing questions such as:

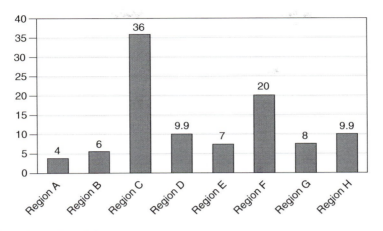

*Figure 11.1* **Sample regional cluster**

▓ *Table 11.3* *Sample aerospace cluster definition (four-digit industry)*

| Aerospace 4 digit NAICS | Industry |
| --- | --- |
| 333200 | Industrial machinery manufacturing |
| 333500 | Metalworking machinery manufacturing |
| 333600 | Engine, turbine, and power transmission equipment manufacturing |
| 336400 | Aerospace product and parts manufacturing |
| 336900 | Other transportation equipment manufacturing |
| 541700 | Scientific research and development services |

- What are the current economic conditions in the community?
- What will the economic and fiscal impact of a proposed investment or policy in our community be (jobs, payroll)?
- What components of the community have been growing and what components have been declining?
- What are the community's options for improving its economic future and which of those options should be pursued first?

Numerous tools and techniques may be used to give some insight into the functioning of a community's economy. This insight is part of the early preparation which must occur before the community can

initiate an effective strategy for change. The following considerations are an important part of any community economic impact analysis:

1 No single number generated by the analysis provides the answer to all of a community's concerns. For instance, when a firm experiences an increase in profits, it is still unknown whether the increase is caused by greater sales or more cost-efficient production. Each of those sources suggests a different strategy by the owner.

2 It is important to make comparisons among communities because the numbers used are not absolute. Just as any owner of a firm would compare his or her success against the industry average or similar firms, it is important for

## BOX 11.1 ASSESSING THE MARKET POTENTIAL OF UNDERSERVED NEIGHBORHOODS

Drive through a high-income neighborhood and you will likely see a plethora of retail businesses providing a variety of products and services ranging from the necessities such as groceries and clothing to the more discretionary such as day spas and art galleries. You will notice quite a difference if you drive through a low- or even moderate-income neighborhood. It would not be surprising to find an absence of jewelry stores and expensive boutique clothing shops, but you may be surprised by the lack of basic retail services we all take for granted. Instead of a large modern grocery store, some poorer neighborhoods are served only by convenience stores with high prices and a limited selection. Lack of retail and commercial services is often a major problem for low- to moderate-income neighborhoods, especially in inner city areas. Residents of these underserved neighborhoods must deal with higher prices and a limited range of products locally, or take time and pay money to travel where better stores are located, often many miles away.

Some local governments, community development corporations, and other interested organizations are working to alleviate this problem and make underserved neighborhoods better places to live. As neighborhoods decline for various reasons (e.g., population movement from city to suburbs), typically wealthier residents move out, crime increases and the housing stock and physical infrastructure deteriorates. As income and retail spending decline, many retailers relocate to better areas or go out of business. Some time later, the neighborhoods may stabilize or even dramatically improve through gentrification. However, retail companies may not immediately see the new potential in these old neighborhoods, which may still look rough.

In these types of situations, communities can utilize a different type of economic analysis to assess the retail potential of the neighborhood and help attract retailers to the area. One way to do this is to compare potential retail "demand" with current retail "supply" in the area. Potential retail demand is what neighborhood residents would spend locally if stores were in the area, rather than drive miles away to better stores. A common way to estimate retail potential is to collect the most recent data on incomes in the study area from the U.S. Census Bureau or a private data vendor. In some cases, community organizations and local governments may conduct their own surveys of the neighborhood to collect the most up-to-date income data (this should be done carefully and anonymously to encourage participation in the survey). Next, data on how much households and individuals at different income levels typically spend on various retail products and services (e.g., clothing, food, pharmaceuticals) is collected, typically from the U.S. Bureau of Labor Statistics' Consumer Expenditure Survey (or collected with a neighborhood survey). By combining the income data and expenditure data and doing a lot of multiplying, retail potential demand can be estimated.

The supply of retail services is usually estimated by surveying and inventorying the types of retail business and their square footage in the study area. Using data on average sales per square foot for different types of retail stores (again, available from data vendors or from local surveys), one can then estimate the total retail sales in an area by type of product or service. Finally, by combining the estimates of potential retail demand with retail supply, one can calculate the excess retail demand in the study area that is presumably being met by residents driving out of the area or simply forgoing purchases. This data can be very valuable in helping recruit retail stores to underserved neighborhoods. Government incentives for retailers to locate in underserved areas and demonstrated community enthusiasm to support local retailers can also help recruit stores.

There are many consultants and organizations that can assist with this kind of retail potential analysis. Here are two:

Social Compact: www.socialcompact.org
Brookings Institution: www.brookings.edu/metro/umi.htm

communities to make similar comparisons. These comparisons need to be with similar communities in order to increase the legitimacy of their analysis.

3   It is important to compare changes over time to sense the direction of community change. This comparison helps to confirm or deny perceptions of current and recent conditions.

4   It is crucial to use a variety of information sources. The steps in local economic analysis in this chapter indicate that a lot of "hard data" can be collected from various agencies as a starting point. It is also important to incorporate into the analysis the insights local citizens have about their community and its economy. If there are differences, then these sets of information need to be challenged to determine which more accurately reflects current conditions.

## Driving forces behind the local economy – the water analogy

Perhaps the easiest way to think of community economic impact analysis is to imagine the community itself as a tub with money flowing into the top as well as draining out of the bottom (Figure 11.2). This analogy represents a number of key concepts. First, any community is intimately linked with the rest of the world through the inflow and outflow of income and goods – the tap and the drain both lead to "everyplace else." Second, the economy (the pool of water in this analogy) uses resources, may be purchased elsewhere (come in through the tap), or may be available locally (be contained in the tub of local resources). Third, the amount of water (the local economy) is determined essentially by the inflow of outside income, the "recycling" of income within the local economy, and the volume of resources used to produce the community's output.

When a community sells or "exports" the goods or services it produces to the next county, state or country, money flows into the local economy from the tap in the form of revenues, profits, and wages. This is

why "export-based" industries – such as manufacturing, advanced services, or even tourism – are so important to a local economy. They bring new money into the community through the tap and raise the level of the water (the local economy). On the other hand, when local consumers spend their money outside of the area (e.g., in a shopping trip to another town), money flows out of the tub and reduces the level of local economic activity. Purely local industries such as barbershops and local restaurants, while important to the quality of life in the local community, typically recirculate money already in the economy and don't increase the overall level of the water. Economic development programs should therefore be based on recruiting, retaining, and starting export-based industries. Local restaurants and grocery stores will open as export-based industries grow.

In a broad sense, economic impact analysis is a systematic examination of this tub; the forces of supply and demand in a local economy. This can be translated into some basic economic development policy questions:

- What are the linkages with the rest of the world?
- What are some ways to increase the potential inflow of income?
- How can the community better use its existing

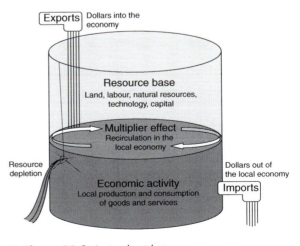

**■ Figure 11.2** A visual guide to your economy

resources and businesses to produce more output, associated jobs, and income?

- How can the community reduce the loss of resources to improve its local income situation?

To stay with the simple "tub economy" analogy, the basic goal of economic development is to collect as much water in the tub as possible. This may be done by:

1 *Decreasing the size of the drain* (import substitution). This strategy involves identifying and attracting industries that are currently not present in the community and whose products are, therefore, imported. One highly controversial example involves smaller communities that try to attract "big box" retailers. This strategy rests in the hope that shoppers will buy locally, keeping their dollars local rather than driving to a larger community to do their shopping. This outcome reduces the outflow, sends their money "down the community's drain," so to speak, and into the neighboring community's tap. This strategy is generally only applicable to the smallest communities and over a narrow range of industries; larger communities generally have all the retail and service activity they would need and manufacturers and other industries are almost never attracted to an entire region because they plan on serving just the local market.

2 *Patching the leaks.* This strategy of preventing local resource depletion can take numerous forms. Cities such as Portland, Oregon, and Boston have attempted with some success to encourage economic growth through the preservation of the natural environment and "green space," making the regions more attractive for hi-tech businesses whose highly paid workers desire a good quality of life. Other communities have campaigned to retain their college graduates and/or recruit "creative workers," artists and the like, for much the same reason.

3 *Increasing the size of the tap* (an "export orientation" approach). This approach involves attempts to directly increase the exports of the regional economy. Remember: exports increase the flow of money into the region from the rest of the country/world, thereby increasing the size of the tap. This may involve recruiting new export industries or working to retain and support such industries already in the region.

For this discussion of economic impact analysis, the focus is on a single, fundamental question: How can a community increase the amount of water in the tub by increasing the amount of water flowing in through the tap? When a community chooses to expend resources to attract firms, the decision has (implicitly, perhaps) been made to promote economic development by increasing the size of the tap.

In most cases (the exception being the import substitution scenario described above), a key characteristic of any firm that is recruited into a community should be that it is primarily an exporter (i.e., it serves a national or international marketplace). As an exporter, the firm is expected to bring money into the community from the rest of the world which means more water in through the tap and more water in the community tub.

Many communities are interested in increasing the amount coming through their tap which means that they are all trying to attract the same kinds of firms. In any given year, depending on the state of the economy, there are literally thousands of communities seeking to recruit the several hundred or so company location projects that exist.

Because of this competition for new investment, communities are more likely to have to pay (or at least provide a competitive baseline of financial incentives) to attract the firm(s) of their choice. From a public policy analysis perspective this is perhaps unfortunate, but it does reinforce the need for high-quality analysis of the economic impact of a proposed project. A quality economic impact analysis of a potential new firm gives the community a good idea of just what they would be getting for their incentive package (what is the return on investment?) or, conversely, gives the economic development official a good idea of how valuable the new firm would really be for the local economy. The goal of the community economic impact analysis, in this situation, is to quantify the degree to which the firm in question would increase the size of the regional economy; or, given the very particular way in which this firm will effect the tap and the nature of the community's tub and drain, what effect would the firm have on the water level?

A common term in any conversation about economic impact analysis is the "multiplier effect." Sticking with the tub analogy, the multiplier effect identifies the following: When the cash flows in from the tap, how many times will it swirl around in the tub before it runs down the drain? The more times it swirls around before draining out, the higher the water level and the bigger the regional economy will be. To see how the multiplier phenomenon works, image that a new manufacturing facility employing 50 local residents has just located in a community. As discussed above, the manufacturing firm sells its products outside of the community and brings in new income in the form of sales revenue, profits, and wages for the workers. The new firm buys some natural resources and services locally, and the workers then spend some, hopefully most, of their wages in the community on groceries, haircuts, clothes, and so on, which increases incomes and jobs in these "secondary" sectors. In turn, the grocers, barbers, and clothing store owners spend some of their increased income locally which promotes a third round of job and income creation. The process repeats itself through several spending cycles. Hence, a dollar of new outside income flowing into the community's tap increases local income and economic activity (the level of the water) by more than just that dollar.

At this point, it is important to dispel several myths about multipliers in order to avoid some very common misunderstandings:

1   There are several different multipliers: employment multipliers, which indicate the total number of jobs created per employee at the new facility; wage multipliers, which measure total additional wages paid in the region per dollar earned by employees at the new facility; and output multipliers, which measure the total of additional sales in the region per dollar of sales at the new facility. These are the most commonly reported multipliers, though several others will be seen from time to time.

2   The employment, output, and wage multipliers for a given project or facility are rarely equal. As a very general rule, employment multipliers for a project tend to be larger than wage multipliers, and output multipliers will tend to be higher than either employment or wage multipliers. An impact analysis that assumes the same multiplier for different applications (e.g., an income multiplier to calculate the increase in the number of jobs) is probably not done correctly.

3   The size of any given multiplier will depend completely on the industry and the region in question; there is not a single regional multiplier that applies to all industries, or a single industry multiplier that applies to all regions. If a report assumes a regional or rule-of-thumb multiplier in the analysis, it is probably not done correctly.

## Information requirements for economic impact models

Economic impact analysis is used to calculate the appropriate multiplier effect(s) in order to determine the total economic impact on the region of the firm in question. An analysis of this sort will be able to determine what the overall impact of a specific new firm will be on the overall size of the economy. While this chapter cannot provide all the tools needed to conduct an impact analysis, it can provide information to help understand how and when to use economic impact analysis.

An impact analysis should be conducted any time a firm requests any incentive concessions from the community to determine exactly what that community is getting for its money and what the return on its investment might be. Whether the analysis is conducted by the community or by a consultant, the more information available about the new firm, the better. The information that a community should obtain to facilitate the best possible impact analysis includes details on:

1 *The industry of the firm in question.* A description of the activity in which the firm is engaged would likely be adequate, but providing the NAICS Code (North American Industry Classification System) or the more dated SIC Code (Standard Industrial Classification) for the industry maximizes the likelihood that the community will be analyzing the industry within the correct framework. These codes are self-reported by companies, so it is possible to use national data sources to identify codes, though firms may report multiple codes. It is important to note that different impact analysis tools operate at different levels of industry detail; some tools can identify only 14 different industry types, while others can identify over 700. When analyzing the economic impact of a new firm on a region, the more industry detail, the better the result.

2 *The region in question.* This can be a bit tricky and involve making some compromises. Good economic impact analysis tools generally use federal government data sources regarding regional economies. Because most of these data sources only collect information down to the county level, the analysis is generally going to be limited to defining a region as a single county or a group of counties. A few modeling tools allow analysis down to the zip code level of detail, but that level of economic data is very limited, so those tools will be using much less reliable information. In addition, some models can conduct true multi-regional analyses that would, for example, allow a regional planning council to determine the impact of a new firm locating in one county, on every individual county in their service area. Others might look only at a single region. The analyst will need to know the precise area(s) of interest when conducting the analysis.

3 *Employment and/or wages and/or output of the firm.* Technically, the analysis could be done with any one of these three numbers, but the more information provided on each of these, the better. Thankfully, the order of importance of these

numbers for the analyst – (1) employment, (2) wages, (3) output – is probably the same order in which the firm is willing to provide the data. Firms are often reluctant to disclose project output (sales) data, so the analysis is often done with employment and wage data. It might also be important to know *when* the employment is expected. Some economic modeling tools allow for multi-year analysis, while others do not. If a community is interested in examining multi-year impacts, or the firm has definitive plans for increasing employment at the facility over time, it is important to use a multi-year modeling tool and to collect as much information as possible about the timing of employment increases.

4 *Construction and investment spending.* An important short-term impact of the new firm is likely to come in the form of these types of spending in the region. The more information provided to the analyst regarding the time, location, and type of expected investment, the better.

5 *The nature of the (potential) incentive package.* An important consideration of any impact analysis is how much a community is granting in incentives to get the firm. If this information can be provided, the analyst should be able to determine the net impact of the new firm including the concession package as well as the gross impact of the firm itself. If these concessions occur over time, then it would be best to provide the year in which those concessions occur (such as tax abatements). Alternatively, one could use the net current value of the recurring incentives. Again, under this type of assumption, it would be best to use multi-year impact tools.

6 *Other factors that might be of interest.* For every project, there may be some interesting bits of information such as the identity of any supply firm that is likely to relocate to the community because of the relocation of the firm in question; a major competitor firm already located in the region; or the kind of new school construction that will be necessary because of additional population moving to the region. If there are some potential special circumstances, or additional

tidbits of data that might be important, the analyst should be informed.

## Conclusion

This chapter provided an introduction to assessing the local economy and using economic impact models. Both of these activities are highly data-driven and, therefore, a community needs to understand where to collect the necessary data. Looking at local industry composition can be a valuable tool in understanding where a community is today and perhaps where it has come from. However, it will not tell us where a community needs to be in the future. This is a policy decision. The same may be said for economic impact models. While there are many such modeling tools available in the market-place, one must be aware of the limitations and challenges presented by each approach. One common concern is that when a model fails to fully account

for the public costs associated with a project, it leads to an overly optimistic assessment of the potential economic impacts. Local economic analysis and impact modeling are useful tools that can be used to support both economic development strategy formulation and decision-making challenges as they occur. Ultimately these tools and techniques just scratch the surface, but when accurately brought to bear on a problem, they can mean the difference between good and poor community decisions.

## Keywords

Economic impact, multipliers, local economy, economic analysis, local industry, industry composition, impact assessment.

---

**CASE STUDY: THE LIMA-ALLEN COUNTY, OHIO CIVIC CENTER**

Lima (Allen County), Ohio used economic impact analysis to help elected officials and policy makers decide whether or not to build a new Civic Center. The Civic Center Board retained the Ohio State University Extension Service to conduct the study. The analysis assessed the amount of expenditures that would be lost to the community due to events not hosted if the Civic Center were not built. The study estimated the total economic impact to the community in terms of direct and indirect employment (due to the multiplier), new local spending due to the facility, and tax revenues to local governments.

The study concluded that there would be a net increase in local spending of five million dollars if the Civic Center were built. Even though the Center itself would directly employ only 22 persons, a total of 119 jobs, direct and indirect, would be created.

The study director explained the value of the analysis as follows: "The Convention and Visitors Bureau and other economic development groups can point to this kind of data and say you will make money here. In the same sense, if you do not build this facility, this money and additional jobs will be lost."

Since the County had committed to fund one-quarter of the Center's annual one million dollar budget, the commissioners wanted to understand their "return on investment." One County commissioner stated: "I think the study shows that the Center is indeed participating in the Allen County economy; however, we need to see greater benefits than have been indicated." In this case, economic impact analysis helped County officials make an informed decision to postpone the Civic Center project.

The Editors

Source: *The Lima News*, August 14, 2007.

## Review questions

1 Why is it important to assess the local economy for community and economic development decisions and policy making?
2 What are some sources of information about the local economy?
3 What are location quotients and industry clusters and why are they useful in assessing the local economy?
4 How is economic impact analysis useful in community and economic development decisions?
5 What are "export-based" industries and why are they important to the local economy?
6 What are economic multipliers and how are they used in community and economic development analyses?

## Bibliography and additional resources

Braddock, D. (1995) "The Use of Regional Economic Values in Conducting Net Present Value Analysis of Development Programs," *International Journal of Public Administration*, 18: 59–82.

Lann, R. and Riall, W. (1999) "LOCI: A Tool for Local Fiscal and Economic Analysis," *Economic Development Review*, 16: 27–33.

Mondello, M.J. and Risne, P. (2004) "Comparative Economic Impact Analyses: Differences Across Cities, Events, and Demographics," *Economic Development Quarterly*, 18: 331.

Okuyama, Y. (2007) "Economic Modeling for Disaster Impact Analysis: Past, Present, and Future," *Economic Systems Research*, 19: 115.

Smith, D.G. and Wheeler, J.R.C. (1986) "How to Determine the Impact of New Business on a Local Economy," *The Journal of Business Forecasting Methods and Systems*, 5: 20.

Throsby, D. (2004) "Assessing the Impacts of a Cultural Industry," *Journal of Arts Management, Law, and Society*, 3: 188–205.

## Websites that contain useful data for assessing a local economy

*Government/free sites*

County Business Patterns. http://www.census.gov/epcd/cbp/view/cbpview.html.
EconData.net. www.econdata.net.
FreeDemographics.com (Census data). www.freedemographics.com.
U.S. Bureau of Economic Analysis. www.commerce.gov/egov.html.
U.S. Bureau of Labor Statistics. www.bls.gov.
U.S. Census Bureau. http://factfinder.census.gov/.

*Fee services*

DemographicsNow. www.demographicsnow.com.
Dun & Bradstreet. www.dnb.com.
Factiva. www.factiva.com.
Harris Infosource. www.harrisinfo.com.
Hoover's Online. www.hoovers.com.

# PART III

# Programming techniques and strategies

# 12 Workforce training for the twenty-first century

## Monieca West

Global competition has fundamentally changed the structure of America's economy. Traditional "high-tech" manufacturing jobs are being outsourced overseas and, internationally, mobile workers are competing everywhere for jobs at all skill levels, even the high-skilled, high-wage jobs that were once considered to be immune from export. Communities and states must focus on raising the skill levels of their resident workers or risk becoming noncompetitive in the global economy and losing jobs to other states and countries. This chapter discusses the kind of workforce development program that states and communities should develop to assure they will have sufficient numbers of trained workers in the twenty-first century. It provides examples of state and local initiatives for reaching this goal. The policy perscriptions for the U.S. discussed in this chapter apply for many countries worldwide.

## Introduction

Workforce development strategies have become increasingly important in recent years as predictions concerning a "new economy" have begun to play out in the lives of individuals, businesses, and communities. Scenarios formulated in the early 1990s predicted that America would lose large numbers of low-skilled jobs to emerging countries. It was thought that the loss of these jobs would be replaced by better jobs created as a result of technology and innovation. As it turns out, these predictions were correct but incomplete. The realization of the predicted impact of technology and globalization on the local economy was constant, often harsh, and with no apparent relief in sight. Now, a decade later, the United States is not only dealing with the expected loss of low-skilled, low-wage jobs, but additionally many desirable high-skilled, high-wage jobs are being lost to countries with well-educated workers such as India and China. The need to develop a workforce that can withstand the pressure will continue as countries such as China – soon to become the number one English-speaking country in the world – become better equipped to challenge U.S. economic dominance.

As the workforces of the United States and other developed countries compete for jobs in an international arena, policy and practice must be revolutionized to meet the demands of this new environment. A public school system designed for the rural, agricultural economy of the 1800s must now prepare students to succeed in a very different twenty-first-century life. Colleges and universities that have not evolved as the world has changed but remained grounded in academic tradition and policy must become more responsive to their role in economic prosperity. Immigration policies that are politically driven must be revamped to permit the flow of needed talent into the country, especially in light of expected shortages due to the aging and retiring U.S. workforce. Federal and state workforce development programs that struggle to keep up with the emerging needs of business and industry in this ultra-competitive environment must be streamlined with policies that easily adapt to changing conditions. Perhaps most important, some observers believe that the United States as a society has not demonstrated an understanding of the value of lifelong learning nor a willingness to demand rigor and excellence of all students at all points along the educational system. All of these issues must be addressed if the economic prosperity Americans have taken for granted is to continue.

The changing economic realities of the twenty-first century cannot be predicted with certainty but communities can prepare with a variety of tools, and a flexible workforce is among the most essential.

> It is not the strongest of the species that will survive, or the most intelligence, but ones most responsive to change.
>
> (Charles Darwin)

## Primary and secondary education

Attention to early education is the most cost-effective strategy to increase the number of skilled workers in the workforce. Primary and secondary education provides students with strong foundations in reading, writing, mathematics, reasoning, and computer skills. It also paves the way for success in higher education and placement in employment. (Judy, D'Amico and Geipel 1997).

Community leaders often mistakenly focus workforce development efforts too narrowly, thinking only in terms of customized training for industrial recruitment or business retention. In reality, holistic workforce development begins in earnest at the secondary level with career exploration, academic preparation for postsecondary education, or technical preparation for entry into apprenticeship programs or job placement. Numerous reports and studies affirm that academic achievement at the high school level is extremely important in preparation for the high-skilled, high-wage, high-demand jobs of the new economy. This is equally true for the "vocational" student who traditionally moved through the education system focusing almost exclusively on trades-based studies, with less emphasis on academics such as math and communications skills that are now required across all levels and types of employment.

To make significant progress, public education can no longer engage in "business as usual." America must consider new approaches to public education commensurate with that required of industry as it faces the challenges of the new economy. The standard approach to K-12 education has changed only marginally since the late 1800s. After decades of "educational reform" and immense increases in funding, far too many high school graduates still leave public education with inadequate skills for the workplace, or require remedial classes to raise their level of academic proficiency for college. High schools must do a better job of preparing students for either entry into the workforce or postsecondary education, as both are equally important to a competitive workforce.

Unfortunately, education reform has been slow to nonexistent, which results in U.S. students lagging behind their peers in countries that compete with American businesses for the best and brightest talent, the currency of the new economy. To demonstrate the significance of this brain drain, consider that 25 percent of the Chinese population with the highest IQs is greater than the total population of the U.S. This means that China has more honors students than the U.S. has children, which certainly positions China to compete for high-skill and high-wage jobs.

Because K-12 education is the foundation of all future learning, a comprehensive workforce development strategy should include support and advocacy for excellence in education, beginning with K-12 and extending through two- and four-year postsecondary institutions. The issues surrounding K-12 reform are very difficult and often controversial but community leaders and community developers must find ways to facilitate a rational public discussion of the issue. Support for educational reform must be included if workforce development efforts are to be effective.

## Postsecondary education

If a community or state is to increase its competitive standing nationally and internationally, more graduating high school seniors and working adults must start or return to college. Postsecondary education is a crucial way that working adults can acquire the skills and credentials necessary to succeed in current and emerging business and industry. Unfortunately, since working adults have full-time jobs and family responsibilities, they often lack the time, money, and flexibility of schedule to fit into traditional higher education models. Recent studies report that working adults get very little financial aid from federal or state sources. Working adults who hold full-time jobs are typically able to attend school on a less than half-time basis, which renders them ineligible for most aid.

The problem is not just the absence of financing but, on the part of most institutions of higher education, the lack of programs and schedules developed to accommodate working adults. Degree attainment and other credential requirements often seem too daunting for a working adult attending school part-time, who may have been out of school for several years and may require some level of basic skills remediation. The pathway connecting education with improved jobs is obstructed for many adults. The states do not effectively provide the financial aid, student support, and basic skills remediation essential for working adult students to advance in their education and career.

Yet, the ability to compete effectively in the new economy is directly related to a community's or state's ability to increase the educational attainment of all its citizens, from those with minimal job skills to those seeking the highest in academic achievement. Numerous national studies have confirmed the direct correlation between academic attainment and earnings potential. The benefits of postsecondary education accrue not only to individuals but to families, communities, states, and to the nation as a whole. For individual citizens, it is well documented in U.S. Census Bureau data and other sources that personal and family "lifetime" earnings increase directly with higher education levels. The Bureau also documents that quality of life factors, such as improved health for individuals and families, are dramatically enhanced by higher levels of education. As quality of life issues are improved for the citizens

of a state, the need for other state programs directed toward health issues and correctional activities is reduced. Policy makers increasingly recognize that improving educational attainment is a key ingredient in their efforts to strengthen economic competitiveness and enhance quality of life.

The role and contributions of community colleges in workforce development initiatives cannot be overstated. These institutions are uniquely positioned as the bridge between high school and university studies, yet are flexible enough to offer classes responsive to general community needs. Community colleges provide developmental, remedial classes to high school graduates without college-level academic preparation, and provide continuing education for non-degree-seeking students. These various approaches to education are important to the overall level of workforce quality, making a strong community college a valuable local asset. Because community colleges are deeply embedded in the local community, they are naturally attuned to the needs of local business and industry. Typically, community colleges are the institution of choice for nontraditional students above the age of 25 who also happen to make up the majority of the local workforce. The U.S. Bureau of Labor Statistics has noted that one-third of the fastest growing occupations nationally are those that require certificates or associate degrees provided by community colleges. Community colleges are extremely important in producing students to fill critical shortages in occupations such as nursing; providing industry certifications such as Oracle and Cisco; offering services that help students transition from secondary to postsecondary education or from postsecondary into the workplace; and providing direct services to local employers and organizations.

Virginia offers a model for comprehensive workforce development. Through its 23 colleges, the Virginia Community College System extends workforce development services and transitional programs into the community, directly serving businesses, employees, and special populations. Programs focus on instructional curricula that prepare incumbent, upcoming, and displaced employees for jobs in current and emerging occupations. In partnership with the office of Workforce Development Services, Virginia's community colleges house the state's Workforce Development Service Centers that provide a single point of entry for job seekers and employers.

## Secondary to postsecondary career pathways

One of the most common barriers to educational attainment is the difficulty encountered by students as they progress through the separate secondary and postsecondary systems. Rare is the occasion when high schools and colleges develop occupational programs of study with curricula that are aligned. Aligned systems allow for articulation agreements; that is, concurrent and dual enrollment programs designed to allow high school students to acquire college credit prior to graduation. To facilitate this alignment, the U.S. Department of Education has established 16 career clusters with 81 separate career pathways.

Career clusters provide a framework for grouping occupations according to common knowledge and skills. Using sequences of coursework required at secondary and postsecondary levels, a career pathway provides a student with a plan of study that leads to specific occupational titles found in his or her career cluster. Rather than training for a specific job, this allows students to acquire skills that are transferable across related occupational fields. For example, the health science career cluster includes pathways for therapeutic services, diagnostic services, health informatics, support services, and biotechnology research and development. A student wishing to become a nurse would develop a plan of study in the therapeutic services pathway, which would include foundational skills that would be transferable to the other pathways in the health sciences career cluster if desired at a later time. This plan of study would map core, technical, and elective coursework beginning with the ninth grade through graduation. If the high

school and the postsecondary institution have aligned coursework using the career cluster framework, the student could also map courses required for the nursing degree and identify opportunities to take college-level work while in high school. This would save tuition dollars and reduce the time between high school and employment. Sample plans of study in all career clusters are available from Career Clusters Institute (www.careerclusters.org)

Virginia has enhanced its career pathways initiatives with career coaches – community college employees who spend most of their time on high school campuses, where they serve as resource specialists for career planning and connect students with businesses and community colleges. To further support the career pathways initiative, Virginia is piloting the "Middle College" concept which will provide an opportunity for high school dropouts between the ages of 18 and 24 to attain a GED and enroll in college coursework that will enhance basic workforce skills as well as count toward a certificate or associate's degree and attainment of a workforce readiness certificate.

Virginia has also established the Workforce Development Academy to train workforce development practitioners and advance the workforce development profession. Professional development opportunities include a graduate Certificate in Workforce Development and a noncredit certificate for the completion of the Workforce Development Professional Competencies course.

Career pathways to improve adult postsecondary attainment initiatives can be designed for secondary to postsecondary transitions. However, career pathways are also needed to enable adult individuals to advance, particularly in high-wage, high-growth careers, and to meet the long-term demands of the new economy. A framework of study must be developed to help an adult student start at the beginning of the career path, if necessary, then move along the path to a certificate or degree. Steps along the path could include basic skills development, earning a GED, learning how to integrate education into work, how to be successful with higher level responsibilities, and how to manage wages. The pathway is not a single program, but a collective framework of multiple programs and services to serve the client at all points along the path.

The Southern Growth Policies Board (SGPB 2002), in its report titled *The Mercedes and the Magnolia*, looked at how the South is preparing for the new economy and offered recommendations that stress the need for graduates to understand and prepare for future career advancement, rather than simply meeting the requirements of today. These recommendations include: creating a seamless workforce system to maximize the client (worker) control over the outcomes; identifying and developing underutilized sources of workers and talents; and creating a self-directed workforce with the attitudes, learning habits, and decision tools necessary for making wise career choices throughout life.

Workforce development has often been aimed at one career path, which is much too narrow to apply in today's economy. Institutions must now focus on the entire client, not just the skill the client wants to learn. This new focus will help clients raise the standard of living for themselves as well as future members of their families. Too often in the past, workforce training programs relied on someone else to focus on the personal and social development needs of the client, which resulted in confusion as to who was responsible.

Therefore, workforce development must now be an entire approach organized with an economic development focus applied to the current and future training of clients. At the same time, it must provide lifelong learning, a wide range of employment support services, and attention to character development and career awareness. Thus, a much broader scope of organizations and individuals must be involved in the design and delivery of the new workforce development program.

This concept has been developed in many states, in many different forms, and is now a reality in the State of Arkansas. The "Arkansas Career Pathways Initiative" (CPI) is a comprehensive project designed to move low-income adults who are eligible for Temporary Assistance for Needy Families (TANIF) into self-sufficiency. CPI provides a series of connected or sequential education courses with an internship or on-the-job experience, as well as enhanced student and academic support services. This combination of structured learning creates achievable stepping-stones for career advancement of adult workers and increases the quality of workers at all levels. As such, career pathways directly link investments made in education with economic advancement.

CPI pathways start with employability skills and adult education, then progress to a certificate of proficiency, a technical certificate, an associate degree and, finally, a baccalaureate degree. Along the way, students are taught to apply knowledge gained in the classroom to local jobs, and to gain work experience through internships, on-the-job training, and shadowing services. These real-world experiences allow students to make connections between education and jobs in a given industry or sector. To increase chances of success, intensive student services such as career assessment, advising, tutoring, job search skills, and job placement assistance are provided, in addition to more flexible class schedules needed by working adults. In some cases extraordinary support, such as assistance with transportation and childcare, is provided.

CPI is a partnership between the state Departments of Higher Education, Workforce Services, and Human Services; the Arkansas Transitional Employment Board; the Arkansas Association of Two Year Colleges; and the Southern Good Faith Fund. CPI programs have been funded for all two-year colleges in the state.

## Developing a demand-driven workforce development system

Designing an effective workforce development system requires a new thought process. Traditionally, workforce training has focused on job training — developing the specific skills necessary for a particular job as well as the human interaction and leadership skills required for the team approaches of the past decade. Unfortunately, this approach may result in limited training and preparation, designed to move individuals into entry-level jobs without consideration of long-term skill development, thereby increasing the number of working poor. Job training must be differentiated from workforce development, as the former is but a single strategy and too narrow for a comprehensive workforce development plan: "The world we have created is a product of our thinking; it cannot be changed without changing our thinking" (Albert Einstein).

The ideal workforce development approach is a demand-driven system of both short- and long-term programs and policies which integrate workers, employers, educators, community and economic developers, and government policies into one seamless system. The system must be coordinated as well as flexible and responsive to changing economic conditions. It must include strategies for incumbent workers, the unemployed and the underemployed, special needs populations, mature workers, and the future workforce. This will require ongoing collaboration among employers, trade associations, and labor organizations, Chambers of Commerce, economic developers, public and private secondary and postsecondary institutions, and community-based nonprofit organizations.

Workforce development strategies for the twenty-first century should be grounded in the community development process and support the economic development needs of the community. An effective demand-driven workforce development system will both bridge the traditionally separate policy domains of education, labor, and economic development, and address both demand- and supply-side issues (Porter 2002). The demand side includes workforce development polices and programs that respond to employer needs in finding and retaining qualified workers, and increasing their skills so that the employer can be more competitive. This may include employer practices that increase on-the-job learning and getting employees to work in teams. On the supply side, workforce development systems that support short- and long-term economic priorities by producing increased skills among current and future workers.

These systems help people find and keep jobs and advance in employment, with programs for out-of-school youth, those already working, those seeking employment, and special populations such as women, minorities, and the disabled.

What sets demand-driven systems apart from other approaches is the focus on lifelong learning and advancement. To achieve this, systemic changes will be required in education policy to improve achievement levels of all students, to offer alternative pathways which facilitate lifelong learning, and to create more flexible credentials and course delivery methods that allow working adults to attain education in small increments over extended periods of time (Porter 2002).

The community development process detailed in this book should be the basis for a workforce development plan because it should be part of a larger overall strategic plan for the community. Because most communities have an established workforce development entity, it may be that a review, redirection, or consolidation of efforts is in order rather than starting from scratch. Common elements of most effective programs include programs that are employer/employee centered, maintain strong networks of diverse partnerships between government- and community-based organizations, contain employers who value training for their employees, and serve a network of clusters in the industry, government, and educational sectors.

In addition to the community development process, the following are points to consider when developing a demand-driven workforce development system:

- Establish a vision for the workforce network supported by an action plan with key indicator targets and performance measures. The vision and action plan should contain labor market policies that both support local and regional economic development goals for job creation and economic growth, and contain financial strategies and incentives that support public and private sector investment in skills development.
- Ensure that the vision and subsequent decision

making are data- and fact-driven to provide an analysis of the labor market including current economic conditions, major forces, and local trends. Routinely evaluate current programs funded under the Workforce Investment Act and other public programs using data to measure progress toward goals and to modify and adjust strategies as conditions warrant. Use data and creativity to replace bureaucracy with innovation.

- Involve all stakeholders in visioning and planning, especially if agencies may be eliminated, reduced, or combined. Develop strategic linkages among stakeholder groups including service providers (secondary and postsecondary educators, state agencies, private companies); clients (employers, unions, current workers, those in the pipeline); government agencies (policy makers, state and federal funding sources); and community-based organizations and individual citizens. In this way, the system will be characterized by responsiveness to economic needs, continuous improvement, and results-based accountability. Because industry representatives are the best source of information for layoff aversion strategies, as well as of advance knowledge related to preparing for emerging jobs, they are critical team members in developing workforce strategies.

- Give strong consideration to regional programs since today's mobile workforce is not necessarily restricted to a specific municipality or county. Because economies are regional in nature, develop workforce development strategies that are also regionally focused. Develop industry-led regional skills alliances to address areas of skills shortages across the region, in order to provide training for industry clusters rather than individual companies.

- Design incumbent worker training programs so that recipient firms are required to develop long-term, work-based training plans before financial assistance is provided. Firms should be required to consider overall training needs rather than providing quick fixes for immediate needs.

- Build a seamless learning system that focuses on K-16 or K-20 systems that produce graduates with marketable skills. To achieve this, education must be viewed as one system rather than separate secondary and postsecondary silos. It also involves advocating for overall high school reform and implementing increased linkages between high school and college.

- Make sure higher education understands the economy and advises students of current and emerging job opportunities so that education is aligned with available employment opportunities. Increase resources of community colleges so that services can be expanded for students pursuing a four-year degree, those seeking technical certifications for beginning employment, and incumbent workers needing to acquire new skills to keep abreast of changes in the workplace. This will help people raise their standards of living through increased educational attainment.

- Create strategies for certifying knowledge and skills that are recognized by employers though gained outside the formal education system, and build pathways for continuing education in the informal and formal learning systems for adults and youth including literacy programs, apprenticeship, vocational training, youth development, technology-enabled learning, and alternative education programs.

## Federal resources

The federal government has demonstrated its interest in workforce development for many years through several legislative acts. The Workforce Investment Act of 1998 (WIA) provides the framework and direction for the bulk of federal funds for workforce development. The Carl D. Perkins Career and Technical Education Act of 2006 provides significant funding for career and technical education at both secondary and postsecondary levels. Other legislation relevant to workforce development includes the Higher Education Act, the previously mentioned Temporary Assistance for Needy Families, the Vocational Rehabilitation Act, and the Adult Education and Family Literacy Act.

## Department of Labor – Employment and Training Administration

The Workforce Investment Act (WIA) is administered by the U.S. Department of Labor's Employment and Training Administration (ETA). The ETA is responsible for federal government job training and worker dislocation programs, federal grants to states for public employment service programs, and unemployment insurance benefits. The ETA provides the majority of its workforce development support through the Office of Workforce Investment, which has operations that focus on one-stop operations, business relations, and services to adults, youth, and workforce systems.

The WIA reformed and consolidated federal job training programs and created a new, comprehensive workforce investment system intended to be customer-focused, to help Americans more effectively plan and manage their careers. It authorized the establishment of state workforce investment boards and state workforce development plans. The purpose of the law was to increase the employment, retention, occupational skill attainment, and earnings of participants.

## Department of Education – Office of Vocational and Adult Education

Funds provided by the Carl D. Perkins Career and Technical Education Act are administered by the Department of Education's Office of Vocational and Adult Education (OVAE). The OVAE oversees programs that prepare students for postsecondary education and careers through high school programs and career and technical education. It also provides opportunities for adults to increase literacy skills, ensures equal access and opportunities for special populations, and promotes the use of technology for access to and delivery of educational instruction.

Perkins focuses on integrating academic and technical study for career and technical education students, strengthening the connections between secondary and postsecondary education, and preparing students for nontraditional occupations. In addition, it focuses on serving the needs of special populations, improving professional development opportunities for educators and administrators, and promoting partnerships between education and industry.

Federal Perkins funds are awarded annually to a state agency designated by its governor. Typically, it is a department of workforce services, a department of education or, in some cases, a department of higher education. Once the award is received by the state, the recipient agency determines the amount of funds to be provided to secondary and postsecondary institutions.

## State and local resources

### State agencies

The delivery of workforce development services at the state level is structured in various ways and often involves several agencies including workforce services, economic development, employment security, general education, higher education, and the workforce investment board. Several states, such as California, have created a cabinet-level interagency team to coordinate all economic development activities including workforce development.

State-level services are provided for employees through occupational and labor market information; skills assessment; job matching and job banks; and special services for dislocated workers, veterans, the disabled, and other special need populations. Services provided to employers include assistance in finding employees; information related to grants, training incentives, tax credits, and state and regional employment data and analysis. Additional services are provided related to job skills profiling and assessment, employee skills gap identification, and related training.

To better compete in the new global economy, several states have developed strategic plans specifically for workforce development. The State of Oregon plan includes increased funding for pre-kindergarten through postsecondary education; a Cluster Investment Fund targeted to demands of employers and clusters of businesses in high demand areas; and a "Skills Up Oregon" fund to both upgrade the skills of unemployed and lower wage workers, and to increase GED and professional certificates among high school dropouts. The plan also includes targeted education investments such as career pathways, the creation of work-readiness certificates, and health care training. The program also focuses on increased math and science skills, engineering and manufacturing research and development, and apprenticeship programs.

Florida has developed initiatives for "First Jobs/First Wages," which targets youth and adults entering the workforce for the first time; "Better Jobs/Better Wages," which helps underemployed workers improve skills to advance to higher wages; and "High Skills/High Wages." The latter acts as a catalyst among industry, economic development organizations, and training providers, in order to identify job skills that are critical to business retention, expansion and recruitment activities.

Customized training is an integral element of workforce development and should be provided to meet the needs of a specific business rather than those of individuals. In most states, this type of training is provided through its economic development agency. For example, in North Carolina, free customized training is offered to businesses in targeted business sectors that create a certain threshold of new jobs annually. Instruction and training are furnished by local colleges, a state network of training specialists, and third-party vendors. The program also provides both extensive continuing education through community colleges for the existing workforce, and a special retraining program for programs with high technical costs.

## State and local workforce investment boards

The Workforce Investment Act established state workforce investment boards as well as local workforce investment boards. The state boards must include the governor and state legislators; representatives of business and labor; chief local elected officials; relevant state agency heads; and representatives of organizations with expertise in workforce and youth activities. The majority of the board must be representatives of business. The board advises the governor on developing the state-wide workforce investment system, the state-wide employment statistics system, performance measures, and a system for allocations. It also develops a state strategic plan which describes the workforce development activities to be undertaken in the state. The plan also describes how the state will enact WIA requirements including those related to special populations, welfare recipients, veterans, and those with other barriers to employment.

Local workforce investment areas are designated by the governor, as well as a local board responsible for planning and overseeing the local program. Like the state board, the local board must also have a majority of business members and include education providers, labor organizations, community-based organizations, economic development agencies, and each of the one-stop partners. A youth council is also established as a subgroup of the local partnership. It is comprised of local board members, youth service organization representatives, local public housing authorities, parents, youth, and other organizations as deemed appropriate locally.

Under the WIA, each local area must establish a one-stop delivery system through which core employment-related services are provided by local one-stop partners, and through which access to other related federal programs is provided. Each local area must have at least one one-stop service center which may be supplemented by a network of affiliated sites.

The law designates certain partners that are required to participate in the one-stop system including those authorized under the following federal legislation: adult, dislocated worker, youth, and vocational rehabilitation activities under the WIA; postsecondary Carl Perkins Act funding; and welfare to work grants. Under the Older Americans Act: veterans employment and training programs; community services block grants; Housing and Urban Development-administered employment and training programs; Trade Adjustment Assistance and NAFTA–TAA (North American Free Trade–Transitional Adjustment Assistance); and unemployment insurance and state employment services required by the Wagner-Peyser Act.

The local board either selects the operator of a one-stop center through a competitive process, or may designate a consortium of no fewer than three one-stop partners to operate the center. An eligible operator may be a public, private, or nonprofit entity or a consortium of such; a postsecondary educational institution; the employment service organization authorized under the Wagner-Peyser Act; another governmental agency; or other organizations.

The cornerstone of the new workforce investment system is the one-stop service delivery, which unifies numerous training, education, and employment programs into a single system in each community for youth, adults, and dislocated workers. The goal is for services from a variety of government programs to be delivered in a seamless manner including customer intake, case management, and job development and placement services. Access is also facilitated to support services that reduce barriers to employment such as transportation providers, childcare services, nonprofit agencies, and other human service providers. Required core services for adult and dislocated workers include: job search and placement assistance; career counseling; labor market information related to job vacancies; skills training for in-demand occupations; skills assessment; information on available services and programs; case management and follow-up services to assist in job retention; development of individual employment plans; and short-term vocational services.

Using the Workforce Investment Board as its base, the State of Maryland has made dramatic changes at all levels of its workforce development system. In so doing, it has created a model, demand-driven workforce development system with partnerships among the three "E" cornerstones: education, employment, and economic development. The system includes all stages of education and training, from K-16 education through retirees. While geared toward meeting the needs of business, the system also provides support services for workers, such as daycare and transportation, and is aligned with the economic development goals of the state.

The Maryland model encourages the following:

- Increased cooperation and collaboration among state agencies to eliminate duplication, reduce costs, unify planning, and maximize resources and organized activities around industry clusters.
- Multiple initiatives that support existing Maryland businesses with retention, growth and development of the workforce including a statewide Web-based career management and job marketing system.
- Refocused postsecondary workforce development initiatives to include an industry sector approach, customized coursework, and degree resources targeted to high-demand industry sectors.

## Labor market intermediaries

Workforce development is a complex system of services and service providers that offer academic and technical skills training, social support services that make it possible for job seekers with extenuating needs to obtain training, and connections to the employers that offer employment. These services are delivered by a variety of labor market intermediaries (LMI) including government agencies, community colleges, nonprofit community-based organizations (CBOs), and more recently, for-profit corporate providers. LMIs may offer services that are industry specific or services that target a specific segment of the workforce development system. They may operate independently or form networks or alliances with other providers.

The scope and mission of LMIs are markedly different in scope and mission from their predecessors, due primarily to welfare reform and the revamping of federal job training programs under the Workforce Investment Act (WIA) of 1998 (Melendez 2004). Prior to these two events, service providers held a "training first" attitude which focused on longer and more intensive education and training. During this training period, unemployed people were eligible for government assistance to support themselves and their families while obtaining skills to find a job. With the passage of the Personal Responsibility and Work Opportunity Reconciliation Act (PRWORA) of 1998 (also referred to as welfare reform), the approach switched to "work first" which favored direct job placement and short-term job-readiness skills. As a result LMIs were forced to become more responsive to employer expectations and were held to increased levels of accountability on performance measures required by federal training programs.

The increased competition for workforce development services contracts challenged many community-based organizations and prompted greater collaboration among agencies and organizations in the public, private, and nonprofit sectors. CBOs found that they could compete more effectively when leveraging the resources and organizational capacities acquired through alliances. As the one-stop centers required by the WIA legislation began coordinating all workforce training services, local CBOs structured various types of arrangements to provide these services. The most successful ones are those that demonstrated flexibility in an evolving environment, were willing to collaborate even if this required organizational or philosophical changes, capitalized themselves through diverse means to have the capacity to pursue and win large WIA contracts, and created reciprocal relationships with employers. Emphasis on building human capacity through partnerships, networking, and collaboration

– hallmarks of the community development process – are key elements of success in the new era of federally funded workforce development.

## Conclusion

An insurance company advertisement from the late 1990s gives good advice for today's community leaders: "You can't predict. You can prepare." To navigate successfully in the new economy, communities must develop an agile workforce in the throes of continual change and uncertainty, characteristics of the new workplace. According to the "New Economy Index," almost one-third of all jobs are now in flux, either being created, or are eliminated every year. The U.S. Department of Labor estimates that today's learners will have 10 to 14 different job assignments by age 38. Former Secretary of Education, Richard Riley, predicts that the 10 jobs most likely to be in demand in 2010 will not have existed in 2004, and that the U.S. must prepare students for jobs that don't yet exist, using technologies that haven't yet been invented in order to solve problems not yet recognized. This churning makes the workplace very unpredictable for employers, employees, and policy makers. Employees must be equipped not only with industry-specific technical know-how, but have both the ability to create, analyze, and transform information that helps solve real-world problems, and to work and communicate effectively with others. Both formal and informal learning must become lifelong in nature, as opposed to stopping after a degree or certificate is attained or employment found. Employers must understand that investing in and facilitating employee training affects productivity and, therefore, determines the company's competitive position. Policy makers must provide a demand-driven workforce development system that is flexible, responsive, and innovative – more than just "business as usual."

## CASE STUDY: COLLABORATING TO DEVELOP A HIGH-TECH WORKFORCE IN TULSA, OKLAHOMA

One of the biggest challenges many communities face today is training and retraining the workforce for high-skilled jobs in technology-based industries. No one program or organization can accomplish such a major workforce development effort – its takes a coordinated, community-wide effort. As such, workforce development is an exercise in community development.

Sleezer et al. (2004) studied how Tulsa, Oklahoma succeeded in creating an advanced workforce development program with participation from key organizations in both the public and private sectors. By 1990, Tulsa had begun to attract a significant number of telecommunications companies. A community assessment revealed that there was a shortage of skilled workers for this industry in the area, and no workforce development program in place to train more. Business, political, and academic leaders began working together to address the issue and concluded that the region needed a Masters of Science degree with an emphasis in telecommunications.

Implementation of this recommendation proved to be a challenge. After some initial cooperation, the branches of the two major state universities located in Tulsa could not agree on how to move forward collaboratively to create a joint program. Each university decided to proceed independently, which would mean fewer resources for each program. In 1999, a change in leadership at one university provided a spark to rekindle the cooperative effort. With input and assistance from business and political leaders, the two universities worked together to create the Center for Excellence in Information Technology and Telecommunications. Since that time, Tulsa has been cited in several studies as having a strong concentration of information technology workers.

Sleezer et al. describe Tulsa's workforce development effort as an exercise in community development as well. Business, political, and academic leaders had somewhat different motives and goals going into the process. For example, faculty members emphasized providing a high-quality curriculum, while business leaders emphasized meeting their firms' educational needs. Through mutual trust and good-faith negotiations, the parties resolved their differences and achieved success. Contributing to this success was the fact that the parties involved in the negotiations had a history of working together on other committees and community efforts. In other words, there was a preexisting degree of social capital or cohesion (see chapter 4) in Tulsa which helped the community plan and act together effectively.

## Keywords

Demand-driven workforce development, demand driven, career pathways, postsecondary education, new economy, workforce investment boards.

## Review questions

1   Why should workforce development strategies be integrated into primary and secondary education?
2   Why are community colleges important in workforce development?
3   What are career pathways and career clusters and how are they related?
4   What is a demand-driven workforce development program and what are its advantages?
5   What are some federal programs for workforce development?
6   What are the state-level programs for workforce development?
7   What are state and local workforce investment boards and how do they operate?

## Bibliography and additional resources

Arkansas Department of Higher Education. Available online at http://www.arpathways.com/ (accessed December 11, 2007).

California Economic Development Partnership. Available online at http://www.labor.ca.gov/cedp/default.htm (accessed December 11, 2007).

California Labor Federation AFL-CIO. Workforce and Economic Development Strategies. High Road Partnerships. Available online at http://www.calaborfed.org/workforce/dev_strat/index.html (accessed December 11, 2007).

Career Clusters Institute. Available online at http://www.careerclusters.org/ (accessed December 11, 2007).

Carl D. Perkins Career and Technical Education Act. Available online at http://www.ed.gov/policy/sectech/leg/perkins/index.html (accessed December 11, 2007).

Commission on the Skills of the American Workforce. Available online at http://www.skillscommission.org/staff.htm (accessed December 11, 2007).

Council for Adult and Experiential Learning. Available online at http://www.cael.org/ (accessed December 11, 2007).

Council for a New Economy Workforce. Available online at http://www.southern.org/cnew/cnew.shtml (accessed December 11, 2007).

Employment and Training Administration. Available online at http://www.doleta.gov/ (accessed December 11, 2007).

Global Workforce in Transition. Available online at http://www.gwit.us/ (accessed December 11, 2007).

Gordon, E.E. (2005) *The 2010 Meltdown: Solving the Impending Jobs Crisis*, Westport, CT: Praeger Publishers.

Government of South Australia (2005) *Better Skills. Better Work. Better State: A Strategy for the Development of South Australia's Workforce to 2010*. Available online at http://www.dfeest.sa.gov.au/dfeest/files/links/wds2005.pdf (accessed December 11, 2007).

Indiana Strategic Skills Initiative. Available online at http://www.in.gov/dwd/employus/ssi.html (accessed December 11, 2007).

Jobs for the Future. Available online at http://www.jff.org/ (accessed December 11, 2007).

Judy, R.W., D'Amico, C. and Geipel, G.L. (1997) *Workforce 2020: Work and Workers in the 21st Century*, Indianapolis, IN: Hudson Institute.

Maryland Department of Labor, Licensing and Regulation. Governor's Workforce Investment Board. Available online at http://www.mdworkforce.com/ (accessed December 11, 2007).

Melendez, E. (2004). *Communities and Workforce Development*, Kalamazoo, MI: W. E. Upjohn Institute for Employment Research.

National Association of State Workforce Agencies. Available online at http://www.workforceatm.org/ (accessed December 11, 2007).

National Center on Education and the Economy. (2003) *Toward a National Workforce Education and Training Policy*. Available online at http://www.ncee.org (accessed December 11, 2007).

National Council for Workforce Education. Available online at http://www.ncwe.org/ (accessed December 11, 2007).

National Governors Association. Center for Best Practices. Available online at www.nga.org (accessed December 11, 2007).

North Carolina Department of Commerce. Available online at http://www.nccommerce.com/en/Workforce Services/ (accessed December 11, 2007).

Office of Vocational and Adult Education. Available online at http://www.ed.gov/about/offices/list/ovae/index.html (accessed December 11, 2007).

Porter, M.E. (2002) *Workforce Development in the Global Economy*. Available online at http://www.gwit.us/global.asp (accessed December 11, 2007).

Progressive Policy Institute (2004) *Economic Development Strategies for the New Economy*. Available online at http://www.ppionline.org/ (accessed December 11, 2007).

Progressive Policy Institute. New Economy Index. Available online at http://www.neweconomyindex.org/index.html (accessed December 11, 2007).

Public/Private Ventures. Available online at http://www.ppv.org/ppv/workforce_development/workforce_development.asp (accessed December 11, 2007).

Sleezer, M. et al. (2004) "Business and Higher Education Partner to Develop a High Skilled Workforce: A Case Study," *Performance Improvement Quarterly*, 17: 65–81.

Southern Growth Policies Board (SGPB) (2002) *The Mercedes and the Magnolia: Preparing the Southern Workforce for the New Economy*. Available online at http://www.southern.org/ (accessed December 11, 2007).

University of Virginia Workforce Development Academy. Available online at http://www.workforcedevelopmentacademy.info/ (accessed December 11, 2007).

U.S. Bureau of Labor Statistics. Available online at http://www.bls.gov/ (accessed December 11, 2007).

U.S. Census Bureau. Available online at http://www.census.gov/ (accessed December 11, 2007).

U.S. Department of Education. Available online at http://www.ed.gov/ (accessed December 11, 2007).

Virginia Community College System Workforce Development Services. Available online at http://www.wdscommunity.vccs.edu/ (accessed December 11, 2007).

Workforce Investment Act. Available online at http://www.doleta.gov/usworkforce/wia/act.cfm (accessed December 11, 2007).

Workforce Florida. Available online at http://www.workforceflorida.com/ (accessed December 11, 2007).

Workforce Oregon. Available online at http://governor.oregon.gov/Gov/workforce/workforce_vision.shtml (accessed December 11, 2007).

Workforce Strategy Center (2001) *Workforce Development: Issues and Opportunities*.

WorkforceUSA.net. Available online at http://www.workforceusa.net/index.php (accessed December 11, 2007).

WorkKeys. Available online at http://www.act.org/workkeys/index.html (accessed December 11, 2007).

# 13 Marketing the community[1]

## Robert H. Pittman

The competition to attract new investment and jobs into communities is fierce. There are thousands of communities competing for business expansion and relocation projects annually. In today's economy, the competition for new investment is increasingly global. Communities must also market themselves to firms already located in their areas (business retention) and to new business start-ups. Marketing is a critical component of a successful community and economic development program. This chapter defines marketing in the community and economic development context, explains the components of a successful industry marketing program, and discusses the key elements in the marketing plan.

## Introduction

Communities market themselves to a variety of entities: new firms, existing firms, nonprofit organizations, tourists, new residents, restaurant chains, and so on. While the principles behind marketing to these various audiences are similar, the specifics of marketing programs vary according to the target audience. The focus of this chapter will not be on marketing to tourists (the subject of Chapter 16). Nor will it be on attracting retail stores, restaurants, or new residents. Instead, it will focus on the marketing of communities in a specific economic development context: that of attracting new firms and retaining and expanding existing ones. In most communities these companies – manufacturing operations, call centers, service firms – form the basis of the local economy, which in turn supports retail stores, restaurants, and new residents.

## Marketing definitions

There are many textbook and lay definitions of marketing. However, what does marketing mean in the specific context of economic development? Let us start with a broad definition of marketing and narrow it down to the economic development field.

*General marketing:* "The performance of business activities that direct the flow of goods and services from producer to consumer" (American Marketing Association 2007). This is a broad, consumer product definition of marketing.

*Societal marketing:* "To determine the needs, wants and interests of target markets and to deliver the desired satisfactions more effectively and efficiently than competitors in a way that preserves or enhances the consumer's and the society's well being" (Kotler 1991: 26). This is getting closer to a definition of marketing in the economic development context. In order to attract new investment that will create jobs and benefit the community at large, communities are trying to meet the needs and wants of their target markets. A company or organization needs a good location for business. This, in turn, is a complex and multifaceted concept including potentially hundreds of factors affecting business operations.

*Economic development marketing:* "Creating an image in the minds of key company executives who rarely make expansion decisions; staying in contact with them so that when the time comes to act, they consider a particular community; and, finally, ensuring that their business location needs are fully satisfied."

For the purposes of this chapter, economic development marketing can be defined in this way. Creating an image is a key to successful community and economic development marketing. Of course, actively pursuing firms with mailings, visits, phone calls, and so on is also critical to a successful economic development campaign. Some observers might refer to this part as "sales" instead of "marketing." However these activities are defined, both will be covered in this chapter.

## Components of marketing

Community marketing may be divided into four broad steps:

- Define the product and message.
- Identify the audience.
- Distribute the message and create awareness.
- Satisfy the needs and wants of the customer.

## Define the product and message

It would be remiss to discuss community marketing without covering two of the key elements on which a marketing program must be based: community SWOT analysis (strengths, weaknesses, opportunities, and threats) and strategic visioning and planning. SWOT analysis (or community assessment) is covered in Chapter 9, while strategic planning is covered in Chapter 6. However, it is worthwhile to review these tools in the context of a marketing program.

A SWOT analysis identifies the economic development strengths and weaknesses of a community. Which community strengths — such as highly productive and trained labor force, good utility cost and service, and supportive local government — will attract industry? Conversely, which weaknesses —

such as inadequate transportation infrastructure, lack of available industrial site and buildings, and permitting difficulties — will hurt the recruitment of new businesses, retention and expansion of existing community, and new business start-ups? All the basic site selection factors should be covered but there will also be community-specific issues that will surface in the analysis such as factions quarrelling over which direction the community should go. Opportunities might include economic development potential from an excellent community college while threats might include the potential closure of a local plant.

It should be obvious why a SWOT analysis is such a critical component of community marketing. If an audit of strengths and weaknesses hasn't been conducted, how can a product (the community) be defined and made attractive to its potential audiences (types of industries)? Many communities claim they are a great place to "live, work, and play." However, what specific set of assets does a community have relative to the thousands of others also vying for new investment? Without a SWOT analysis this question cannot be answered, the product cannot be defined, nor can an effective marketing image and campaign be created. A SWOT analysis also identifies the weaknesses and threats that need to be addressed in the short and long term.

The SWOT analysis gives the community a foundation on which to base effective community strategic planning and visioning. Once community leaders and citizens compare their economic development strengths and weaknesses with those of other communities, they can more effectively chart a direction for future growth and development. So many communities want to attract the next automobile assembly plant or company headquarters, yet have a completely unrealistic assessment of their chances for attracting such large projects. Perhaps they should focus instead on smaller advanced manufacturing companies or customer support centers. It is great to dream about a future in which the community is growing and attracting high-paying jobs, but if a dream is unrealistic it will likely never happen.

Another component of strategic planning and visioning involves what the community wants to be in the future. Does it want to focus on attracting manufacturing and service firms, or does it want to become a "bedroom" community and thus rely more on tourism and retail and jobs in surrounding communities to support its residents? Again, it is easy to understand why strategic planning and visioning is critical to an effective marketing campaign. Without this element based on SWOT analysis a community does not even know to whom it wants to market. Imagine that a person wakes up one day and decides to get in their car and go somewhere. However, they are not sure where to go or why, so they drive around aimlessly and return two hours later, happy they made good time on the highway. Their neighbors and family would certainly question the person's sanity. Community marketing without a SWOT analysis and strategic plan is just as pointless.

*Marketing image*

A fundamental part of defining the product and marketing message is to create an effective marketing image. A community marketing image may be defined as: "The sum of beliefs, ideas and impressions that people (residents, target audience, outside public, etc.) have of a place" (Kotler et al. 1993: 141). Another term often used for image is the "market brand" of a community. The image or brand of a community or place can profoundly affect companies' location decisions. As an example, consider North Dakota. The state has an image in many people's minds of being too cold, remote, and snowy for effective business operations. In reality, North Dakota boasts a number of leading national and international companies such as Great Plains Software (acquired by Microsoft) and the Melroe Company (manufacturer of Bobcat construction equipment). The state has worked to counteract this negative image by running advertisements featuring testimonials from North Dakota executives. In magazines and other publications, these testimonials tout the state as a profitable business location with productive labor and few down days due to weather.

As another example, consider the image boost Alabama enjoyed when Mercedes-Benz (now DaimlerChrysler) decided in the early 1990s to locate its first North American assembly plant in the state. That decision sent a message around the world that a leading global company found Alabama to be a good location for its operations. Any negative stereotypes in the minds of corporate executives (e.g., that Alabama was a poor, agricultural state with a lack of skilled labor) were refuted by that one location decision.

A state's or a community's image often determines whether it is on the initial facility location search list or not. Once the short list of locations has been selected, it also often has an intangible but strong effect on the final location decision. Most of the time, when the decision is down to the final two or three communities, the profit and cost profiles of the final communities are similar. The final decision is affected by a number of intangible factors including image. For example, a company executive might be afraid to recommend a location with a negative image for fear his management and board of directors would think he was making a mistake.

A marketing image should be:

- valid,
- believable,
- simple,
- appealing,
- distinctive (not just another great place to live, work and play),
- related to the target audience.

If the marketing image does not meet these criteria it can do more harm than good to a community.

*Slogans and logos*

Slogans and logos may be used to communicate a community's image and rise above the marketing clutter of hundreds or thousands of other communities. Slogans are briefly stated ideas, themes, or "catchphrases" used to convey a community's or area's image (see examples in Box 13.1). Logos are graphic or pictorial images used to help convey a community's image. They can provide a visual unifying theme for marketing materials such as brochures; letterheads; and promotional giveaways such as key chains, pens, coasters, and so on. Again, themes and logos must clearly reflect a community's carefully crafted marketing image. It has been said that we live in a "sound-bite" world today and slogans and logos can be effective in instantly communicating an image.

---

**BOX 13.1 EXAMPLES OF ECONOMIC DEVELOPMENT MARKETING SLOGANS**

What image do these slogans create in your mind?

- Mississippi: Yeah, We Can Do That!
- Michigan: The Future is Now
- Seattle: Leading Center of the Pacific Northwest – The Alternative to California
- Fairfax County, VA: The Nation's Second Most Important Address
- Atlanta, Georgia: Center of the New South
- Henderson County, NC: Smart Business Location With a Metropolitan Blue Ridge Lifestyle

A slogan can also effectively communicate a community's geographic location to corporate executives worldwide. For example, two hours north of Dallas, the southern Oklahoma City of Ardmore uses the slogan "Dallas is Coming Our Way." This slogan successfully communicates two facts. One is that some businesses are choosing Ardmore instead of Dallas as a location. Another is that Ardmore is close to Dallas, because corporate executives use the better known Texas city as a point of reference.

---

## Identify the audience

After the community image and message have been crafted, the next step is to identify the audience for the development marketing campaign. That market is certainly broad but, again, this chapter focuses on marketing to companies. This audience includes:

- Outside companies, site selection consultants, industrial real estate companies, and others involved in corporate expansion (recruitment).
- Lead-generating economic development organizations (e.g., state and regional economic development organizations, utilities).
- Existing businesses already in the community (business retention and expansion).
- Entrepreneurs (new business start-ups).

This group may be referred to as an *external* marketing audience: those organizations and individuals that can help a community attract investment and create jobs. There is another marketing audience that can be referred to as *internal*:

- Community stakeholders (e.g., elected officials, board members, sponsors).
- Media (newspapers, TV, radio).
- The general public.

Community and economic development organizations need to communicate with this internal audience so that they may continue to secure strong cooperation and support in the overall development effort. Many economic development marketing programs have died premature deaths because not enough attention was given to ensuring that a strong community coalition stayed together to support the overall effort. As discussed in Chapter 1, this is one area where community development and economic development have a strong overlap.

The external development marketing audience is still quite large and successful, and communities identify their prime prospects by drilling down to specific sectors and organizations in the marketplace. This is critical because communities have limited marketing budgets and cannot spread their resources too thinly by marketing to a broad audience that may not even be interested in their message. The challenge is to identify those industries and companies that would be most likely to view a community as a potential location match. In order to do this, communities need to segment their markets.

Community economic development market segments include:

- *Sector/industry.* Manufacturing, service, professional, wholesale/distribution, and so on. Within each sector (e.g., manufacturing), which industries would make the best targets (e.g., auto parts, plastics, electronics)?
- *Geography.* For a community in rural Arkansas, for example, it would probably be more fruitful to try to recruit companies in higher cost locations such as Chicago, Dallas, or another urban area rather than similar rural areas nearby.
- *Type of company/organization.* Smaller communities may be better off targeting smaller, privately owned firms. Larger communities can target these as well, but also may want to target larger, publicly traded companies. Again, it is all about the best match with the community.

Substantial effort should be given to the target industry analysis because this often makes the difference between a successful and unsuccessful marketing campaign. Target industry selection criteria include:

- A match between the location needs of the industries and the strengths/weaknesses of the community (from the SWOT analysis).
- Historical and potential growth rates of industries.
- Skill levels and wage rates of industries.
- Diversification potential of the industries for the local economy.

- Other community-specific considerations such as environmental friendliness, industry image, and so on.

Once the target industries are selected, the next step is to identify companies in those industries that will make likely targets for recruiting. The community should look for companies that are most likely to expand and that also meet the above mentioned criteria of geography, company size, and ownership.

The target industry and company selection process must be based on sound research. Large amounts of data on growth rates, location criteria, and patterns should be collected and analyzed in order to select the best targets. Many communities hire professional consultants or rely on local universities to assist with this complicated procedure. If the community undertakes the analysis itself, there are numerous sources of information on industries and companies that are available through government and private organizations. So much of this is now only a few clicks away on the Internet.

## Distribute message and create awareness

After creating the message and identifying the target audience, the third basic step in community marketing is to distribute the message and create awareness. There are different means a community can use to distribute its marketing message and promote itself. The mix of marketing elements should be chosen very carefully and with much research in order to make the best use of limited marketing dollars. Ways to promote and market a community for economic development include:

- *Advertising.* Media choices include television, radio, and print. Types of print media for economic development advertising include national business publications such as the *Wall Street Journal* or *Business Week*. However, advertising in these national publications is quite expensive and probably not feasible for smaller communities. Other print media include industry trade publications oriented toward industries such as aerospace or plastics, and site selection/development magazines such as *Expansion Management, Area Development*, and *Site Selection* (see Figure 13.1).

Industry trade publications are read by executives within specific industries and thus may be used to reach a community's target industries. Site selection and development magazines target corporate executives that are typically involved in corporate expansion decisions but which cover a broad range of industries. There are other niche print publications, such as local business journals and Chamber of Commerce magazines, that communities can explore as potential advertising options.

- *Direct mail.* We are all regularly bombarded with "junk mail" and 99 percent of it winds up unopened in the trash. However, many communities operate successful direct mail campaigns aimed specifically at the industries and companies they have carefully selected, as

described above. Research shows that repeated mailing to a select audience has a much higher response rate than general blanket mailings. If a community sends out 1000 targeted mailings, gets 10 company prospects, and lands one new company, the benefit will be tremendous even though the batting average is low.

- *Email.* Many communities use email to market themselves. Some email messages are direct "sales pitches" while others are electronic copies of newsletters, announcements, etc. Targeted email can be effective but random mass e-mailings are usually treated as just more spam.

- *Trade shows.* Many communities successfully market themselves to their target industries at trade shows. There may be hundreds of executives from a target industry there, so attendance can be a very cost-effective way to reach the target audience. Many communities buy or share booths at the trade show and stock them with marketing materials. Others just "walk the floor" and look for opportunities to meet executives from key companies face-to-face.

**■ *Figure 13.1* An economic development advertisement**

- *Personal contact.* This is one of the best ways to market a community but it is often difficult and expensive to get a personal contact opportunity with a corporate executive. Trade shows are one way to do this, as is the telemarketing follow-up which is part of the direct mail campaign discussed above. Another personal contact method is to visit prospects, but this is expensive and is usually done after a prospect has expressed interest in a community.
- *Networking.* Many leads come to communities through networking with other economic development officials and agencies. Networking contacts should include the state economic development agency, utility companies, real estate brokers and developers, site selection consultants, existing employers in the community, railroads, and other organizations that may know of expanding companies.
- *Public relations.* While all of the above activities may be considered public relations, to get their message out to their target audience and general public, successful communities have even more proactive programs at their disposal. Local, regional, and state news channels (television, radio, and print) should be regularly sent press releases and otherwise contacted to get the community's name in print and publicize significant events.
- *Websites.* The Internet is now a primary way to gather information, and site selection data is no exception. It is almost taken for granted today that a community will have a website with relevant economic development information. Not having one sends a negative message to potential investors. In addition to containing basic economic and demographic information about a community, a website can contain testimonials from local executives, preferably in audio or video format; community pictures; and interactive video "virtual tours" of the community (see Figure 13.1).

---

**BOX 13.2 SUGGESTIONS FOR EFFECTIVE ECONOMIC DEVELOPMENT WEBSITES**

- Don't bury the economic development website in a municipal, chamber, or other website. At most, locate the economic development page one link away from the home page.
- Orient your economic development website toward external users (prospects). Don't make users wade through chamber banquet announcements.
- Keep the information up-to-date. Old data give a bad impression and may contribute to eliminating your community from consideration.
- Liven it up with graphics, pictures, and videos; however, keep it accessible to lower speed connections.
- In marketing materials, encourage people to visit the website. Good marketing materials are essential to effectively promote a community.

---

Types of marketing materials include:

- *Brochures.* One of the best formats is a six-to-eight-page four-color professional brochure highlighting the economic development assets and livability of the community. However, these can be expensive so many communities opt for "desktop" versions.
- *Postcards and other brief mailings.* As discussed above, these brief reminder pieces sent out quarterly or semi-annually help keep a community in the minds of target company executives (see Figure 13.2).
- *Community profiles.* These are more in-depth fact books about communities. Often 15 to 20 or more pages in length, they are really more

valuable as a follow-up tool once an executive's interest in a community has been piqued. Brochures or other shorter pieces should be used to create the initial interest.

- *Audiovisual presentations.* This category includes CDs, DVDs, computer presentations, and the like (websites are covered above). They can be expensive to develop and are being replaced in many instances by websites that can do the same thing over the internet at the click of a mouse. However, these presentations can be effective as a downstream selling tool once initial interest and specific contacts have been established. Mass mailings of these media are almost never effective.

- *Promotional items.* This category includes T-shirts, pens, tote bags, and many, many other creative items. As mentioned above, these promotional gifts can help with community branding and imaging.

- *Newsletters.* Short newsletters mailed to prospects can be effective marketing tools. However, newsletters mailed to the *external* audience should not contain purely local information such as the date of the next city council meeting. Instead,

newsletters sent to the external audience should briefly highlight recent positive developments, company locations, and expansions in the community.

## Satisfying the needs and wants of the customer

The final basic step in marketing the community is satisfying the needs and wants of the customer. To his definitions of marketing, Kotler (1991: xxii) adds: "Marketing ... calls upon everyone in the organization to think (like) and serve the customer." This is a critical but often overlooked distinction in community marketing. Development organizations must keep their focus on attracting and satisfying the needs and wants of the companies, tourists, restaurants, nonprofits, or other groups they are trying to attract. When making a location decision, most companies refer to a "must-have" (or needs) list of key location drivers for the particular project. These might include a skilled labor force, low tax rates, low transportation costs, high-quality utilities, or any number of factors. In addition to the needs

*Figure 13.2* **An economic development marketing postcard**

list, they usually have a "want" list that might include a low cost of living, good education, recreational opportunities, and so on. It should be stressed that all company location projects are unique. There is no "universal" list of top location factors. Obviously the location criteria for a steel mill are different from those of a corporate research and development center or nonprofit organization.

Attracting companies and new investment often requires communities to be flexible and make accommodations for companies such as improving water and sewer infrastructure, helping with labor training, or granting incentives. Of course it is up to the community to decide which accommodations to make, if any, based on the economic benefits of new corporate investment and on the desires and values of local residents.

## The marketing plan

A marketing plan is the roadmap communities follow for successful promotion. It should be a written document widely agreed upon by all elements of the community. The marketing plan should contain:

- A mission statement for the marketing plan and a vision statement for the community.
- A situation analysis (SWOT analysis summary).
- A description of the target audience (industries and companies).
- Marketing goals and objectives.
- Strategic action items to achieve each objective.
- Budget and resource requirements.
- Clearly defined staff requirements and positions.
- Clearly defined responsibilities of participating organizations and stakeholders.

Marketing goals are what a community or other entity has set out to accomplish. For example, a major goal of a marketing plan could be to "Attract new business and industry to the county through an aggressive targeted marketing campaign." Marketing objectives generally describe what needs to be done to achieve the goal. An example of an objective could be to "Convince prospects with an active project to visit the county." Strategic action items are specific, measurable activities designed to accomplish an objective. An example could be to "Invite corporate prospects as expense-paid guests to enjoy the county's Fall Apple Festival and local quail hunting."

Marketing plans may be 10 or 100 pages long, depending on the community and level of detail desired. Too little detail causes the marketing plan to be viewed as just another general statement of goals and objectives; too much detail can discourage local stakeholders from reading the plan and buying into it. Detailed marketing plans are best and they can include an executive summary for the general public's benefit.

Marketing plans should be dynamic. No one can predict with certainty how the various elements of the plan will work in practice. As the community learns about the elements of the marketing program that work best, and as external conditions such as the economic situation change, the marketing plan should be revisited and modified at least yearly.

## The regional approach to marketing

Marketing a community can be a very expensive activity. Advertising, travel to trade shows, developing promotional materials and other marketing activities often require significant budgets which may be beyond the means of smaller communities. For example, a booth at a major trade show can cost tens of thousands of dollars. Yet in a global economy with increased competition for investment, tourism, and other economic activities, it is even more important to have an effective marketing program to "get on the radar screen."

In the past several years there has been a trend toward regional marketing whereby adjacent communities join together to develop and support a joint marketing program. In this way, smaller communities can pool resources and compete with

**BOX 13.3 SOME *SUCCESS* FACTORS FOR DEVELOPMENT MARKETING**

- Unify your community vision and strategic plan;
- Understand the product you are marketing;
- Know which audience wants the product;
- Understand the target audience;
- Write a marketing plan including:
  - how to get the marketing message to the target audience
  - adequate budget, resources, and staff
- Have committed parties to stay the course to ensure that:
  - board members, elected officials, the private sector, and other citizens understand the plan and process
  - public and private concerns cooperate and form partnerships.

larger communities. This joint approach has many added benefits including better geographical recognition (e.g., people know where Northwest Louisiana is, but not necessarily where Shreveport, Louisiana is), and encouraging adjacent communities to cooperate in other ways such as joint infrastructure development.

Larger communities can also benefit from regional marketing. For example, Memphis, Tennessee joined forces with Tunica, Mississippi, which has become a major casino gaming destination, to jointly market the southwest Tennessee–northwest Mississippi region. Even though the cities are one hour apart in separate states, surveys showed that visitors to one often traveled to the other, and they realized that marketing the joint Memphis–Tunica experience would be more appealing to tourists (Gillette 2008).

Regardless of whether a community takes a regional approach or a "go-it-alone" approach, careful consideration must be given to developing the right kind of marketing program to ensure positive returns on invested resources. Many communities fall prey to the "ideal community syndrome" (see Box 13.4) and throw money at ineffective marketing campaigns that portray them as the best place in the world to locate any business. To be effective, as discussed in this chapter, marketing campaigns should be targeted, focused, and credible. Targeting

involves segmenting the market and selecting the right industries and companies (or tourists or other targets), being focused means selecting the right media for the message, and being credible means crafting the right image and message that is believable and resonates with the marketing audience.

## Conclusion

Community marketing is a broad topic. This chapter has focused on economic development marketing and specifically on recruiting new firms to a community. However, the marketing elements discussed in this chapter have broad applicability throughout the field of community and economic development. They are particularly relevant to business retention and expansion, and new business start-ups. Research shows that most new jobs in communities are created by retention and expansion of existing companies and business start-ups as opposed to recruitment of new companies. Marketing plans should include goals, objectives, and strategic action steps for all of these important economic development activities.

Even the smallest, most isolated communities can achieve marketing success through applying the principles described in this chapter. For a metropolitan area, marketing success may be attracting a new

**BOX 13.4 SOME *FAILURE* FACTORS FOR DEVELOPMENT MARKETING**

- *Ideal community syndrome.* "This is the best place in the world to live, work and play. Why wouldn't anybody in his right mind want to put a facility here?" The fallacy is that there are thousands of "ideal communities" competing for a relatively few number of corporate investments. Often, residents cannot or choose not to view their community objectively from an outside investment standpoint. This is one reason why a SWOT analysis is a key part of the marketing process.

- *If you build it they will come.* This relates to the ideal community syndrome above. If a community just develops a site or industrial building and waits for the phone to ring, the chances are slim that it will be successful. In addition to developing the product, it must be marketed.

- *If it ain't broke, don't fix it.* "We've been doing fine for years so why should we change things now?" What if "doing fine" means that the local textile mill has not laid off yet? If communities do not actively assess their economic well-being and industrial base on a regular basis (as in Chapter 11) nor have development plans to move forward, they will inevitably move backwards.

- *Throwing the baby out with the bathwater.* "We didn't meet our goal of recruiting new jobs last year so we must change the program or quit wasting money." Marketing success is dependent on a number of external influences beyond the control of the community such as national economic conditions and corporate spending cycles. While success goals should be set, programs should not be solely judged on "hitting the numbers." Real economic development success comes from establishing a good marketing plan, evaluating it regularly, making midcourse corrections as necessary, and staying the course.

Fortune 500 company with 1000 local employees. For a small rural community, marketing success may be as basic as helping a start-up company with five employees or attracting a McDonald's restaurant.

The key point to remember is that manna rarely falls from heaven. Success is more likely to come to communities that develop solid marketing programs and stay the course.

**CASE STUDY: BUFFALO NIAGARA REGION**

Located on Lake Erie in far Western New York State, Buffalo originally thrived as a Great Lakes port and manufacturing center. In the late twentieth century, as shipping routes and technologies changed and U.S. manufacturing growth abated, Buffalo's economic development fortunes declined. In 1999, Buffalo Niagara Enterprise (BNE) was founded. A regional economic development marketing organization, its mission was to market the eight-county area and attract new corporate investment.

In the minds of some corporate executives, Buffalo is well known for being featured in the video highlights on weather reports. With its high average snowfall, Western New York gets a lot of "lake effect" snow, and storms can be quite intense. To change their image from a place known for harsh weather to a place recognized as a good business location, BNE undertook a systematic, well-funded marketing campaign widely supported by the region's stakeholders. Their marketing program included:

- *Advertising to modify the region's image.* BNE became a National Public Radio sponsor using the slogan "Buffalo Niagara: Home of Snow-white Winters." This is a good example of directly tackling a perceived marketing weakness and turning it into a positive. The phrase "snow-white winters" evokes a more pleasing image of winter in the Buffalo region than blizzard footage shown on weather reports. It also appeals to winter sports enthusiasts.
- *Target industry research.* BNE spent a year carefully identifying industries that might be well suited to the economic development strengths and weaknesses of the region and are growing on a national basis. The target industries identified included primary life sciences, agribusiness, and back-office operations. BNE developed a list of 600 companies in these industries for direct marketing activities.
- *Call trips to prospects.* To personally explain the benefits of locating business operations in the Buffalo Niagara region, BNE began making trips to visit high-level executives in these target companies. Taking advantage of its location, BNE paid visits to numerous Canadian companies, touting the Buffalo Niagara region as a good place for Canadian companies to establish U.S. operations.

While it is sometimes difficult to directly link marketing activity with recruiting success, there is no doubt that the Buffalo Niagara region has been successful in attracting corporate investment. For BNE's 2006 to 2007 fiscal year, it announced 21 new or expanded projects worth more than $1.8 billion in private sector investment, 1052 new jobs, and 1337 retained jobs. Companies that created new jobs in the region included Astronics Corporation, 100 jobs; Multisorb Technologies, 95 jobs; and Citicorp, 500 jobs.

Source: Author discussions with BNE, the BNE website (www.buffaloniagara.org), and Glynn (2007).

## Keywords

Marketing, recruitment, business retention and expansion, business retention, business expansion, image, market brand, slogan, target industry, marketing plan.

## Review questions

1 How is community and economic development marketing different from general or "societal" marketing?
2 What are the four major components of community marketing?
3 Why is a SWOT analysis important in community marketing?
4 What is a marketing image and why is it important?
5 What are a community's external and internal marketing audiences?
6 What are some economic development market segments?
7 What are some ways to distribute a community's marketing message and create awareness?
8 What are some key components of a community's marketing plan?

## Note

1 The author wishes to express appreciation to David Kolzow, Rhonda Phillips, and Jennifer Tanner. Together, we have developed and taught many of the concepts in this chapter. However, the author assumes responsibility for any errors or omissions.

## Bibliography and additional resources

American Marketing Association. Available online at www.marketingpower.com (accessed December 5, 2007).

Finkle, J.A. (2002) *Introduction to Economic Development*, Washington, DC: International Economic Development Council.

Gillette, B. (2008) "Regional Approach Maximizing Impact of Marketing Campaigns," *The Mississippi Business Journal*, 30(9): 14.

Glynn, M. (2007) "Buffalo Niagara Enterprise Returns Its Focus to Marketing," *Knight Ridder Tribune Business News*, October 4.

Koepke, R.L. (ed.) (1996) *Practicing Economic Development*, Rosemont, IL: American Economic Development Council.

Kotler, P. (1991) *Marketing Management*, 3rd edn, Englewood Cliffs, NJ: Prentice-Hall.

—— et al. (1993) *Marketing Places*, New York: The Free Press.

Shively, R. (2004) *Economic Development for Small Communities*, Washington, DC: National Center for Small Communities.

# 14 Retaining and expanding existing businesses in the community

## Robert H. Pittman and Richard T. Roberts

While the recruitment of a new business into a community usually garners front-page headlines, the retention or expansion of an existing business rarely does. Communities tend to take their existing businesses for granted, which can be a big mistake. Industries choose to expand in different communities from where they are located (or even worse, relocate entirely to another community) for many reasons. Some of those reasons are readily discovered and prevented through a proactive existing business retention and expansion program. In most communities, it is a lot easier to create 25, 50, or 500 jobs through the retention and expansion of existing industry than to spend the time and money it takes to recruit new industry.

## Introduction

When people hear the term "economic development," they often think about recruiting new businesses into their community or region, sometimes visualizing large projects such as automobile assembly plants, high-tech research facilities, or corporate headquarters. These types of projects make headlines and can transform a regional economy almost overnight. Because of the publicity and attention these large projects command, many communities make the mistake of pursuing the "Holy Grail," overlooking the small and medium-size businesses that create most new jobs. According to the Organization for Economic Cooperation and Development (OECD), small and medium-size enterprises (fewer than 250 employees) account for 95 percent of the firms and over half of private sector employment in most OECD countries (OECD 2002).

As discussed in Chapter 1, the three legs of the economic development stool are recruiting new busi-

nesses, retaining and expanding existing businesses in the community, and facilitating new business start-ups. Research has shown that retention and expansion and new business start-ups account for the majority of new jobs created in most areas. Consider these findings:

- In pioneering research into job creation in the United States, David Birch found that 60 to 80 percent of all new jobs created in most areas come from retention and expansion of existing businesses and new small businesses (Birch 1987).
- Data compiled by the consulting firm Arthur D. Little revealed that between 31 and 72 percent of jobs in six states surveyed were created by business retention and expansion alone (Phillips 1996).
- *Site Selection* magazine found that 57 percent of manufacturing capacity increases in the U.S. in 2006 (50 or more jobs or at least 20,000 square feet) were expansions of existing facilities (*Site Selection* 2007).

- *Southern Business and Development* magazine reported that, in manufacturing and nonmanufacturing companies, 55 percent of all new projects of 2005 that created at least 200 jobs in the South, were expansions of existing facilities (*Southern Business and Development* 2006).

Additional data compiled by *Site Selection* magazine shows that, whether new facilities or the expansion of existing facilities, there was a total of 3076 company projects in the U.S. in 2006 (*Site Selection* 2007). Competition for these projects is still fierce and the odds of a given community landing one are low. It has been estimated that there are over 15,000 economic development organizations in the U.S. actively recruiting new businesses. To increase their chances of success, smart communities figure out the types of industries and size of businesses most suitable for their area and concentrate on recruiting those projects.

While this is certainly a better approach than pursuing large "trophy" projects, many of these communities still place too much emphasis on recruiting new industry. In the heyday of postwar manufacturing in the U.S., this was referred to as "smokestack chasing." Recruiting new firms, even small to medium-size businesses, is extremely expensive and time-consuming. To stand out from the dozens or hundreds of competitors vying for the same new facility, communities must market themselves. They must advertise, attend trade shows, call on companies in distant cities or countries, and engage in a host of other expensive activities.

Companies do not simply knock on doors and ask to locate in a community. Many communities and citizens do not understand how intensely competitive it is to recruit new businesses, and fall prey to what may be referred to as the perfect community syndrome, to wit: "My community is a great place to live, work and play. I've raised a family here, operated a successful business, and enjoyed the quality of life. Why wouldn't anyone in their right mind want to locate a new corporate facility here?" There are two major problems with this attitude. First, citizens often do not see the shortcomings of their community from a business location standpoint and,

second, there are thousands of self-perceived "perfect" communities competing for every project.

If recruiting new companies – large or small – is difficult and competitive, especially in the era of globalization (see Chapter 22), what should communities do to create jobs, increase incomes, and improve the quality of life? The answer is to adopt a balanced approach to local economic development. It is ironic that when a new firm moves into town it makes the front page of the local newspaper, even if only a few jobs are created. However, if a business already located in and contributing to the community announces an expansion that creates hundreds of new jobs, the announcement is often buried in the back of Section C. In short, many citizens take their existing industries for granted. They do not realize that local companies, especially branch facilities of national firms, regularly consider other communities for their expansions. They may even decide to relocate or consolidate all operations in a different area. Often, one community's loss is another's gain. What elected official or citizen wants to lose an existing business and/or its expansion to another community?

Even thriving local economies may be subject to sudden downturns from unforeseen events such as the closure of a manufacturing plant, call center, or other local facility. In such cases there is a downward economic multiplier effect. The closure of a local facility has a ripple effect throughout the economy. Primary job and income losses translate into less consumer and business spending in an area, a loss that hurts the retail industry and local consumer and business service industries. A loss of 100 primary jobs can translate into a total job loss of 200 or more.

Recovering from such a shock to the local economy can take years. Surely no one in a community wants such an economic shock to occur, regardless of the strength of the local economy. However, in today's extremely competitive and fluid global economy, economic dislocation is becoming more common. Industries are moving offshore; product lines and therefore production facilities are rapidly changing; and mergers and acquisitions are occurring more frequently. If there is agreement that

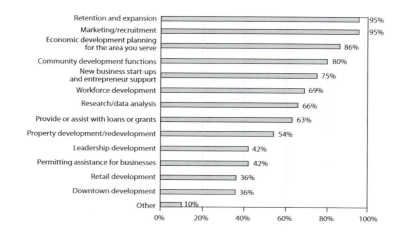

**■ *Figure 14.1* Economic development activities performed by SEDC members (source: Southern Economic Development Council Member Survey 2005)**

these economic shocks are undesirable, then communities should do what they can to prevent them in order to maintain and grow local industries. Fortunately, most economic developers understand the importance of business retention and expansion. As shown in Figure 14.1, a recent survey by the Southern Economic Development Council, a professional association of economic developers in 17 southern states, indicated that 95 percent of the members were engaged in some sort business retention and expansion activities (SEDC 2006).

Companies may decide to relocate or expand away from where they are currently located due to a variety of external factors that are completely outside the control of the community such as mergers, acquisitions, moving offshore, and changing product mix. However, factors that are within the control of the local community often enter into a company's decision to relocate or expand elsewhere. Some examples include:

- Inadequate services and infrastructure (e.g., water/sewer, electricity, roads, telecommunications).
- Poor business climate (e.g., the community's permitting and regulatory policies and procedures don't support business).

- Inadequate labor force.
- Lack of available sites and buildings.
- Poor education system and workforce training.

Seemingly insignificant issues that the community can easily address may contribute to a business's decision to relocate or expand in a different area (see Box 14.1).

## What does business retention and expansion encompass?

Business retention and expansion (BRE) covers a wide spectrum of activities, but a general definition is to work with existing businesses in an area to increase the likelihood that they will remain and expand there instead of relocating or expanding elsewhere. BRE programs should be tailored to the specific needs of a community – one size does not fit all. However, there are several components that are usually included in BRE programs. They include business visitations and surveys to identify problems and issues; troubleshooting community-related problems that companies might have; and industry recognition/appreciation programs.

## BOX 14.1 CAN POTHOLES CAUSE LOCAL BUSINESSES TO RELOCATE?

In the course of doing an economic development assessment for a town in the South Central U.S., a consultant interviewed the manager of a large plant that employed almost 500 persons, about the strengths and weaknesses of the community as a business location. The manager did not point to any major problems – the labor force was suitable, the community's location on a major highway was good, and the local regulations were not onerous. As the manager escorted the consultant out of the front door after the session, one of the company's trucks pulling into the plant hit a large pothole, shaking the truck and potentially damaging the cargo. The plant manager's smile turned into a frown as he said, "I've been trying to get the City to fix that pothole for over a year, and nobody does anything about it." That pothole, an insignificant issue to the City, was a sign to the manager of the City's lack of support and concern for local businesses. If the City won't fix the pothole, how can it be trusted to upgrade water and sewer infrastructure and improve local roads to support the growth of local businesses?

Factors like the pothole repair issue can influence a business location decision. Imagine – as often happens – an executive visiting the community to screen it as a potential location hearing about this event. What impression would he take away concerning the City's support for local businesses? Rarely does a company move to a community without first talking to local executives about their experiences in that community. The local executive with the pothole problem would certainly not give the City a glowing reference.

Source: Pittman and Harris (2007)

Regular visits to businesses and interviews with executives are important components of a BRE program. While BRE programs should be tailored to fit a community's unique situation, they should generally include the following components:

- Interviews and surveys with local businesses to detect trends, problems or issues. It is preferable to conduct personal interviews with key executives from local businesses. This demonstrates the value the community puts on these businesses and facilitates better communication and information gathering. Surveys may also be used to gather data and they are relatively easy to create and administer, especially if they are Internet-based. Often a survey followed by a personal interview is the best approach. Interviews and surveys give local businesses a feeling that the community cares, and if there is a problem an executive will likely identify it in the interview

or survey. It is critical, however, that any problems identified be addressed immediately or the business will not fill out a survey or consent to an interview again. All local businesses should be visited and/or surveyed at least once a year. This can be a major undertaking, especially in larger communities. Economic development agencies often solicit the assistance of board members and volunteers to help conduct the interviews and surveys.

- Industry appreciation functions such as annual banquets. These typically provide a casual setting in which to interact with local businesses and provide an excellent opportunity for businesses to network and discuss issues of mutual interest.
- Initiatives to improve the general business climate. This might include streamlining permitting processes, making government agencies customer-friendly, and so on.
- Sponsorship of meetings and forums to help companies deal with issues of concern to them.

One example would be roundtable discussions with local operations or human resource managers. These work most effectively when the host provides talking points to get the discussion started, possibly over breakfast or lunch. However, there should be plenty of time for unstructured discussion to allow issues to bubble to the surface.

- An area industry directory listing all businesses in a community including manufacturing, service, and retail. These directories can help businesses find local products and services they need to operate effectively. Information for each business should include key executives, contact information, a brief description of the product or service, and the North American Industry Classification System (NAICS) number.
- Labor and wage surveys to help local businesses gauge their pay scales.

Some BRE programs even offer business consulting services. These services may include assistance in identifying export markets for local products or sponsoring informative sessions about market trends and specific topics of interest. One of the most valuable BRE services is helping local companies attain "lean and agile" production status to help them remain competitive in the global economy (see Case study, p. 218).

## Benefits and advantages of BRE

As a part of the overall community and economic development program, the benefits and advantages of business retention and expansion are significant.

- *BRE is a cost-effective way to create jobs and keep local economies strong.* Existing firms are already located in the community. There is no need to incur travel expense, spend staff time, and mount large marketing campaigns as required when recruiting new firms. Economic development, just as any sales-related activity, is influenced by personal relationships. Companies want to feel comfortable that local officials and citizens support local businesses. Personal relationships and trust are important. Economic developers, committee members, and many other local stakeholders can more easily establish these relationships with local managers than with out-of-town firms. They have the opportunity to do so because they live and work in close proximity.

  Furthermore, when new companies move in, communities incur new expenses. Often new infrastructure must be developed, site preparation may be necessary, and cash incentives may be granted. The expansion of an existing business in its current location is usually a less expensive proposition.

- *Relationships with local executives can be leveraged to create even more jobs.* As noted above, key components of a good BRE program are regular visits and contacts with local businesses and executives. Through these visits, relationships and trust are developed. Local executives who feel supported and appreciated think of themselves as members of the economic development team. There are many examples of communities and local executives working together to entice companies that supply local firms, or with whom they have other business relationships, to move to the community. This strengthens the supply chain relationship, makes both firms more competitive, and helps the local economy.

  Many communities leverage their relationships with local firms to "move up the food chain" and gain access to decision makers at parent company headquarters located in different cities or countries. If a company already has a successful branch operation in a community, it knows that community is a good place for business and may be more inclined to expand existing facilities or locate new facilities there.

- *BRE can facilitate the recruitment of new industry.* Despite the difficulty and expense of recruiting new firms, it is still a critical component of the overall economic development process for most communities. Very often, local business executives are a community's best recruiters. In almost all cases, before a company makes a decision to locate a new facility in a community, representatives from that company or their consultants will talk at length with executives of local firms about operating a business there. A local executive who feels that the community supports his business and has benefited from a BRE program will make a much better salesperson for the community. On the other hand, a disgruntled executive (such as the one with the pothole problem in the example above), will not be a good community advocate. All too often, unfortunately the latter is the case.

Companies make location decisions very carefully based on myriad factors – some are quantifiable (e.g., wage rates, transportation costs) and some are not as easily quantifiable (e.g., local work ethic, labor/management relations). Facilities are long-term investments for companies. The last thing they want to do is close them prematurely after a few years because of problems with the local business climate or global economic events. A strong BRE program sends a signal to companies making location decisions, that a community will help support their profitability as conditions change in the future. With global markets and competition, product lifetimes are shorter and facilities become obsolete sooner. The willingness of a community to "partner" with local firms and help them meet their changing resource needs as products and markets change increases the likelihood that facilities can be retooled to produce different products or services and avoid closing down. It is important for anyone involved in the economic development process to question what existing businesses would have to say when asked about the community. If "I don't know" is the answer, the community is likely in need of a better BRE program.

---

## BOX 14.2 EARLY WARNING SIGNS OF A POSSIBLE BUSINESS RELOCATION, DOWNSIZING OR CLOSURE

- Declining sales and employment: Are they having issues with cost of production, struggling with supply chain issues, or is employment declining just because their processes are becoming more automated?
- Obsolete buildings and/or equipment: Why aren't they spending money on the facility to maintain buildings and equipment?
- Ownership change – merger, buyout, acquisition: What plans do the new owners have for the local facility?
- Older industries and product lines: Do they need help with advancing their technology or will they require a complete retooling due to obsolete products?
- Excess production capacity: If they have not gone through a lean initiative (see Case study, p. 218), allowing them to produce more efficiently, why are they utilizing less of their production capacity?
- Complaints about labor supply, local business climate, infrastructure, and services. Have their complaints fallen on deaf ears, or is someone communicating the concerns to the right people?
- Expansions in other locations: Why aren't they choosing to expand in the local community? Are they landlocked or did they receive incentives to start a new facility elsewhere?

## BRE is a team effort

Successful BRE programs are usually team initiatives. It is not just the job of the Chamber of Commerce or economic development professional to operate the program. Issues affecting local businesses are varied and complex, often requiring special expertise. There are many specialty areas – such as utilities, permitting and regulation, government services, transportation – that may require attention from local government agencies or other private organizations.

BRE is most effective when undertaken by a team composed of representatives from economic development organizations, local government agencies, educational institutions, and other fields. Knowledgable specialists from utilities, transportation departments, and so on should be on the team as necessary to help resolve technical issues. The multidisciplinary BRE team should work together and coordinate communications and activities. All members should assist in BRE activities, such as business visitations and surveys, and the team leaders should know whom to contact to handle a specific situation. The team leaders should always make sure that any problems or concerns reach the appropriate experts on the BRE team. This is often where the ball is dropped and, as a result, the issues and concerns fall on deaf ears.

## How successful are BRE programs?

As shown in Figure 14.1, the vast majority of economic development professionals report that their organizations are involved in some facet of business retention and expansion. Most of them would undoubtedly offer stories and anecdotes on the success of BRE programs if asked. In a survey of economic developers, Morse and Ha (1997) gathered information on the benefits and best practices of BRE programs. They found that 30 percent of the plans adopted in BRE written reports had been substantially or completely implemented and another 35 percent of the plans were being actively pursued. Among the organizations reporting implementation or pursuit thereof, the most common programs included (percentage reporting on each program in parentheses):

- publicize the area's strengths (72 percent);
- share information on business programs (66 percent);
- continuous BRE program (64 percent);
- labor training (53 percent); and
- informing politicians of concerns (51 percent).

The survey respondents reported that the following factors, in their opinions, facilitated success in BRE programs:

- adhere closely to the BRE written plan;
- share the results with the community;
- assure adequate funding and personnel for the program;
- specify responsibility for program implementation;
- concentrate on key aspects of the BRE program and implementation rather than dilute the effort with too many efforts and initiatives.

Because of the many benefits and high return on the investment of resources and time, a BRE program should be one of the initial and primary functions in a community's overall economic development program.

## BOX 14.3 HOW TO CONDUCT BRE INTERVIEWS AND SURVEYS

With regard to their operations and potential for relocation, downsizing or expansion, obtaining sensitive information from local companies requires a diplomatic approach. This is why it is important to establish a relationship of trust with local executives. It is best to obtain BRE information in a personal interview or meeting as opposed to a mail survey. However, a survey may be used to gather routine statistical data before a personal interview, and to gather information from a broader cross-section of local companies.

A good BRE program should track the health of local companies. Sample questions could be: How much are employment and revenues at the local facility fluctuating year after year? How much are they fluctuating at corporate headquarters, if applicable? In either or both cases, are they growing or declining? To avoid asking repetitive questions and to show they have done their homework, the BRE team should do thorough research into a local firm before initiating discussions with them. A good BRE interview is a conversation, not an interrogation. When setting up the interview, the BRE team should request 30 to 45 minutes of the interviewee's time and confirm that this is acceptable. The interview should never go overtime unless the interviewee prolongs it.

The interview is composed of three parts: opening, main body, and closing. The opening is the ice breaker, in which the interviewer introduces himself, explains the purpose of the visit, and establishes rapport with the interviewee. Always assure the interviewee that the information shared will be kept confidential within the BRE team, and ask permission to take notes. The main body of the interview is the information-gathering portion. Rather than simply read questions and jot down answers, the interview should be conversational. The interviewer should know the questions so well that he can ask them at the appropriate place in the interview (not always necessarily in the same order) without having to read them. Often, interviewees have one or two key issues they want to "get off their chest." If the interviewee seems anxious to get on to the next question, let that person lead the discussion and get to what is really important to them. In the closing, the interviewer thanks the interviewee for their time, once again gives assurances of confidentiality, and asks them to call if the BRE team or local economic development agency can ever be of assistance.

Below are sample BRE interview and survey questions. While each situation is unique, these are the types of questions that will uncover clues on company issues and potential future actions.

- What are the current levels of employment and production at this facility? How do these compare with previous levels?
- Is your facility adequate for current and anticipated future operations?
- How old is this facility and how long has it been in operation?
- What products do you make? Do you anticipate phasing out this/these products or adding new ones in the near future?
- What/where are your principal market areas? Where do you ship your product or provide your service?
- Is the company locally owned or do you have a parent company (who and where)?
- What do you consider to be the advantages of doing business in this area (e.g., labor, transportation, business climate, utilities)? The disadvantages?
- Are there local issues or problems affecting the success of your operations? How can we help resolve them?
- Do you anticipate any downsizing, closure, or expansion of this facility in the near future? Why and when?
- Do you have any suppliers or customers that might consider locating in our community?

- If a new company were looking to relocate to the area and asked you for information on the business climate in the community, what would you tell them?
- May we speak to or visit your corporate headquarters to express our appreciation for the economic contribution you make? Whom should we contact?
- Are there any other issues you would like to discuss? Feel free to contact us any time if issues arise.

Some of these questions may be sensitive and must be asked in a diplomatic way with the assurance of confidentiality. Some companies will be comfortable answering all the questions, while others will decline to answer some in spite of confidentiality. Any information obtained will be useful. For more guidance on appropriate questions for BRE interviews, see Sell and Leistritz (1997) and Canada (1999).

---

## CASE STUDY: THE CENTER FOR CONTINUOUS IMPROVEMENT

In the course of conducting an economic development assessment in the Athens, Georgia region, a consultant identified a powerful example of a successful business retention and expansion program. During an interview at a local plant that employed several hundred persons, the manager stated that corporate headquarters had almost decided two years ago to shut down the facility due to low productivity. He went on to say that, if not for the Center for Continuous Improvement's help with increasing productivity, the axe would have fallen.

Hearing such a strong testimonial, the consultant investigated and discovered that the Center for Continuous Improvement was founded at the Athens Technical Institute in response to the needs of several local companies. These needs were discovered in the course of BRE interviews. The main symptom was that productivity was lagging at many of these facilities. In both manufacturing and service sectors, these companies' managers knew they needed to improve productivity and quality but lacked the knowledge and/or resources to implement an improvement program by themselves. At the suggestion of Athens Tech, several local companies agreed to combine resources and establish the Center for Continuous Improvement. The Center would work jointly with the companies to improve productivity and quality.

The Center became a research, training, and implementation resource for Total Quality Management (TQM), also called "lean" operations. Lean may be described as the minimization of waste while adding value to the product or process. Lean distinguishes between *value-adding* activities that transform information and materials into products and services that the customer wants, and *non-value-adding* activities that consume resources but don't directly contribute to the product or service. In today's global economy, lean operations are key to remaining competitive and profitable (Stamm and Pittman 2007).

Each member company working with the Center was able to accomplish the following:

- instill team culture throughout the facility;
- create a vision and values;
- perform cycle time studies resulting in improved quality, less waste, and higher productivity;
- implement ergonomic practices to reduce injuries and cycle time;

- improve customer service;
- improve purchasing and material management techniques.

TQM or lean transformation is not just a one-time event; it is an ongoing process. The Center continues to serve as a focal point for research and training in the field. Member companies help each other, sharing ideas and visiting each other's facilities. This example demonstrates that productivity and quality can be dramatically improved in local companies by adopting TQM or lean operations, increasing the likelihood that companies will remain competitive, stay, and grow in a community. The revolution in management and production techniques required to achieve this can take place using local resources. Through BRE programs, economic development agencies can be a catalyst to make this happen. For more information on this case study, see Ford and Pittman (1997).

## Keywords

Business retention and expansion, business interviews and surveys, business climate, industry directories, labor and wage surveys, early warning signs.

## Review questions

1   What are some common reasons companies decide not to expand in their present location or to relocate to another community?
2   What are the key components of BRE programs?
3   What are the advantages of BRE programs?
4   What are some key success factors in BRE programs?
5   What are some early warning signs that a company may be considering closing or relocating away from a community?
6   What are some important questions to ask in a BRE interview?

## Bibliography and additional resources

Birch, D. (1987) *Job Creation in America*, New York: The Free Press.

Canada, E. (1999) "Rocketing Out of the Twilight Zone: Gaining Strategic Insights from Business Retention," *Economic Development Review*, 16: 15–19.

Ford, S. and Pittman, R. (1997) "Economic Development Through Quality Improvement: A Case Study in Northeast Georgia," *Economic Development Review*, 15: 33–35.

Morse, G. and Ha, I. (1997) "How Successful Are Business Retention and Expansion Implementation Efforts?," *Economic Development Review*, 15: 8–12.

Organization for Economic Cooperation and Development (OECD) (2002) *OECD Small and Medium Enterprise Outlook*, Paris: OECD.

Phillips, P.D. (1996) "Business Retention and Expansion: Theory and an Example in Practice," *Economic Development Review*, 14: 19–24.

Pittman, R. and Harris, M. (2007) "The Role of Utilities in Business Retention and Expansion," *Management Quarterly*, 48(1): 14–29.

Sell, R.S. and Leistritz, F.L. (1997) "Asking the Right Questions in Business Retention and Expansion Surveys," *Economic Development Review*, 15: 14–18.

*Site Selection* Online (2007) *New Corporate Facilities and Expansion*. Available online at www.siteselection.com/issues/2007/mar/cover/pg02.htm (accessed November 1, 2007).

*Southern Business and Development* (2006) Available online at www.sb-d.com (accessed November 1, 2007).

Southern Economic Development Council (SEDC) (2006) *2006 Member Survey*. Available online at www.sedc.org (accessed December 11, 2007).

Stamm, D. J. and Pittman, R. (2007) "Lean is a Competitive Imperative," *Business Expansion Journal*, March. Available online at http://www.bxjonline.com/ (accessed December 11, 2007).

# 15 Entrepreneurship as a community development strategy[1]

## John Gruidl and Deborah M. Markley

## Introduction

A profound change in community economic development strategy over the past decade has been the emergence of entrepreneurship. Now, as never before, community developers recognize that entrepreneurship is critical to the vitality of the local economy. This emerging strategic change is due to several factors. A primary reason is the impact of globalization in driving many manufacturing jobs to overseas locations and, thus, reducing the effectiveness of using industrial recruitment as a strategy.

Another factor leading to the rise of the entrepreneurship development strategy is the evidence that entrepreneurs are driving economic growth and job creation throughout the world. For example, the U.S. National Commission on Entrepreneurship reports that small entrepreneurs are responsible for 67 percent of inventions and 95 percent of radical innovations since World War II (National Commission on Entrepreneurship 2001). On the international level, studies demonstrate the close connection between entrepreneurship and economic development. According to the Global Entrepreneurship Monitor Project, there is not a single country with a high level of entrepreneurship that is coping with low economic growth (Reynolds 2000).

Entrepreneurship has also succeeded in revitalizing many communities. For example, in Fairfield, Iowa, support from the Fairfield Entrepreneurs Association has sparked equity investments of more than $250 million in more than 50 start-up companies since 1990, generating more than 3000 high-paying jobs (Chojnowski 2006). Another example is Littleton, Colorado, which pioneered the concept of "economic gardening." The City of Littleton provides entrepreneurs, many in high-tech enterprises, with extensive market information, either free or at low cost, and arranges networking opportunities with other entrepreneurs, trade associations, universities, and think-tanks for the latest innovations. Littleton created twice as many jobs in its first seven years of "economic gardening" as were created in the previous 14 years (Markley et al. 2005a). It is clear from these examples that an entrepreneurial strategy can lead to significant increases in new businesses, jobs, and private investment in a community.

The goal of this chapter is to provide the fundamentals for implementing a strategy of supporting entrepreneurs and creating a supportive environment for growing these businesses. Although this chapter is about new enterprises, it is not intended to provide a primer on how to start a business or prepare a business plan. Instead, it is written from the perspective of the community developer or leader who seeks to expand the number of enterprise start-ups in the community and to create a culture of entrepreneurship among community residents. The main premise of the chapter is that community support must be focused on the needs and wants of entrepreneurs if it is to succeed.

## Who is an entrepreneur anyway?

The word "entrepreneur" is in frequent use, but it is one of those words which evoke different images for different people: high-growth, high-tech wizards like Bill Gates of Microsoft or Michael Dell of Dell Computing; a local artisan running a micro-enterprise; "mom and pop" struggling to survive in their main street store.

Yet, in order for our communities to effectively support entrepreneurs, we must have a deeper understanding of entrepreneurs and what they need. Let us start by defining what we mean by entrepreneur. A useful definition of entrepreneur is "a person who creates and grows an enterprise" (Markley et al. 2005b). This definition is simple, yet it helps clarify who an entrepreneur is. First, it reminds us that our focus is the person, not the venture itself. This is an elite group of people as only about one in ten American adults is currently an entrepreneur; that is, actively engaged in the process of starting an enter-

prise (Markley et al. 2005b). Beth Strube, whose story appears in Box 15.1, provides one example of a successful business entrepreneur.

Second, not all small business owners are entrepreneurs. Ewing Marion Kauffman (founder of the Kauffman Foundation) often identified the difference between a business owner and an entrepreneur in the following way: "A business owner works 'in' the business while an entrepreneur works 'on' the business." The difference is profound. Revitalizing, growing, and reinventing a business are inherently entrepreneurial.

There is also a difference in the level of financial risk between an entrepreneur and an owner/ manager. The level of risk associated with entrepreneurs is generally greater than that of managing an existing small business. Purchasing a franchise or an existing retail store may be an option for those who do not want to take large financial risk, yet still want to own their own business. However, it may also be a first step for entrepreneurs who go on to start new ventures.

### BOX 15.1 MEET A BUSINESS ENTREPRENEUR

#### Beth Strube, Dickinson, North Dakota

Beth Strube wanted to plant her roots in rural America. Her home is Dickinson, a small town in North Dakota. Her typical day runs the gambit from editing a preschool lesson on animal ABCs to reading the *Wall Street Journal*. Strube is President of Funshine Express, Inc., which originates and distributes preschool curriculums. Her business serves small children and their caretakers, but running a business grossing one million dollars in sales annually is hardly child's play. From downtown Dickinson, North Dakota, she manages 14 employees and serves a customer base of 3000 to 4000 – an impressive increase from her start in 1995 when she had two employees, 30 customers and a copy machine in her basement.

Beth illustrates the style of many successful entrepreneurs. She has a passion (kids) and saw a need to fulfill (better preschool curricula). She networked and acquired the skills she needed to take an evolving dream and grow it into an exceptional business. Dickinson, which some refer to this community as North Dakota's entrepreneurial community, gave Beth its full support. It embraced her idea for a unique business and treated her venture as a more traditional business development opportunity, making the full range of economic development resources, including financing, available to her.

Source: Markley et al. (2005b).

Third, the definition is intended to include civic entrepreneurs. Civic entrepreneurs create programs and resources that benefit our communities and our lives. They develop children's museums, organize a new chapter of Big Brothers Big Sisters, provide public health care, and build new playgrounds and parks. Often, this work is done through a nonprofit organization or informal community or neighborhood association. Civic entrepreneurs need skills in planning their enterprise, marketing their product or service, earning revenues or obtaining funding to keep the organization financially solvent, and creating value. As with business entrepreneurs, they perceive and act upon opportunities. Maxine Moul, whose story appears in Box 15.2, is an example of a woman who excels at both business and civic entrepreneurship.

To better understand entrepreneurs, it is worthwhile to examine the ingredients of the entrepreneurial spirit: creativity, innovation, motivation, and capacity. Creativity is characterized by originality, expressiveness, and imagination. Entrepreneurs are often not the same folks who develop the new idea, product, services, or approach. Rather, they are the ones who see value in new ideas, are able to take an abstract idea, and make it concrete (Markley et al. 2005b).

Innovation occurs when something new is created. Innovation may not be high tech, or lifesaving. Often, innovation revolutionizes how a process works. For example, the invention of the self-polishing cast steel plow by John Deere cut through the heavy Midwestern soil much better than previous plows and completely changed farming practices. At the heart of the entrepreneurial process, entrepreneurs perceive opportunity and transform ideas into commercial products or civic services that people want and are willing to pay for. New wealth is generally created during this process of commercialization. While this innovation may make some entrepreneurs involved in the early commercialization phase wealthy, it also has the potential to bring much needed change and wealth to a community.

The motivations that drive entrepreneurial behavior are wide-ranging. We might assume that most entrepreneurs are motivated by greed – making money. However, like most myths, this one is largely untrue. While entrepreneurs enjoy and appreciate making money, it is not usually their most important, driving motivation. Smilor (2001), in his book *Daring Visionaries*, talks about the soul of the entrepreneur and uses the word "passion." If

## BOX 15.2 MEET A CIVIC ENTREPRENEUR

### Maxine Moul, Lincoln, Nebraska

Maxine and her husband Francis are journalists. They had a dream of owning their own newspaper. The chance presented itself and they used "the three Fs" (family, friends, and fools) to raise the funds necessary to purchase the a weekly newspaper in Syracuse, Nebraska (population 1762). They not only succeeded with this venture, but also created the "Penny Press," a multi-state want ad publication. By all standards, Francis and Maxine Moul were highly successful entrepreneurs and business-people, and those family, friends, and fools who took a risk with the Mouls did really well when their media company was sold. Maxine went on to become Nebraska's Lieutenant Governor and Director of Nebraska's State Economic Development Agency.

However, our story on Maxine is not only about her business successes – it is about her role as a civic entrepreneur. Her true legacy is in the creation of the Nebraska Community Foundation (NCF). Maxine played a central role in its early beginnings and steady growth. NCF has become one of America's most successful and innovative rural community foundations.

Source: Markley et al. (2005b).

there is a common thread running through all entrepreneurs, this may be it. All entrepreneurs do indeed have some kind of a passion – to live a certain lifestyle; to grow a globally competitive company employing thousands; to provide a service to meet some critical community need; to prove (sometimes over and over again) their creativity and innovative spirit.

However, bringing together creativity, innovation, and motivation is not enough. Successful entrepreneurs acquire the capacity they need to put their ideas into action. Most entrepreneurs do not start with all the knowledge, skills, and insights necessary to create thriving ventures. They acquire these skills, develop them, and employ them consistently and effectively. The RUPRI Center for Rural Entrepreneurship has identified five capacities as being most important (Markley et al. 2005b).

1  *Ability to perceive opportunities.* First and foremost, entrepreneurs develop a heightened ability to perceive opportunities. They see how an abstract idea can be made practical and brought to the market. However, discipline and judgment are essential to this capacity. An entrepreneur must not only perceive opportunities but assess them and determine if they are a good fit with his evolving game plan.

2  *Ability to assess and manage risks.* There is a myth that entrepreneurs are reckless. In fact, most entrepreneurs hate risk. They become very skilled at risk identification, assessment, and management. Entrepreneurs understand that too much risk can kill an idea. The ability to deal effectively with risk distinguishes successful from less successful entrepreneurs.

3  *Ability to build a team.* Ernesto Sirolli (1999), the founder of Enterprise Facilitation™, believes that no single individual has all the skills necessary to succeed in starting a business. He argues that to succeed, a new venture must have capability in production, marketing and finance, the three legs of the stool. However, Sirolli doubts that any single person is highly skilled in all three functions. Therefore, successful entrepreneurs often bring in partners who have skills that complement theirs. In fact, entrepreneurs can become highly skilled at team building. They learn what kinds of team members they need and figure out how to assemble the right human resources. Great teams build great ventures.

4  *Ability to mobilize resources.* Another myth is that successful entrepreneurs have deep-pocketed investors. The reality is that most successful entrepreneurial growth companies start like everyone else – with too few resources and dependency upon family and friends. Here again is an attribute that distinguishes highly successful from less successful entrepreneurs. Skilled entrepreneurs learn how to mobilize resources. They can mobilize not only necessary investment capital, but also strategic partners, people, facilities, and whatever else is necessary for success.

5  *Ability to sustain creativity.* Finally, many entrepreneurs are not happy becoming managers of successful growing companies. They hire chief financial officers to ensure good management. Skilled entrepreneurs learn how to keep growing companies creative. They learn how to develop environments and processes that allow companies to be re-created.

## What are the needs of entrepreneurs?

Entrepreneurs are the creative, passionate people who drive the start-up of enterprises. However, entrepreneurs are not all the same. At the RUPRI Center for Rural Entrepreneurship, we use the concept of "entrepreneurial talent" as a way for leaders to better understand the kinds of entrepreneurs in their communities. Every community has a range of entrepreneurial talent and, once you have identified this talent, you can begin to develop the resources and strategies that each entrepreneur needs to be successful. The three main types of entrepreneurial talent are potential (those who may become entrepreneurs), existing business owners (some of whom may be entrepreneurs), and entrepreneurs (including those with growth and even high growth potential). For more information on a tool to help you understand and identify the entrepreneurial talent in your community, go to: http://www.energizingentrepreneurs.org/content/chapter_4/tools/1_0 00220.pdf.

We will focus on a few examples of entrepreneurial talent and then consider the resources and assistance that each type of entrepreneur is likely to need. We will also focus specifically on one pool of potential entrepreneurial talent that is critically important to creating a culture of entrepreneurship in communities – youth (Markley et al. 2005b).

## Potential – aspiring and start-up

Aspiring and start-up entrepreneurs are early in the venture creation process. Aspiring entrepreneurs are motivated toward making a life change and are actively considering crossing the bridge and starting an enterprise. Aspiring entrepreneurs are likely to be researching and developing their business idea. They may even be testing it out in informal ways. They may be attending business workshops and networking with potential customers, related businesses, and other entrepreneurs.

Aspiring entrepreneurs need guidance in assessing how good their idea really is. Will it meet a market test? Will someone pay for the good or service they want to offer? Is it really feasible for them to create this business? They need someone to ask the tough questions so that they can make an informed "go – no-go" decision about starting a business.

Start-up entrepreneurs, on the other hand, have crossed the bridge and made the decision to start their business. They have given thought to the business model, although there may still be some gaps. Some may have a formal business plan, although most do not. Some have the necessary skills to launch their enterprise, perhaps with the assistance of other team members. However, it is likely that they may need to acquire more skills to create a successful business.

Like aspiring entrepreneurs, start-up entrepreneurs need someone who can help them move from ideas to a solid game plan. They have already made a decision to start a business – now they need help making sure all the pieces are in place. Is the management team strong? Is there capital to start the venture? Are markets clearly identified and strategies for tapping them tested? Can the team acquire the necessary skills quickly enough to succeed with the venture?

The specific needs of aspiring and start-up entrepreneurs are closely related and can be met with moral support, networking and mentoring, business counseling, and entrepreneurship training.

### Networking and mentoring

Entrepreneurial networks that serve aspiring and start-up entrepreneurs can be formal or informal. An example of a formal network would be monthly forums sponsored by the Chamber of Commerce that offer an opportunity for entrepreneurs to meet their peers and share information about service providers, markets, or frustrations about doing business. A network is a great place for an aspiring entrepreneur to get moral support as he speaks with others who have already traveled the path to a successful business.

However, networking does not have to be formal. Entrepreneurs will network whenever they come together in one place. An informal breakfast meeting for young entrepreneurs, or a friendly Friday happy hour for entrepreneurs on the main street, can provide opportunities for entrepreneurs to develop connections and identify the resources and support they need from their peers. In many cases, simply providing a venue for entrepreneurs to come together is all the work the community needs to do – the entrepreneurs often take it from there.

Mentoring programs can be effective in strategically linking an experienced entrepreneur with an aspiring or start-up entrepreneur. Mentoring can happen organically. For example, experienced Hispanic restaurant owners in Hendersonville, North Carolina "adopted" new immigrants who were interested in starting restaurants or catering businesses to help them learn the ins and outs of the sector. Mentoring programs can also be established by creating a pool of experienced entrepreneurs who are willing to work with new entrepreneurs in sectors where they have expertise (Markley et al. 2005b).

### Entrepreneurship training and business counseling programs

Aspiring and start-up entrepreneurs can often benefit from participation in training and counseling programs, either one-on-one or with other entrepreneurs. Small Business Development Centers (SBDC) are an important resource in providing both training and counseling opportunities to aspiring and start-up entrepreneurs. Furthermore, a number of well-tested "how-to" training programs take entrepreneurs through the process of starting their own businesses. FastTrac and NxLeveL are examples of programs that provide an excellent curriculum for entrepreneurs who are in the early stage of building their ventures. Community developers and leaders need to be closely linked with the SBDC and other regional providers of training and counseling. Microenterprise programs often combine training with micro-lending programs, providing opportunities for entrepreneurs who are self-employed or have very few employees to get the assistance they need.

Counseling programs may be more appropriate for entrepreneurs who have already developed a business plan but need assistance with specific aspects of the business. For example, an entrepreneur might need assistance in accessing export markets or understanding the licensing requirements for operating a commercial kitchen. These types of questions are best addressed through the services of a business counselor who works one-on-one with the entrepreneur.

Whatever training or counseling programs are used, community developers should keep several things in mind:

- Entrepreneurs need to be able to understand where to get these services so they don't become frustrated as they try to get help. Communities can help by having a "one-stop shop" for entrepreneurs where they can obtain a directory with information about service providers including an explanation of the specific services offered by each provider.
- Training and counseling programs should be user-friendly. Aspiring and start-up entrepreneurs may still have their "day jobs" and need to be able to take a class or meet with a counselor in the evenings, on weekends, or even online.
- A community developer needs to know enough about an entrepreneur and her skills to steer her toward the training program or business counselor that can best meet her needs.

## Potential – youth

Youth represent a large population of potential entrepreneurs. In rural communities, where retaining young people is a vital concern, entrepreneurship offers a promising tool. National surveys indicate that most high school students want to start their own business but do not feel prepared to do so (Markley et al. 2005b).

Since the necessary motivation is present for many young people, the community leader's task is to provide the potential entrepreneur with opportunities to learn new entrepreneurial skills. An important first step is engaging young people in conversations about their dreams and what businesses they are interested in. Youth have at least three specific needs as potential entrepreneurs. Those needs are to become engaged with the community, to learn what the business world is like, and to develop the skills set to start a new enterprise. Civic entrepreneurship, networking and mentoring, and entrepreneurial training are support mechanisms that meet these needs.

### Civic entrepreneurship

Many young people have a desire to serve and be proud of their community. Engaging youth as civic entrepreneurs is an excellent way to build their loyalty to the community and to give them practical experience as civic leaders. For example, in Miner County, South Dakota, Future Business Leaders of America (FBLA) students conducted a "community cash flow study" of local residents' spending habits and attitudes toward local business. The youth were able to help community members understand the impact of buying locally and to inform local merchants about what people wanted. The county enjoyed a 41.1 percent increase in sales in the year following the study! Young people remain active in community development through the Miner County Community Revitalization organization and other local groups (Markley et al. 2005a).

### Networking and mentoring

How can young people learn about business opportunities? Networking with entrepreneurs and business owners is an excellent way to open the door to entrepreneurship. A mentoring or apprenticeship program which matches a business owner to a young person in a one-on-one mentoring relationship gives the potential entrepreneur the chance to see how business is created in the real world. This, at least in part, may provide the young person with the insight and encouragement for choosing the entrepreneurial path. A first step in getting a mentoring program going in your community might be to invite community entrepreneurs into the classroom. Entrepreneurs who tell their stories in the schools are providing the role models these young people need to envision an entrepreneurial future for themselves.

### Entrepreneurial training and experience

How can youth acquire the skills they need to become successful business entrepreneurs? Traditionally, K-12 education has focused on preparing students to be good workers rather than successful entrepreneurs. However, this is beginning to change as more schools are offering activities, such as Junior Achievement, that provide youth with the knowledge and practice of business. Some schools have gone even further by incorporating notable entrepreneurship training programs into the curriculum. A successful example is the Rural Entrepreneurship through Action Learning (REAL) program which is now being used in 43 states and foreign countries. In the REAL program, youth create viable businesses that generate on average 2.2 jobs and sometimes move out to a permanent site in the community (Corporation for Enterprise Development). Another well-established curriculum is the National Foundation for Teaching Entrepreneurship's BizTech program, offered in high schools throughout the county (www.nfte.com).

Youth entrepreneurship camps are another approach to bringing youth together with experienced businesspeople for intensive training and guidance in starting a new business. For example, the youth entrepreneurship camp sponsored by the University of Wisconsin Small Business Development Center includes 40 hours of training in real-world business skills, team building, leadership development, financial management, verbal communication, and business etiquette. Participants also learn how to successfully negotiate for business materials, set goals, and recognize real business opportunities (University of Wisconsin).

## Business owners – lifestyle

These entrepreneurs may have successful ventures, whether a main street café or a family medical practice, but they often do not have the motivation or capacity to grow. However, it is important for community leaders to know who these entrepreneurs are and to identify whether some existing business owners develop the motivation to grow or change their business in some way. Often the motivations of lifestyle entrepreneurs may change unexpectedly when a son or daughter returns to town and wants to be involved in the business; when the business owner is faced with an opportunity such as the potential to expand into an adjacent store front on the main street, or to take the business "online."

Business owners who are motivated to grow their businesses need many of the same support services as start-up entrepreneurs: networking opportunities; training to build their skill sets; one-on-one assistance with specific business issues such as creating a new website, developing e-commerce tools, or tapping new markets. Community leaders need to stay in touch with business owners throughout the community so that they can identify entrepreneurial talent and provide the support for that talent to thrive.

## Entrepreneurs – growth

Growth-oriented entrepreneurs are already succeeding and are driven to grow their ventures. They see new markets, profit centers, products, or services to be created. They hold the promise of growth with the associated benefits of profits and job creation. Exceptional, even among the growth-oriented ventures, are the high-growth businesses. They represent just 4 to 5 percent of all American businesses. High-growth companies are achieving sustained growth rates of 15 percent or more each year and are doubling in size every five years.

Obviously, growth-oriented entrepreneurs, especially high growth, are gold to a community. Not only are they providing jobs and income to the community now but their success will bring an expansion each year into the future. Community developers want to provide them with the best support available. What support do the growth-oriented entrepreneurs need? They are past the discovery and start-up stages and do not need to sit through a 12-week course on how to start a business.

Since growth-oriented entrepreneurs are often interested in developing a new product or cultivating a new market, they may need knowledge of capital sources or assistance with marketing or expanding production. They may need to expand the management team to encompass new skills. As their success has grown, their support needs have changed from broad forms of general assistance to very specific, targeted business information needs. To effectively support these growth-oriented entrepreneurs,

support services should focus on customized assistance, higher order assistance, and networking.

### Customized assistance

As the growth-oriented entrepreneurs' needs become more targeted and specific, the support must evolve as well. Customized assistance provides one-on-one assistance in response to very specific questions from the entrepreneur. The service provider might provide market research to help the entrepreneur better understand a new market (such as that provided through "economic gardening"). A university textile lab might develop a prototype product for a manufacturing entrepreneur to market to prospective customers. This assistance is designed to respond to the specific needs of the entrepreneur as he actively grows the business.

### Higher order assistance

General practice attorneys and business counselors often have the qualifications needed to help a start-up entrepreneur with legal or financial questions. However, in order to support growth-oriented entrepreneurs, higher order services may be required. The growth-oriented entrepreneur who is developing a new product may need a patent attorney. A rapidly growing entrepreneurial venture may need the capital and expertise that a venture capitalist can provide. If these services are not available locally, community developers need to develop a network of external service providers, often in nearby urban areas, which can be tapped to help these growth-oriented entrepreneurs.

### Networking

Whether you are assisting aspiring, start-up, or growth-oriented entrepreneurs, the importance of networking doesn't change. The sophistication of the network may increase as the growth-oriented entrepreneur seeks specialized knowledge to help the business expand. The informal breakfast group may give rise to a more formalized network, perhaps organized by the Chamber of Commerce or business association. One of the best examples of a sophisticated entrepreneurial network exists in North Carolina – the Council for Entrepreneurial Development (www.cednc.org). A good resource as you think about building an entrepreneurial network is the guide *Hello, My Business Name Is ... A Guide to Building Entrepreneurial Networks in North Carolina* (Pages 2006).

*Expanding/strengthening the management team*

Growth-oriented entrepreneurs often begin to face time and ability constraints in the management of a growing enterprise. A family member serving as part-time bookkeeper might work well for a start-up but a growing venture may need a chief financial officer. The entrepreneur himself may be able to manage employees when the business is family run but a growing venture may need a human resources director. Building this new management team can often be a challenge for entrepreneurs. It is hard to share power in an organization, especially one you have built from the ground up. However, it is even harder to be successful without an experienced team. These entrepreneurs need help recognizing their staffing needs and finding the right people to fill those needs. For example, Kentucky Highlands, an entrepreneurial support organization, helped one of their entrepreneurs find a CFO, even to the point of conducting interviews and recommending candidates (Markley et al. 2005b).

## Entrepreneurs – serial

Serial entrepreneurs make up another category of entrepreneurial talent. These are individuals who start, grow, and often sell one business and then begin the creative process all over again in another business venture. Serial entrepreneurs often start five or more businesses in their lifetime. Although they make up only 2 to 3 percent of all entrepreneurs,

they have a unique experience and are important to identify in your community.

Finally, we have already mentioned civic entrepreneurs. They identify opportunities for improving the community or addressing a problem such as the need for recreational opportunities for youth. They share similar motivations as business entrepreneurs but they live by a different bottom line. Research by the Center for Rural Entrepreneurship indicates that communities with high rates of civic entrepreneurs are the kinds of communities that also create a high quality of life. These are the environments in which business entrepreneurs can thrive (Markley et al. 2005b).

It is important for community developers and leaders to keep in mind that community support makes a difference. The more supportive the community, the larger the number of prospective entrepreneurs who will move ahead to start new enterprises. Furthermore, the more advanced the system of entrepreneurial support, the more likely growth businesses are to expand and create more local jobs. Fairfield, Iowa is an excellent example of how even a small community (population 9500) can energize its entrepreneurs (see Case study).

## How can your community become "entrepreneur-friendly"?

There is no single set of actions that a community should adopt to become entrepreneur-friendly. However, the RUPRI Center for Rural Entrepreneurship (Markley et al. 2005b) has identified three levels of support – basic, advanced, and high performing – that can help you begin to think about creating an environment in your community that is supportive of entrepreneurs.

## Basic support

Investment in a basic support package is the starting point to building a broader and more sophisticated

## CASE STUDY: FAIRFIELD, IOWA: AN ENTREPRENEURIAL SUCCESS STORY

In 1989, a group of entrepreneurial businesspeople formed the Fairfield Entrepreneurs Association (FEA). The FEA was designed to increase the success rate of start-up companies and nurture companies in the second stage of development; that is, after they had generated over $500,000 in annual revenues. The FEA supports entrepreneurs through recognition and awards, acceleration of second-stage companies, and mentoring and networking activities for entrepreneurs. Fairfield is characterized by extensive sharing of information on the "how-to's" of business start-up financing and marketing. It has developed a pool of shared wisdom and experience and a culture of guarded openness about business ideas. The result is great synergy among entrepreneurs, overlapping, and sometimes copy-cat business models.

For example, an entrepreneur named Earl Kappan, after several other start-up ventures, started a company that became known as "Books Are Fun." The company sold bestselling hardcover books via book fairs in schools, hospitals, and businesses across the United States. With financing from an outside investor, the company grew rapidly and developed a network of sales representatives across the country. A synergy occurred when local author Marci Shimoff proposed a collaborative venture to Kappan to explore the potential for distributing a book through Books Are Fun. The book was *Chicken Soup for the Mother's Soul.* Books Are Fun tested the book and started buying the title in lots of tens of thousands, propelling the book to the top of bestseller lists.

Other Fairfield residents developed Chicken Soup topics that went on to become *Chicken Soup for the Pet-lover's Soul, Chicken Soup for the Gardener's Soul, Chicken Soup for the Veteran's Soul,* and more. Approximately 13 Chicken Soup authors live, or formerly lived, in Fairfield. Another resident designed covers for the Chicken Soup series and has become a sought-after book jacket designer. Another Fairfield resident is a leader in self-publishing enterprises. The original enterprise Books are Fun also did well. When it grew to more than 500 employees and $400 million in annual revenue, it was purchased by Reader's Digest for $380 million.

There are many more examples of success in Fairfield among financial services, e-commerce, telecommunications, and art-based businesses. Burt Chojnowski, President of the Fairfield Entrepreneurial Association, attributes much of the success to the networking and sharing among businesses. Chojnowski states that "having access to this specialized knowledge, especially about funding opportunities, was clearly a competitive advantage for entrepreneurs and improved the financial literacy and sophistication of Fairfield entrepreneurs" (2006: 3–4).

The FEA, in conjunction with other Fairfield organizations, offers additional support to entrepreneurs. Entrepreneurial boot camps and an up-to-date library are available, including specialized information for "art-preneurs," civic entrepreneurs, and "food-preneurs." Entrepreneurship training has been expanded to include workshops for youth entrepreneurs. The results have been amazing for a small town. Since 1990, equity investments of more than $250 million have been made in more than 500 start-up companies, generating more than 3000 jobs.

Source: Chojnowski (2006).

community support system for entrepreneurs. To provide a basic level of support in your community, leaders need to:

- Address any issues related to creating a positive entrepreneurial climate such as favorable attitudes toward entrepreneurship and good quality of life, and strong infrastructure such as roads, utilities, and telecommunications. The greatest entrepreneurship development program operating in a weak climate with poor infrastructure will come up short.
- Take stock of your current access to appropriate business services (e.g., legal, marketing, production, financial, accounting). Access to the right services is important. Remember: if these services are not available within the local community, as is sometimes the case in rural areas, they may be accessed over long distances using today's technology.
- Create a focus on entrepreneurs, both business and civic. Creating a focus on entrepreneurs might include raising the awareness level of community residents and leaders about the role of entrepreneurship within the community. Going a bit further, a community might identify entrepreneurs and provide periodic recognition for their contributions to the community.
- Provide regular opportunities for networking and mentoring. Entrepreneurs themselves indicate that the most important support they can receive is a combination of networking with other entrepreneurs and access to mentors.

## Advanced support

Once the basic elements of a support system are in place, a community can consider a number of advanced activities to further energize entrepreneurs. Remember: more advanced support doesn't mean that things should become more complicated for the entrepreneur. Massive directories and complicated pathways for entrepreneurs to access support can be counterproductive. We urge communities at this level to create some kind of simple organization

(probably using existing organizations) to ensure that entrepreneurial support efforts are understandable, easy to access, and seamless.

Leaders providing advanced support typically:

- Link closely to the regional Small Business Development Center office or other provider of entrepreneurial training and counseling. Another option is to offer an entrepreneurial training resource such as FastTrac, NxLeveL, or REAL (Rural Entrepreneurship through Action Learning) if these programs are not being provided by others in the community. These programs are particularly helpful to start-up and early stage businesses.
- Ensure that an entrepreneur has access to appropriate financial capital beyond that provided by local banking institutions. An important step may be the formation of micro-lending programs for smaller start-up entrepreneurs and revolving loan programs for growing and restructuring businesses.
- Implement programs that increase local entrepreneurs' awareness of and access to new markets. In rural areas particularly, entrepreneurs may need assistance to develop strong skills in identifying market opportunities and assessing the commercial feasibility of various opportunities. Sending delegations to conferences, trade shows, and trade missions are all good ways to increase market awareness.
- Encourage programming that introduces youth (the younger the better) to entrepreneurship. Young people are a driving cultural force in our nation and communities. Sooner or later, these same young people will form the backbone of our economies and communities. Creating opportunities for young people to engage in venture and community building is critically important.

## High-performing support

To be a high-performing community that is optimally supporting entrepreneurs requires considerable community commitment and investment. High-performing communities are characterized by:

- Using strategies that offer customized help to the full range of local entrepreneurs. This might involve entrepreneurial coaches to work one-on-one with aspiring or start-up entrepreneurs. It might involve the city and library providing extensive and specialized market information to growth-oriented entrepreneurs, as the City of Littleton, Colorado does. This customized help may require significant investment by local community organizations.
- Building on current financing resources by creating area-based "angel" investment networks and pathways to more traditional venture capital resources (which may be external to the community). Sooner or later, growing ventures need more sophisticated forms of capital including access to equity capital. As entrepreneurial deals emerge and grow, the ability to help these ventures meet their capital needs is critical to keeping these businesses within the community.
- Integrating entrepreneurial opportunities into the core curricula of their K-16 educational systems. Trying to engage youth in entrepreneurship via extracurricular activities usually brings only marginal benefits.
- High-capacity organizations dedicated to supporting entrepreneurs. These entrepreneurial support organizations, such as the Fairfield Entrepreneurial Association, are rooted in communities and provide a comprehensive and sophisticated package of support that energizes start-up entrepreneurs and develops entrepreneurial growth companies.

Relatively few communities meet the standards for a high-performing support environment. Places like Fairfield, Iowa, and Littleton, Colorado, come close. Many more rural communities are providing advanced support to their entrepreneurs and even more have the basic elements of support in place.

## Minority entrepreneurship

The desire to start a business extends across racial and ethnic lines. In fact, there has been a recent surge in entrepreneurship among minorities in the United States, particularly among well-educated African Americans. Research shows that blacks are 50 percent more likely to engage in start-up activities than whites. Hispanic men are about 20 percent more likely than white men to be involved with a start-up, although the difference isn't statistically significant.

Education significantly predicts nascent entrepreneurship, particularly for blacks and Hispanics. Approximately 26 out of every 100 black men and 20 out of every 100 Hispanic men with graduate education experience report efforts to start a new business. This compares to 10 out of every 100 white men with graduate education experience (Ewing Marion Kauffman Foundation 2002).

Venture capital investing in minority enterprises is very profitable. An analysis of 24 venture capital funds making 117 minority-oriented investments found the average investment per firm was $562,400; the average gross yield per firm was $1,623,900 generating an average net return of $1,061,500. (Bates and Bradford 2003).

Communities should be aware of the high propensity of minorities, especially recent immigrants, to start new businesses. This presents an opportunity to communities, but only if community leaders understand and serve their minority entrepreneurs. Many ethnic minority entrepreneurs try to serve their own ethnic market. This is likely the market that they know best, but it can be a limiting factor. The community entrepreneurship team should provide minority entrepreneurs with the information to serve a broader market if they so desire. This can be done by engaging with existing minority networks and by building links between them and the traditional business networks. Again, the importance of building broad networks is important to success.

## What is the next step for your community?

If there is little current support for entrepreneurs in your community, you can start the ball rolling by forming a team of leaders and interested citizens to lead the entrepreneurial effort. It may be that an existing organization, such as the local economic development group, has an interest in entrepreneurship and could be the umbrella organization for the new team.

Once the team has come together, one of its first actions would be to examine what the community is currently doing to foster entrepreneurship. The RUPRI Center for Rural Entrepreneurship suggests using the *Rate Your Community's Support for Entrepreneurs* tool to get a sense of how your community is currently supporting entrepreneurs (see Box 15.3). It is unlikely that a community will score high on every category but the tool may suggest what it is doing well and areas where it can improve.

Next, the team should identify the entrepreneurial talent in your community. You can classify entrepreneurial individuals into the categories that we discussed previously (e.g., potential, business owners, entrepreneurs). The RUPRI Center for Rural Entrepreneurship has excellent tools for identifying entrepreneurs and determining which type(s) of entrepreneurs you want to focus on (www.energizingentrepreneurs.org).

Once your group has focused on the type of entrepreneurial talent you will be supporting, plan visits to each and every entrepreneur on your list. It is critical to build your initiative on the needs and wants of entrepreneurs. Conversations with entrepreneurs may reveal common needs and concerns, and clarify specific actions for the community team to take.

Next, it is important to identify the assets and resources in your community that can help entrepreneurs. Assets may be broken down into three categories: programs for entrepreneurs, business services, and capital (Markley et al. 2005b). Try to be specific and identify known resources that fit into the categories. Ultimately, these identified resources may be organized into a resource directory that the community can use to trigger assistance for entrepreneurs, once needs and opportunities are identified.

Finally, take action! Keep in mind that entrepreneurs need better networks, not simply more programs. It is vital to establish relationships with entrepreneurs and to be responsive to their needs. Adding more complexity and more layers of service is not going to lead to success. Instead, having a clear pathway to available services and helping to establish networks will bring your community closer to being entrepreneur-friendly.

The references at the end of this chapter provide useful resources as you take these next steps in your community.

## Conclusion

Communities are recognizing that entrepreneurship is an important strategy of community economic development. However, as has been seen in this chapter, entrepreneurs are not all the same. Communities must be flexible and responsive to meet the varying needs of entrepreneurs. It is essential that support be tailored to fit the needs and wants of entrepreneurs, rather than based on satisfying an external agency or funding source.

All types of entrepreneurs benefit from networking and mentoring opportunities. Indeed, other entrepreneurs may be the most valuable source of information and ideas! The example of Fairfield, Iowa illustrates the power of networking and mentoring in igniting an entrepreneurial economy.

As the chapter discusses, the support that communities choose to provide entrepreneurs covers a wide spectrum. Basic support provides a positive climate and infrastructure.

Advanced support provides training, capital and access to new markets. Finally, the highest order of support promotes customized assistance, "angel" investment networks, and entrepreneurial curricula in local schools.

Entrepreneurship has always been the engine that brings new vitality to places in the U.S. and around the world. Yet it has been largely overlooked as a

## BOX 15.3 RATE YOUR COMMUNITY SUPPORT FOR ENTREPRENEURS

*Directions:* Use this worksheet to explore and rate the ways that your town helps to identify, support and nurture entrepreneurs.

| | Not at all | | Somewhat | | Very strong |
|---|---|---|---|---|---|
| Community clubs or school activities that promote entrepreneurship. | 1 | 2 | 3 | 4 | 5 |
| A Chamber of Commerce or development corporation helps local businesses get started and supports existing business expansion. | 1 | 2 | 3 | 4 | 5 |
| Public recognition or acknowledgment for business achievement. | 1 | 2 | 3 | 4 | 5 |
| A program to identify and recognize entrepreneurs in the area. | 1 | 2 | 3 | 4 | 5 |
| Inter-generational mentoring by business owners and managers. | 1 | 2 | 3 | 4 | 5 |
| Internship opportunities for local youth and young adults returning from college. | 1 | 2 | 3 | 4 | 5 |
| Networks linking entrepreneurs to capital, new employees, and trategic partners. | 1 | 2 | 3 | 4 | 5 |
| An environment that values and supports young people who are starting new businesses. | 1 | 2 | 3 | 4 | 5 |
| Entrepreneurial education as part of the K-12 curriculum. | 1 | 2 | 3 | 4 | 5 |
| An information resource center or person to help entrepreneurs develop their enterprises. | 1 | 2 | 3 | 4 | 5 |
| Access to affordable and professional. legal, accounting and consulting services. | 1 | 2 | 3 | 4 | 5 |
| Participation in a business expansion and retention program. | 1 | 2 | 3 | 4 | 5 |
| Access to financing resources supporting start-ups and expansions. | 1 | 2 | 3 | 4 | 5 |
| Locally available entrepreneurship training. | 1 | 2 | 3 | 4 | 5 |
| A micro-enterprise development program. | 1 | 2 | 3 | 4 | 5 |

Source: Markley et al. (2005a).

development strategy. Community developers will be surprised at the potential of local entrepreneurs. As the community learns to support and honor entrepreneurship, remarkable and surprising things may happen.

## Keywords

Entrepreneur, civic entrepreneur, aspiring and start-up entrepreneur, potential youth entrepreneurs, potential entrepreneurs, youth entrepreneurs, business owners, lifestyle entrepreneurs, innovation, small business development centers, economic gardening, micro-enterprise.

## Review questions

1 How would you define "entrepreneur?"
2 What are some of the characteristics of the entrepreneurial "spirit?"
3 What are the different types of entrepreneurs?
4 What are the needs of these different types of entrepreneurs that the community can provide to support them?

**CASE STUDY: USING COMMUNITY DEVELOPMENT PRINCIPLES TO INCREASE ENTREPRENURIAL SUPPORT: THE NORRTALJE PROJECT IN SWEDEN**

Research on a collaborative initiative in Norrtalje, Sweden, a municipality of approximately 50,000 residents one hour north of Stockholm, illustrates how community development principles can be used to improve the local business climate for entrepreneurs. Professor Amy Olsson of the Royal Institute of Technology examined the effectiveness of collaborative dialogue in building interpersonal and inter-organizational networks to increase the availability of financing and professional support services to small businesses in Norrtalje.

Professor Olsson's case study draws on the extensive literature of collaborative planning (sometimes referred to as collaborative dialogue) from community planning theory. Collaborative planning is defined by its emphasis on common exploratory learning achieved through "authentic dialogue" among diverse and independent parties. The characteristics of authentic dialogue include no hidden agendas or withholding of key information, no misleading or manipulating other parties, open and honest discussions and sharing of information, and non-structured, free-flowing conversations. Often an impartial, trusted facilitator can help with the process. Collaborative dialogue promotes social capacity building as participants develop trust and social bonds as well as a shared sense of purpose and accomplishment.

Community and economic developers and stakeholders in Norrtalje not only desired to improve the availability of small business financing through local commercial banks, but also to strengthen relationships among small businesses and banks so that the latter could serve as a referral source or even a mentor for small businesses seeking assistance. There were two general types of barriers working against this goal: structural/organizational issues (e.g., banks couldn't grant low-collateral loans) and social/relational issues (e.g., bankers and entrepreneurs came from different backgrounds and didn't "speak the same language").

The situation in Norrtalje attracted the attention of several university and government organizations, which teamed up with local stakeholders to further the collaborative dialogue process and to create and fund new local facilitating organizations to mentor small businesses and help increase the availability of local support services. More effective collaborative dialogue among key individuals and organizations and actions by the newly created facilitating organizations helped ameliorate some of the structural/organizational barriers (for example, a collateral-free "growth loan" financial instrument was created). They also helped bridge some of the social/relational issues. Through extensive personal interviews, Professor Olsson documented that the relevant parties, including entrepreneurs, banks, and professional service providers achieved a higher propensity to cooperate, an increased level of trust, and a better understanding of common issues.

The Editors

Source: Olsson, A.R. (2008) "Collaboration to Improve Local Business Services," *International Journal of Bank Marketing*, 26(1): 57–72.

5  What are ways a community can become entrepreneur-friendly?

## Note

1  This chapter draws heavily from research and fieldwork by the RUPRI Center for Rural Entrepreneurship (www.energizingentrepreneurs.org), and their book, *Energizing Entrepreneurs: Charting a Course for Rural Communities*. The Center's goal is to be the focal point for efforts to stimulate and support entrepreneurship development in rural U.S. communities.

## Bibliography and additional resources

Bates, T. and Bradford, W. (2003). *Minorities and Venture Capital: New Wave in American Business*, Ewing Marion Kauffman Foundation. Available online at http://www.kauffman.org/pdf/minorities_vc_report.pdf (accessed April 28, 2008).

Chojnowski, B. (2006) "Open-source Rural Entrepreneurship Development," *Rural Research Report*, Illinois Institute for Rural Affairs, Western Illinois University, 17 (2). Available online at http://www.iira.org/pubsnew/ publications/IIRA_RRR_660.pdf (accessed September 27, 2006).

Corporation for Enterprise Development, *Real Entrepreneurial Education*. Available online at http://www.cfed.org/focus.m?parentid=32&siteid=341&id=341 (accessed September 26, 2006).

Ewing Marion Kauffman Foundation (2002) *The Entrepreneur Next Door; Characteristics of Individuals Starting Companies*. Available online at http://www.kauffman.org/pdf/psed_brochure.pdf (accessed April 28, 2008).

Markley, D., Dabson, K. and Macke, D. (2005a) *Energizing an Entrepreneurial Economy: A Guide for County Leaders*, RUPRI Center for Rural Entrepreneurship National Policy Brief. Available online at http://www.energizingentrepreneurs.org/content/chapter_8/supporting_materials/1_000123.pdf (accessed September 26, 2006).

Markley, D., Macke, D. and Luther, V. (2005b) *Energizing Entrepreneurs: Charting a Course for Rural Communities*. RUPRI Center for Rural Entrepreneurship. Available online at http://www.energizingentrepreneurs.org/content/chapter_3/tools/1_000027.pdf (accessed December 4, 2007).

National Commission on Entrepreneurship (2001) *High Growth Companies: Mapping America's Entrepreneurial Landscape*. Available online at www.publicforuminstitute.org/nde/sources/reports/2001-high-growth.pdf (accessed September 26, 2006).

Pages, E. (2006) *Hello, My Business Name Is ... A Guide to Building Entrepreneurial Networks in North Carolina*, Council for Entrepreneurial Development. Available online at http://www.entreworks.net/whatsnew/06/HelloMyBusinessName.pdf (accessed October 1, 2006).

Reynolds, P.D. et al. (2000) *GEM 2000 Global Executive Report*. http://www.gemconsortium.org/about.aspx?page=global_reports_2000 (accessed September 26, 2006).

Rural Policy Research Institute (RUPRI) Center for Rural Entrepreneurship (2006) "Energizing Your Economy Through Entrepreneurship," *Assessment Series: Understanding Entrepreneurial Talent*. Available online at http://www.energizingentrepreneurs.org/content/chapter4/tools/1_000220.pdf (accessed May, 2006).

Sirolli, E. (1999) *Ripples from the Zambezi: Passion, Entrepreneurship, and the Rebirth of Local Economies*, Gabriola Island, BC: New Society Publishers.

Smilor, R.W. (2001) *Daring Visionaries: How Entrepreneurs Build Companies, Inspire Allegiance, and Create Wealth*, Cincinnati, OH: Adams Media Corporation.

University of Wisconsin School of Business. Small Business Development Center. Available online at http://exed.wisc.edu/sbdc/ (accessed September 26, 2006).

# 16 Tourism-based development

## Deepak Chhabra and Rhonda Phillips

Tourism has long been recognized as a community and economic development strategy to bring in revenues. As one of the fastest growing industries in the world, many communities are seeking ways to tap into this vast and productive industry to capture local community and economic development benefits. This chapter focuses on presenting models and processes of tourism-based development along with applied approaches such as the Main Street program, popular culture, heritage, and other specialty approaches.

## Introduction

The tourism industry has grown to become one of the largest economic activities in the world with an estimated 200 million jobs worldwide and accounting for over 10 percent of global gross domestic product (World Travel and Tourism Council 2005). This is "big business" – in the US and Canada alone, the industry is worth over one trillion dollars and second only to health care in the US as the largest industry (Nickerson 1996). Tourism also accounts for nearly 12 percent of all consumer spending (Smith 1995). In both developed and developing countries, it is one of the fastest growing industries in the world for all sizes of communities on the continuum from rural to urban.

What exactly is the tourism industry? According to Nickerson's *Foundations of Tourism* (1996), it is the action and activities of people taking trips to a place or places outside their home community for *any purpose* except daily commuting to and from work and includes business travel as well as travel for pleasure. It comprises many different organizations from both the public and private sectors including:

- *Transportation providers:* Scheduled and charter air services; maritime services; ground service (auto, bus, rail transport).

- *Attractions:* Recreational, cultural, historic, scenic (may be any type from festivals to state parks and museums to entire communities).
- *Food services:* Restaurants and food stores.
- *Accommodation:* Ranges from hotels/motels and bed and breakfast inns to campgrounds and private homes.
- *Travel distributors:* Travel agents, tour operators, wholesalers, ticket agents.
- *Tourism promoters:* State tourism offices promote economic development through marketing and promotion as do chambers of commerce and convention and visitor bureaus at the regional and local levels.
- *Supporting infrastructure:* In addition to transportation, other necessary infrastructure including utilities and other services.
- *Land managing agencies:* Federal, state, and local parks, recreational area and forest area management (Inskeep 1991; Nickerson 1996).

**Tourism industry:**
The mix of interdependent businesses and organizations directly or indirectly serving the traveling public (Nickerson 1996).

Why would a community want to pursue a tourism-based development approach? There are several compelling reasons: (1) tourism can provide both direct and indirect economic benefits; (2) tourism can generate various social/cultural benefits; and (3) tourism can help achieve environmental conservation objectives (Inskeep 1991). On the other hand, tourism can negatively impact communities on all three of these fronts – economic, social/cultural, and environmental. Projects or programs that are poorly designed and implemented without proper planning and consideration of impact result in negative outcomes. Tourism-based development that is too successful may result in undesirable outcomes such as stressing infrastructure limits or causing conflicts between visitor and resident populations. Yet despite these concerns, tourism can be a beneficial development strategy for communities when approached correctly.

While many may think of tourism as synonymous with marketing, it goes far beyond that as the prior list illustrates. Marketing is absolutely essential, yet building the tourism industry in a community or region requires much more. Planning, development, and policy issues transcend the concerns of marketers – for example, tourism as a method of community development demands better information and tools to assist with product development, planning, and assessment of impacts (Smith 1995). Given that the tourism industry comprises numerous players and is fairly complex in terms of including marketing as well as planning, development and policy concerns, among others, how do communities tap into this vast and productive industry to capture local community and economic development benefits? Presented below is information on tourism models that may be used as a basis for considering tourism-based development followed by a discussion of strategic approaches that may be implemented by communities.

## A review of tourism planning models

A conventional approach to tourism planning involves encouraging the introduction of a few hotels, ensuring transportation links, and developing a promotional campaign, but with the advent of mass tourism and its visible social and environmental impacts, long-term planning and controlled tourism development has become crucial. Inskeep (1998) provides several reasons for the need to optimize tourism benefits while preventing or mitigating the externalities resulting from tourism. These include lack of experience at both the governmental and private sector level, the need to integrate the multi-sectoral and fragmented components of tourism, the careful match of tourism markets and products with minimum impact on the local environment, sustainable use of natural and cultural resources, development of new tourism areas and specialized training facilities, and the need for comprehensive and integrated planning that can be easily related to tourism policy. Careful planning will "allow for future flexibility of new development and revitalization of older tourism areas" (Inskeep 1998: 17).

Successful planning strategy "goes far beyond schemes to maximize profits" and, therefore, should include a "detailed, 'on the ground' outline so as to determine how each of the factors affecting the success of a tourism destination should be developed" (Goeldner and Ritchie 2006: 438). In brief, previous literature suggests the following objectives for sustainable tourism development: formulate a framework to enhance local resident quality of life; develop multiple-use infrastructure and recreational facilities which cater to locals and tourists alike; ensure appropriate developments that are reflective and sensitive to the features unique to the area; take the cultural, social, and economic values of the host community into consideration; and optimize visitor satisfaction (Goeldner and Ritchie 2006; Inskeep 1998; Mason 2003).

As Butler (1991) has suggested, tourist destinations evolve and go through life cycles and eventually decline if preventive and sustained measures are not taken for their rejuvenation. Appropriate planning can help anticipate evolution of tourist destinations and decline can be prevented through careful marketing and management techniques. Hence sustained development and protection of cultural, economic, and natural environments call for tourism stakeholders to coordinate and harmonize their activities. Planning follows policy formulation, and both are required to ensure that a destination retains its competitive and sustainable traits (Goeldner and Ritchie 2006).

## Types of tourism planning models

Several approaches to the planning of tourism are discussed in the tourism literature. These include the (1) sustainable development approach; (2) systems approach; (3) community approach; (4) integrated planning approach; (5) comprehensive planning approach; (6) continuous and flexible approach; and (7) functional tourism systems approach. A brief description of the first six are provided as follows, with more details of the last model presented.

1 *Sustainable development approach:* This approach evolved in the 1970s and sought to improve quality of life and meet human needs and wants. Development emphasis was to enhance the living conditions of people and involved addressing concerns such as life expectancy, nutritional status, educational achievement, and spiritual benefits (Pearce et al. 1990). Further, the sustainable development approach suggests strategizing tourism development with a long-term focus in such a way that it benefits future generations (Fyall and Garrod 1998). Today, under the umbrella of sustainable development, tourism planners usually assemble elements from different approaches to meet their specific needs. Despite vast application of sustainable tourism planning in tourism literature in both public and private sectors, much of it is theoretical (Holden 2000; Mason 2003). A sobering observation is made by Butler (1991) in his argument that all tourism tends to lead to mass tourism regardless of its origin and aim, and suggests that it is very

important to plan sustainable tourism in a way that it "withstands the test of time" (Fennell 1999).

2 *Systems approach:* A system has generally been defined as a set of elements interrelated to each other. "It is like a spider's web; you touch one part of it and reverberation will be felt throughout" (Mill and Morrison 1985: 19). According to Murphy (1985), the systems approach takes a broad view and provides flexibility by formulating appropriate strategies as applicable to different levels or elements of tourism. Another advantage suggested by Gravel (1979: 123) was "programmed learning and continuous improvement" by examining, defining, and synthesizing using a holistic perspective.

It is interesting to note that the tourism system has been viewed differently by various authors. For example, Leiper (1990) divided tourism into three elements: tourists, geographical elements, and the tourism industry, with the tourist as an active actor. Leiper's (1990) model consisted of the destination of origin, transit route regions which provide the transportation link, and the tourism destination. This model represented the whole environment in its "human, socio-cultural, economical, technological, physical, political, and legal" form (Tosun and Jenkins 1998). Mill and Morrison (1985) suggested four main components: market, travel, destination, and marketing. According to Harssel (1994), demand and supply components of tourism constitute the tourism system, and these are linked by transportation and promotion. Gunn (1994) also proposed a demand- and supply-based tourism system and argued that the supply side is dynamic because it is subject to multiple external influences.

3 *Community approach:* This focuses on decentralization and facilitation of coordination between different stakeholders of tourism. This approach stemmed from the increasing emphasis on democratization and gained significance when political power shifted from the central govern-

ment to states, cities, towns, and neighborhoods, thereby giving voice and empowerment to local communities to address their own problems and find appropriate solutions. This approach calls for better participation between the tourism industry and the local residents. It is hoped that the involvement of local residents in decision-making processes will facilitate better working partnerships between the host communities and the travel and tourism industry.

4  *Integrated planning approach to tourism development:* This approach requires inputs from multiple stakeholders and aims to "facilitate integration of tourism into overall sub-national, national, and international tourism markets" (Tosun and Jenkins 1998: 105). According to Tosun and Jenkins, this approach strives to achieve a balance between supply and demand. The authors suggest two important implications of this approach: integrating various components of the tourism industry, and taking a macrosystem perspective by taking socio-cultural, economic, political, and environmental factors into consideration.

5  *Comprehensive planning:* This approach calls for a master plan to guide tourism development. This view focuses on the overall tourism situation and takes all components of tourism into consideration to promote sustainable planning. The main purpose is to facilitate coordination among all sectors relevant to the development of tourism (Bannon 1976).

6  *Continuous and flexible approach:* Continuity here refers to "ongoing research and feedback and flexibility implying adaptable planning and being responsive to rapidly changing environments" (Tosun and Jenkins 1998: 106). This view is supported by Getz (1986: 32) in his statement "constant evolution and reassessment of directions will make the planning process more adaptable to changes in the tourism system, and will lead to greater ability to predict such changes."

7  *Functional tourism system model:* The underlying basis of this model is the functioning tourism system with demand and supply as the main driving mechanisms. According to Gunn and Var (2002), the demand and supply components are both complex and bring forth multiple issues that require the attention of all planners and developers for successful tourism to happen. In other words, the tourist industry is a system of major interdependent components that are closely linked together. It is not just airlines or hotels – it is a combination of many other vital elements such as attractions, information centers, and other modes of transportation, including pedestrianism. The key to correct tourism development is the ultimate match between supply and demand. Figure 16.1 provides a brief overview of Gunn's tourism system model.

Gunn defines demand in terms of market segmentation on the basis of motivation, financial ability, time, and physical ability of a tourist. The supply side comprises all goods and services that are needed to meet the demand. According to Gunn and Var (2002), supply consists of five main components. The first component is attractions which make up the "energizing power unit of the system" because they "entice, lure, and stimulate"

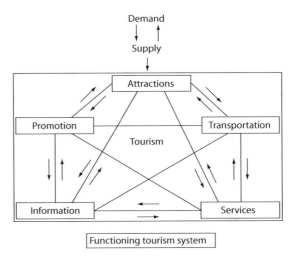

■ *Figure 16.1* **Tourism system model (source: Gunn and Var 2002: 34)**

interest in travel (Gunn and Var 2002: 41). Attractions are often classified by ownership, resource foundation, and trip duration. Attractions can have both rural and urban characteristics, and they gain by clustering. Linkage between attractions and services is important. The service sector, on the other hand, needs to have local as well as non-local markets for long-term viability. Because tourism businesses depend on urban infrastructure and accessibility, services also gain from clustering. The service businesses are highly dependent on attractions. The transportation component includes both tourist and local travel in the transport sector plans, in addition to an emphasis on promoting pedestrianism (Gunn and Var 2002).

Next, the information sector should be represented by a local tourist bureau together with other sources that provide traffic, weather, and religious information. It is important for information centers to provide hospitality training to local residents. Information systems should not be promotion centered, and visitor centers are essential to provide a hub for information and to serve effectively as a distribution channel. Information is needed on several topics such as weather conditions, physical demands, customs, social contact, host privacy, foods, etiquette, religious beliefs, history, politics, communication, facilities, services, and health. Finally, Gunn and Var (2002) include a promotion sector component and call for an integrated promotional plan undertaken either by the visitor center or a combination of businesses.

According to Gunn and Var (2002), several external factors influence the supply components of the tourism system such as resources, organization leadership, finance, labor, entrepreneurship, community, competition, and governmental policies. "The causes of travel to a destination are grounded in the destination's resources, natural and cultural, and the attractions that relate to them" (Gunn and Var 2002: 59). Entrepreneurs are needed due to the dynamic nature of tourism so that they can foresee new developments and creative ways to manage existing developments. Next, capital and labor capacity are required for both public and private tourism development. Finally, competition exercises a significant influence on the tourism system, and community is a highly significant influence through its perceptions and support. Market performance, market share, and market growth are important factors in competitiveness, but they cannot determine or maintain longevity unless sustainable measures in the context of the local community and environment are pursued. "Political, environmental, religious, cultural, ethnic, and other groups in an area can make or break the proper functioning of the tourism system" (Gunn and Var 2002: 67).

Gunn's model of a functioning tourism system has been used extensively for tourism planning by several destination marketing organizations. Several examples of destination planning cases are provided in the academic literature such as the Hopi in Arizona, Bouctouche Bay in New Brunswick, Canada, and Sanibel Island in Florida. The Hopi in Arizona aimed to achieve a balance between economic development through tourism planning and the integrity of their culture. They created their own code of ethics that imposed several limitations on tourists such as restricting the recording and/or publication of tourist observations, providing guidelines regarding appropriate tourist clothing and behavior, and respect for community restricted areas and events. In their effort to strive toward cooperative solutions to issues related to tourism development, the community has been considered successful.

The Bouctouche Bay community in New Brunswick was considered successful in its tourism planning because it considered community participation as integral to the development process. The Bouctouche Bay Ecotourism Project was considered successful worldwide in its efforts to achieve public–private partnerships with its emphasis on resource protection and tourism interpretation. This project promoted educational workshops that involved entrepreneurial/community groups and included leadership by professionals as well as field trips and the sharing of ideas for the future (Gunn and Var 2002: 277). The local stakeholders, community groups, government, industry, and aboriginals were all encouraged to participate in the planning process so that a shared vision and plan could be created.

The example of Sanibel Island is also noteworthy because it illustrates an ideal participatory planning process involving the public such as the local residents, government officers, consultants, traffic specialists, attorneys, and economic specialists. Community involvement was one of the key components of the plan – this is one of the main requisites of sustainable planning and development.

## Starting the tourism planning process

As demonstrated in the models, community participation is a crucial factor for the long-term viability of tourism. In an effort to counter friction resulting from tourism's negative impacts, many researchers are suggesting that tourism-dominated/interested communities should plan their evolution more systematically, thereby taking into account residents' attitudes and perceptions about its growth at the outset (Reid et al. 2004: 624). The significance of community involvement is to provide a voice for those involved in or impacted by tourism, to make sound decisions concerning use of local knowledge, and to reduce possible conflicts between tourists and members of the host community. The emphasis in the tourism planning models on the need for coordination and collaboration among all sectors involved leads to the need for planning. Simply put, planning provides the opportunity to envision what a community wants and how to get there. Without it, there is no direction to achieve desirable outcomes – the community will have to accept whatever comes its way. Planning, including project and program development coordination, is crucial to ensure that all elements are developed in an integrated manner to serve both tourist needs and the desires of the host community (Inskeep 1991). All planning processes, whatever the focus, always begin with an inventory or research phase and

cycles to an evaluation or monitoring of outcomes phase. Note that the process is continuous and is never "over" because change is the only constant! Further, planning is not an end in itself but rather a necessary means for achieving outcomes that people value in their communities (Bunnell 2002). The following process is recommended to start a community on its way to achieving tourism-based development:

1  *What do we have?* Inventory assets (people; organizations; cultural/heritage, natural, financial, and built resources) and contexts (political, economic, social, environmental) of the community. This is the research phase and may include a variety of sources and tools such as surveys, focus groups, asset mapping, and so on.
2  *What do we want?* At this point, the all-important vision as a guide to seeing what could happen is crafted by stakeholders – those in the community that have an interest in helping achieve a more desirable future. Belief is a powerful tool and can inspire a community to achieve remarkable outcomes. The vision should be bold enough to inspire and realistic enough to attain.
3  *How do we get there?* This stage is about developing the plan that is a guide with specifics for achieving the vision and includes goal statements and actions. Most importantly, it selects the strategies or approaches desired – will a program such as Main Street or a theme-based approach be used? What trends should be looked at and which markets are sought? It also identifies which organizations or groups of collaborators will tackle the tasks and action items. Collaborative efforts typically work best, but in some cases it takes a "champion" to start the efforts, and others will join in later.
4  *What have we done, and what do we need to do now?* Monitoring is critical to see if the above steps are working; if not, then adjustments and revisions are needed. Because the nature of this process is continuous, it provides feedback for refining ongoing activities as well as starting new initiatives.

## Getting from here to there: selecting approaches

One of the most critical components of selecting approaches is determining who is going to carry it forward. Many communities develop a collaborative arrangement with both public and private entities; others focus on one organization to serve as the umbrella agency for the community's efforts. Here are examples of approaches that a community can use to implement tourism-based development. Some like to use a combination or create an entirely unique approach with the bottom line being: *whatever works and reflects the desires of the community!*

## Main Street approach

In 1980, the National Trust for Historic Preservation established the Main Street Program to focus on traditional downtown revitalization. It transcends historic preservation and includes community and economic development, infrastructure, and marketing elements. The approach includes not only aspects of encouraging tourism in the form of visitors to shop and spend downtown, but also more broad-scale community development outcomes. The Program is successful, with an average return of $35 reinvested for every $1 spent on revitalization. While not all can be a designated Main Street community, there is much to be learned from looking at their development strategy, the Main Street Four-point Approach™:

1 *Organization* involves getting everyone working toward the same goal and assembling the appropriate human and financial resources to implement a Main Street revitalization program. A governing board and standing committees make up the fundamental organizational structure of the volunteer-driven program. Volunteers are coordinated and supported by a paid program director as well. This structure not only divides the workload and clearly delineates responsibili-

ties, but also builds consensus and cooperation among the various stakeholders.

2 *Promotion* sells a positive image of the commercial district and encourages consumers and investors to live, work, shop, play, and invest in the Main Street district. By marketing a district's unique characteristics to residents, investors, business owners, and visitors, an effective promotional strategy forges a positive image through advertising, retail promotional activity, special events, and marketing campaigns carried out by local volunteers. These activities improve consumer and investor confidence in the district and encourage commercial activity and investment in the area.

3 *Design* means getting Main Street into top physical shape. Capitalizing on its best assets — such as historic buildings and pedestrian-oriented streets — is just part of the story. An inviting atmosphere, created through attractive window displays, parking areas, building improvements, street furniture, signs, sidewalks, street lights, and landscaping, conveys a positive visual message about the commercial district and what it has to offer. Design activities also include instilling good maintenance practices in the commercial district, enhancing the physical appearance of the commercial district by rehabilitating historic buildings, encouraging appropriate new construction, developing sensitive design management systems, and long-term planning.

4 *Economic restructuring* strengthens a community's existing economic assets while expanding and diversifying its economic base. The Main Street program helps sharpen the competitiveness of existing business owners and recruits compatible new businesses and new economic uses to build a commercial district that responds to today's consumers' needs. Converting unused or underused commercial space into economically productive property also helps boost the profitability of the district.

(National Trust for Historic Preservation 2005: 47)

Many of the Main Street applications have been met with success, and often tourism is a vital component of that success. The Program offers many sources of information and help to communities and should be a first stop for a community embarking on revitalization efforts. The Program has worked with over 1600 communities since its inception. Mansfield, Ohio; Danville, Kentucky; Elkader, Iowa; and Enid, Oklahoma are just a few examples of Main Street communities.

A similar program operates in the UK, The Townscape Heritage Initiative (THI) which combines both heritage resources oftentimes with the goal of increasing tourism for community development outcomes. It is a programme of the Heritage Lottery Fund focused on sustainable conversation and beneficial reuse of heritage resources at the community level. By providing funding and support services, THI serves as a catalyst for social, cultural, and economic regeneration – helping make communities more livable and citizens enjoy a higher quality of life. It operates on principles such as those of the Main Street program – bringing together residents to explore how best to use

---

## BOX 16.1 ELEMENTS FOR ATTRACTING HERITAGE TOURISTS

### Historical and archaeological resources

How many of these historic resources are in your community?

- Museums
- Historic properties listed on the National Register of Historic Places or otherwise designated landmarks
- Historic neighborhoods, districts, or even entire towns or villages
- Depots, county courthouses, or other buildings that have historic significance because of architectural or engineering features, people associated with them, or contribution to historic events
- Bridges and barns, battlefields or parks
- Fountains, sculptures, or monuments

### Ethnic and cultural resources

In addition to historic and archaeological sites, consider the ethnic or cultural amenities of your community through exploring the traditions and indigenous products by:

- artists
- craftspeople
- folklorists
- other entertainers such as singers or storytellers
- galleries, theaters
- ethnic restaurants or centers
- special events like re-enactments or ethnic fairs
- farming and commercial fishing and other traditional lifestyles.

Source: National Trust for Historic Preservation (1999: 22).

resources for achieving desirable community outcomes.

## Heritage tourism

Whether it is ethnic activities or styles of architecture, using heritage as a basis for building tourism-based development can be very rewarding. Preserving heritage and tourism have not always been congruent ideas, but in the recent past it has become one of the most popular forms of tourism with heritage travelers typically staying longer and spending more than any other type of tourist. The benefits of this approach are numerous including new opportunities for preserving and conserving an area's heritage while giving the visitor a learning and enriching experience. It can begin with using the community's built heritage, such as the case with Cape May, New Jersey, or Eureka Springs, Arkansas. Other communities use their ethnicity to develop their approach, as with Solvang, California, with a population of just over 5000 and over 2 million visitors per year. How did they do it? Solvang had a rich Danish history and parlayed this into becoming the "Danish Capital of America" complete with festivals and other special events (Phillips 2002). This Danish culture mecca for tourists encourages new buildings and rehabilitations to use ethnic-style architecture for the built environment as well.

## Natural/recreational tourism

Many communities or regions have a bounty of natural resources that lend themselves as a basis for tourism. The U.S. national park system and the individual states' park systems are major destinations for natural and recreational tourists each year. However, at the community level this type of tourism can yield benefits as well. While the scale may be different, the appeal is still high. For example, are there landscapes or even transportation features with unique recreational opportunities such as a canal or

railroad corridor? The Rails to Trails program has become extremely popular and can attract numerous recreational tourists. Combining trails with venues at community locations is a successful approach. Three small towns in south Florida recently did just this – by combining efforts and invoking the help of the U.S. Army Corps of Engineers, a walking/biking trail has been constructed around the levy of a large lake that borders the towns. Visitors have increased to nearly 300,000 on the trail and represent an opportunity to visit the towns when venues are offered.

## Popular culture approaches

Popular culture runs the gamut from visual arts and music to filmmaking. Using these elements as a basis for building a tourism-based development strategy is speculative yet very exciting. The entertainment industry can play an important role in stimulating development outcomes. This approach is not without risks though, since some communities find that the intensity of energy and resources required may be too high to sustain a "popular" appeal. Others find it has worked very well. There are three major popular culture strategies that have been used with success.

### 1 Arts-based

Some view the arts as a powerful catalyst for rebuilding *all* aspects of a community, not just the economic sphere, in terms of venues to attract tourists. Thus the benefits of this strategy can be far-reaching beyond tourism alone. The number and types of tourist venues based on art are tremendously diverse. Some communities use themselves as the palette for the venue, painting murals on the walls of their buildings or incorporating public art on a major scale into the community – Stuebenville, Ohio; Toppenish, Washington; and Loveland, Colorado, are such examples. Care must be taken with this approach so that the commercialization or theming aspect does not threaten community ambience. In addition to the community as a visual arts palette, others attract artists to live and display their work, often in arts districts including converted warehouses or other obsolete industrial facilities. This strategy is not limited to only large urban areas. Small communities, such as Bellows Falls, Vermont (population 3700), have used its downtown structures to develop artist live-work space, attracting visitors to the galleries for shows and to shop. The following considerations are offered to communities interested in pursuing an arts-based strategy:

- *General support for the arts.* Citizens and local government officials need to recognize that a healthy arts presence is a vital part of community infrastructure and is important in terms of community development. Citizen participation approaches in community decision making should be used to further build support.
- *Seek out untapped resources.* Local governments may have more resources than direct funding that can be used to support arts-based businesses and other activities which in turn attract visitors to the community. Examples include rent-free facilities from a variety of sources such as school classrooms and auditoriums, commercial warehouses, conference centers, or vacant retail spaces.
- *Integrate the support of the arts with community development benefits.* Whenever possible, the community should strive to link benefits with arts-based activities. For example, artisans could participate in programs such as placing their art in public venues.
- *Maximize resources through community sharing.* The centralization of facilities and resources is a significant factor in the success of arts-based programs. A centralized facility, such as a production studio, gallery, office or retail space, may be used by numerous groups to provide cost savings.
- *Adopt a flexible approach to arts support.* All artists are different and need different kinds of support and assistance. Business management assistance to arts entrepreneurs is usually a critical need in communities (Phillips 2004).

Building an arts-based development strategy will result in increased tourism, particularly for shopping in galleries and visiting during special venues such as public art exhibits.

## 2 Music

Music has long been used as a basis for attracting tourists. One only has to look as far as Nashville or New Orleans to see the widespread impact that music and related activities can have on an area. Branson, Missouri is often cited as an example of how entertainment and particularly the music side of the business can be used as a basis for building a tourist dependent economy. One strategy that seems to work particularly well is when ethnic or heritage music is coupled with a tourism development strategy. Mountain View, Arkansas, with a population of less than 3000, found that its remote location in the Ozark Mountains has not prohibited it from becoming a tourist destination because of its music heritage. The "Folk Music Capital," Mountain View holds the Arkansas Folk Festival as well as offering other venues and special events. Eunice, Louisiana is now known as the "Prairie Cajun Capital" owing to its Zydeco and Cajun music heritage that is kept alive and well with special venues and a regular television and radio show, the Rendez-Vous Des Cajuns – the Cajun version of the Grand Ole Opry. Eunice attracts many visitors to its town and to its renovated Liberty Theater as a spectacular venue for music offerings. Its visitors are not limited to the region – the British Broadcasting Company has visited several times to film the shows which has prompted many British visitors to journey over to see the live performances.

## 3 Filmmaking

While filmmaking and television production can be an attractive strategy for economic development outcomes, it can have mixed results as a tourism venue. Typically after filming, the community attracts tourists for a period of time to see the venues. However, the attraction may fade after time if the community does not offer complementary venues. Winterset, Iowa, the location of the film *Madison County*, experienced a great influx of tourists for several years after the film was shot. While the initial crowds have gone, the community has incorporated several other venues to attract tourists over the long term such as the Covered Bridges Festival since the county is home to six remaining bridges. If a community can continue to attract periodic filming activities, then the tourist interest stays steadier.

## Corporate culture approaches

Corporate culture has long fascinated Americans. A fast-growing tourist venue are factory tours and corporate museums, with currently over 300 such attractions in the U.S. and more expected. When Binney and Smith Inc.'s Crayola Factory museum opened in Easton, Pennsylvania, over 300,000 visitors came in the first year. Take a look at corporate communities such as Hershey, Pennsylvania – "Chocolate Town, USA" – or Austin, Minnesota – "Spam Town, USA" – to see the appeal. These venues are mostly private sector initiatives, but when combined with integrated community efforts as a development strategy, they can be very effective. The drawbacks are the risks of tying a community's image to one corporation, the issue of who has control over or influence on major decisions affecting the venues, and the resulting community outcomes. Yet the benefits include bringing in external revenues and visitors that may otherwise not have ventured to the community.

## Surrealistic approaches

This category is in a world of its own. This type of strategy is of an incongruous nature that defies or exceeds common expectations. It emerges when a community does not have inherent natural, cultural, ethnic, or built resources to use as the basis for its tourism development approach, so it creates them with energy and imagination. In other instances, a community may already have the genesis of a resource and then take it to a different level altogether. The results can be astoundingly successful in some cases or a flop. It takes a high measure of courage (and sometimes audaciousness) to embark on this strategy, but the flip side is that the rewards can be very high if done well. There are three elements to consider in this strategy:

1 *Shock value.* Is the approach completely unnatural to the point that visitors are not attracted? Or is it workable because of the delight of the visitors when encountering the unexpected?
2 *Scale.* This concerns the ability to develop the created environment (the venues) to a plausible level so that there is enough to see to draw visitors. One venue, depending upon its size, may not do it. Combining venues and carrying the theme throughout can work better.
3 *Scope.* Communities must find the strategy or approach that incorporates activities and venues appealing to a market. In other words, it needs to be targeted to an audience.

A case in point: tiny Cassadaga, Florida has found its niche in its surrealistic strategy by building its reputation as a center for providing palm readings and other mystic services. Its shock value is seeing an entire community that revolves around these services in a rather removed location; its scale is appropriate as most businesses in town have some elements of the theme; and its scope is focused on tourists who enjoy a dip into the mystical side of life. Perhaps the best case to illustrate this approach though is Helen, Georgia. Located in the north Georgia mountains, this small community developed itself as an alpine village with strict design standards, venues, and special events year round. Transforming itself from a dying town with only six businesses open to now over 200 and over a million visitors per year, Helen is a great story of how creativity can lead to a total transformation of an economy.

## Conclusion: bringing it all together

Tourism-based development represents an excellent opportunity for communities to capture benefits of one of the largest and fastest-growing industries in the world. This chapter has presented reasons why tourism is an appropriate strategy as well as some of the drawbacks. The major models for tourism were presented along with processes for planning tourism-based development with a focus on selection of the approaches that a community may use. Identifying, designing, and implementing a tourism-based development process is complex, multifaceted, and requires a large measure of energy, resources, and commitment. It necessitates that communities go beyond marketing efforts to consider all facets – from planning and project and program management to gauging outcomes and adjusting strategies as needs and desires change through time. The stakes are high because tourism-based development has the potential to dramatically change a community in many ways and aspects. This is particularly important for communities with heritage assets. Issues of authenticity should be considered as well as protecting fragile resources. On the other hand, tourism-based development can induce positive changes, and with the correct approach and well-designed strategies, the results can be tremendously beneficial.

## Keywords

Tourism planning, arts-based development, heritage and cultural resources, eco-tourism, community concept and place marketing.

## Review questions

1 What is the "tourism industry," and why is it important to community development?
2 Describe at least two major tourism planning models.

3 Select a community with which you are familiar, and outline how one or more of the approaches could be applied in that context.

## Bibliography

Bannon, J. (1976) *Leisure Resources: Its Comprehensive Planning*, Englewood Cliffs, NJ: Prentice-Hall.

Bunnell, G. (2002) *Making Places Special, Stories of Real Places Made Better by Planning*, Chicago, IL: American Planning Association.

Butler, R. (1991) "Tourism, Environment, and Sustainable Development," *Environmental Conservation*, 18: 201–209.

Fennell, D. (1999) *Ecotourism: An Introduction*, London: Routledge.

Fyall, A. and Garrod, B. (1998) "Heritage Tourism: At What Price?," *Managing Leisure*, 3: 213–228.

Getz, D. (1986) "Models in Tourism Planning: Towards the Integration of Theory and Practice," *Tourism Management*, 7(1): 21–32.

Goeldner, C. and Ritchie, J. (2006) *Tourism Principles, Practices, Philosophies*, New Jersey: John Wiley & Sons.

Gravel, J. (1979) "Tourism and Recreational Planning: A Methodological Approach to the Valuation and Calibration of Tourist Activities," in W.T. Perks and I.M. Robinson (eds) *Urban and Regional Planning in a Federal State: The Canadian Experience*, Stoudsburg: Penn, Dowden, Hutchinson & Ross, pp. 122–134.

Gunn, C. (1994) *Tourism Planning*, 3rd edn, London: Taylor & Francis.

Gunn, C. and Var, T. (2002) *Tourism Planning*, 4th edn, London: Taylor & Francis.

Harssel, J. (1994) *Tourism: An Exploration*, Englewood Cliffs, NJ: Prentice-Hall.

Holden, A. (2000) *Environment and Tourism*, London: Routledge.

Inskeep, E. (1991) *Tourism Planning, An Integrated and Sustainable Approach*, New York: John Wiley & Sons.

Inskeep, E. (1998) *Tourism Planning. An Integrated and Sustainable Development Approach*, New York: John Wiley & Sons.

Leiper, N. (1990) "Tourism System," *Occasional Paper 2*, Auckland, NZ: Massey University.

Mason, P. (2003) *Tourism Impacts, Planning and Management*, Amsterdam: Butterworth Heinemann.

Mill, R. and Morrison, A. (1985) *The Tourism System*, Englewood Cliffs: Prentice-Hall.

Murphy, P. (1985) *Tourism: A Community Approach*, New York: Methuen.

National Trust for Historic Preservation (1999) *Getting Started, How to Succeed in Heritage Tourism*, Washington, DC: National Trust.

National Trust for Historic Preservation (2005) "The Main Street Four-Point Approach™ to commercial district revitalization." Available online at http://www.main-street.org/content.aspx?page=47andsection=2 (accessed June 22, 2006).

Nickerson, N.P. (1996) *Foundations of Tourism*, Upper Saddle River, NJ: Prentice-Hall.

Pearce, D., Barbier, E. and Markandya, A. (1990) *Sustainable Development, Economics, and Environment in the Third World*, Aldershot: Edward Elgar.

Phillips, R. (2002) *Concept Marketing for Communities: Capitalizing on Underutilized Resources to Generate Growth and Development*, Westport, CT: Praeger Publishers.

Phillips, R. (2004) "Artful Business: Using the Arts for Community Economic Development," *Community Development Journal*, 39(2): 112–122.

Reid, D., Mair, H. and George, W. (2004) "Community Tourism Planning of Self-assessment Instrument," *Annals of Tourism Research*, 31(3): 623–639.

Smith, S.L.J. (1995) *Tourism Analysis*, London: Longman Group.

Tosun, C. and Jenkins, C. (1998) "The Evolution of Tourism Planning in Third-World Countries: A Critique," *Progress in Tourism and Hospitality Research*, 4: 101–114.

Webb, S. (2005) "Strategic Partnerships for Sport Tourism Destinations," in J. Higham (ed.) *Sport Tourism Destinations*, Oxford: Butterworth-Heinemann: pp. 136–150.

World Travel and Tourism Council (2005) "Blueprint for New Tourism," *Media and Resource Center*. Available online at www.wttc.org/news (accessed June 16, 2006).

## Suggested further reading and sources of information

Americans for the Arts. Available online at http://www. americansforthearts.org/issues/comdev/index.asp. Great resource! Books such as the *Field Guide* list arts organizations throughout the US See also the economic impact calculator feature – "Arts and Economic Prosperity Calculator" – under the Community Development section.

Community Arts Network. Available online at www.communityarts.net/links. See the "Links for Arts and Community Development" for examples of projects and activities throughout the US.

*Downside Up* is an inspiring film about how art can change a community and serve as the basis for revitalization. The National Endowment for the Arts has partnered with New Day Films to bring this documentary to each state by loaning the film for viewing in your community. To obtain the film, go to www.downsideupthe-movie.com.

*Incubating the Arts: Establishing a Program to Help Artists and Arts Organizations Become Viable Businesses* by Ellen Gerl, National Business Incubation Association. Available online at www.nbia.org.

International Downtown Association. Good resources by topic for members. Available online at www.ida-downtown.org.

National Assembly of Local Arts Agencies (1996) *The Arts Build Communities*, Washington, DC: NALAA.

National Association for County, Community and Economic Development. Various resources and links. Available online at www.ncced.org.

National Endowment for the Arts. Available online at http://arts.endow.gov/. Various link, resources, and information on grants.

National Main Street Program of the National Trust for Historic Preservation. See "Success Stories" as sometimes these are themed in their redevelopment strategies. Available online at www.mainstreet.org.

*Niche Strategies for Downtown Revitalization* by David Milder (1997), New York: Downtown Research and Development Center.

Partners for Livable Communities. Available online at www.livable.com. Information on cultural-based community development.

*The 100 Best Art Towns in America*, 4th edn, by John Villani (2005), Woostock, VT: Countryman Press.

Travel and Tourism Research Association (TTRA). Numerous publications for travel and tourism information. Available online at http://www.ttra.com/.

"Turning Around Downtown: Twelve Steps to Revitalization" by Christopher Leinberger (2005) in *Research Briefs*, The Brookings Institution, pp. 1–21.

Urban Land Institute (2002) *Powerhouse Arts District*, Washington, DC: ULI.

World Tourism Organization (WTO). Resources for promoting tourism-based development. Available online at http://www.unwto.org/index.php.

# 17 Housing and community planning

## Joseli Macedo

This chapter discusses housing and its role in community planning. A few definitions introduce the topic and clarify common terms used by practitioners and researchers. An explanation of housing typology, density, and affordability provides a basic understanding of how these three issues affect housing and community planning. Housing and community planning varies widely worldwide; however, a review of the role of government in housing and a brief historic overview of housing provision in the US serves to provide one context. This chapter provides a narrative of housing delivery systems and assistance programs including public housing. It concludes with a discussion about future trends in housing and community planning.

## Introduction

Housing is the life of a community. Residential land use is the most prevalent land-use designation in most cities; thus, housing and community planning play a very important role in the development of an urban area. Communities need a variety of housing types. The housing stock of a community should comprise housing types that may be made available to different socioeconomic groups. In addition to housing typology, density and affordability are also important aspects that characterize communities and sometimes entire cities.

Government has played a role in facilitating and providing shelter, particularly to those individuals and families who need assistance. In recent times, nonprofits and the private sector have also gotten involved in housing provision. Whether government-sponsored or a result of public–private partnerships, housing assistance programs are available to fulfill the needs of low-income families.

## The current state of housing

The 2000 U.S. Census found the national homeownership rate to be 66.2 percent, which represented a 2 percent increase over the prior ten years in which the country experienced an unprecedented economic expansion. During this period, the nation's housing stock increased 13.3 percent. Most of this housing stock was in metropolitan areas, which also contained the largest number of rental units. Although 45.5 percent of all rental units were in the central cities of these metropolitan areas, 2000 marked the first year that the homeownership rate in central cities exceeded 50 percent.

According to the American Housing Survey, there are about 124 million housing units in the U.S., 15 million of which are vacant or for seasonal use. The typology of the housing stock presented in Table 17.1 shows that the prevalent housing type is the single-family, detached home (62.5 percent), followed by the multifamily unit (24.9 percent).

**Table 17.1** *Composition of the housing stock by type – 2005*

| Housing unit type | Number of units | Percentage |
|---|---|---|
| Single-family, detached | 77,703,000 | 62.5 |
| Single-family, attached | 7,046,000 | 5.7 |
| Multifamily | 30,999,000 | 24.9 |
| Manufactured | 8,630,000 | 6.9 |
| *Total housing units* | 124,377,000 | 100.0 |
| Vacant or seasonal | 15,505,000 | 12.5 |

Source: American Housing Survey 2005 (tabulated by author).

## Housing typology

Different types of housing are developed at different points in time. Usually, housing types are determined by demographics and markets. In the case of demographics, for instance, the detached, single-family housing type dominated developments in the 1950s and 1960s when the "nuclear family" was the prevailing type of family. More recently, the number of single-person households has increased, and smaller, often attached units are becoming a common type of housing in urban areas. Nonetheless, the prevalent housing type in the U.S. is still the single-family, detached home.

Housing trends usually reveal societal changes. In the past half century, household size has decreased considerably: from 3.33 persons per household in 1960 to 2.57 in 2005. Family size has also changed and families have gotten smaller. The average number of persons per family in 1960 was 3.67 and in 2005 it was 3.13. While both household and family size have decreased in the past 50 years, the percentage of one- and two-person households has increased. In 1960, only 13 percent of households contained one person; that percentage had doubled by 2000 (see Table 17.2).

Changes in demographics and family structure bring about changes in housing needs; people demand housing types that better fit their lifestyles and incomes. Box 17.1 lists the most common types of housing.

## Housing density

Density is an important aspect of urbanism; it affects planning in that it has urban design, urban management, and legal implications. Density can be expressed as demographic density – number of people within an area (i.e., persons per acre), or as housing density – number of housing units within an area (i.e., units per acre).

Rural areas have lower densities than urban areas. Within urban areas, there can be low-medium- and high-density neighborhoods. As with several aspects of community planning, density has advantages and disadvantages, some of which are listed in Box 17.2.

Some of the factors that affect urban density are: availability of land, development layout, lot size, street standards, infrastructure standards, housing typology, family size, and urban legislation.

**Table 17.2** *Demographic characteristics of households and families in the U.S., 1960 to 2005*

| | 1960 | 1970 | 1980 | 1990 | 2000 | 2005 |
|---|---|---|---|---|---|---|
| Persons per household | 3.33 | 3.14 | 2.76 | 2.63 | 2.62 | 2.57 |
| Persons per family | 3.67 | 3.58 | 3.29 | 3.17 | 3.17 | 3.13 |
| Percentage of one person households | 13 | 17 | 23 | 25 | 26 | 26 |

Source: U.S. Census Bureau (tabulated by author).

## BOX 17.1 HOUSING TYPOLOGY

Conventional, constructed housing
- Single-family, detached
- Single-family, attached
  - Row house
  - Town house
- Multi-family
  - Low-rise (1–4 stories)
  - Medium-rise (5–9 stories)
  - High-rise (ten or more stories)

Factory-built housing
- Manufactured
- Modulae
- Mobile homes and trailers

Group quarters
- Commercial
  - Dormitories and rooming houses
  - Hotels and motels
- Organizational
  - Fraternities and sororities
  - Seminaries and convents
  - Shelters
- Institutional
  - Boarding schools
  - Military bases
  - Jails

## BOX 17.2 ADVANTAGES AND DISADVANTAGES RELATED TO DENSITY

High density

Efficient infrastructure
Economies of scale
Consumer access
More social control
Efficient land use
Higher revenues
Urban vitality
Access to employment

Overburdened infrastructure
Criminal activity
Pollution
Greater risk of environmental degradation
Congestion and saturation

Disadvantages

Advantages

Less accessibility to services
High cost for service maintenance
Little interaction and social control
Less access and higher cost of transit
High rate of consumption of urban land and infrastructure

Low cost of basic sanitation
Less pollution
Peace and quiet

Low density

## Housing affordability

Housing needs to be affordable to every person in society; even families with higher incomes have to choose a home based on what they can afford. Sometimes market-rate housing is not affordable to some families. Ideally, a community should provide opportunities to families that cannot afford market-rate housing including middle- and low-income housing. Affordable housing no longer equates with low-income and public housing. With the gap between income level and housing costs growing wider, housing affordability has become an issue that affects almost all income levels, and is no longer an issue that only concerns the "poor."

Across the board, incomes have not kept pace with housing costs. Both renters and owners have had to resign themselves to the fact that the quality and size of house they are able to afford has declined. Otherwise, they have had to either supplement their incomes or reduce other expenditures to maintain their lifestyles. The need for government-assisted housing has increased in recent years because there are many gainfully employed families who cannot afford adequate housing in the U.S. today.

## Housing adequacy

Being able to afford a home is not the only problem American families face. The condition of affordable housing is also important. There are three problems related to housing adequacy: physical condition, overcrowding, and cost burden. Beyond adequacy,

the problem of homelessness is probably the most difficult to define and quantify.

Physical characteristics that determine whether a housing unit is in poor condition include not only the state of the basic structure of the house (walls, roof, windows), but other basic amenities for the exclusive use of the unit's occupants such as piped hot and cold water, flush toilet, and bathtub or shower. Housing in poor physical condition is usually referred to as "substandard." Attention to housing quality has broadened in recent years to consider neighborhood as well as individual unit characteristics. Neighborhood conditions may be the most serious problem of housing quality in the U.S. today. According to the 2005 American Housing Survey, 14 percent of all housing units presented some type of deficiency such as cracks and holes, faulty wiring, and so on. In addition, 37 percent of homeowners and 24 percent of renters reported some type of deficiency such as street noise, crime, odors, and other problems in their neighborhoods.

The Census Bureau defines a household as all the people occupying a housing unit but does not enumerate the number of families sharing a housing unit with relatives with whom they would prefer not to live. Overcrowding occurs when multiple families occupy the same housing unit and is more common among low-income families, particularly renters.

Cost burden is the proportion of income spent on housing. According to the 2005 American Housing Survey, the median housing cost burden was 20.7 percent. The median monthly housing cost was higher for owners than for renters and varied among regions, the lowest being in the South and the

**Table 17.3** Monthly housing expenditure as percentage of income, owners and renters, 2005

| Housing costs as percent of income | Owner occupied | | Renter occupied | |
|---|---|---|---|---|
| | With mortgage (%) | Without mortgage (%) | All renters (%) | Unsubsidized renters (%) |
| Less than 30 | 66.9 | 79.4 | 44.8 | 45.2 |
| Less than 50 | 19.3 | 9.2 | 25.2 | 24.5 |
| More than 50 | 13.8 | 11.4 | 29.9 | 30.3 |

Source: American Housing Survey 2006 (tabulated by author).

highest in the West. Although owners spent more on housing than did renters, renters had a higher cost burden than owners. While owners spent 28.4 percent of their incomes on housing, renters spent 19.6 percent (Table 17.3).

Homelessness is very difficult to quantify because the homeless comprise a range of household types, ages, and races. Different from common perception, the homeless are not all alcoholics and derelicts; some have serious health problems and no support from governments or social services. Causes of homelessness include poverty, unemployment, deinstitutionalization of the mentally ill, reduction of government programs, and lack of low-rent housing. The number of homeless people is significant; however, given the nature of homelessness, it is very difficult to know how many people are living in those conditions.

## The role of government in housing

A brief historical overview of the governmental role in housing follows. One attempt to shelter the poor in the US was the implementation of public housing projects. "Model tenements" were first suggested by Jacob Riis in *How the Other Half Lives* (1890). Prior to World War II, public housing was considered a success. Units were clean and safe and maintenance rules were very strict. There were long waiting lists, single parents and welfare recipients were not accepted, constituents were carefully screened, and most were gainfully employed. As part of the effort to overcome the Great Depression, the National Housing Act of 1934 created the Federal Housing Administration (FHA) with the goal of improving housing standards and providing a mutual mortgage insurance system. The public housing program was created by the Housing Act of 1937. In the late 1940s, a radical change in federal rules occurred – tenants with jobs had to leave the enormous government-subsidized high-rise projects which then housed the poor.

Housing supply increased rapidly after World War II and the rate of construction surpassed population growth. In the 1950s and 1960s, African-American migrants from the south streamed into big cities and were contained within these projects, thus avoiding residential integration. The danger in concentrating 100 percent of subsidized units, where people with similar problems lived in close proximity, is that critical mass could result in social upheaval. Guided by federal policies, local housing authorities were responsible for building and operating the disastrous large-scale urban projects.

In 1968, congress banned the construction of high-rise projects for only the poor. The first public housing buildings to be demolished were those of Pruitt-Igoe in St. Louis in 1972. In the mid-1990s, the Department of Housing and Urban Development (HUD) demolished 30,000 units in the worst high-rises built during the heyday of housing projects. The initial goal was to demolish 100,000 failing units. Unfortunately, HUD had, as it still does, an ever-shrinking budget so even the most laudable efforts were hindered by lack of funds. More recently, government has partnered with nonprofits and private corporations in an effort to achieve the goal of making housing accessible and affordable.

## Current approaches

Government can act to guarantee that households, similarly situated in social and economic terms, have equal opportunities to become homeowners. In Great Britain, Germany, France, and the Netherlands, where providing affordable housing does not mean serving primarily the very poor, more than 20 percent of the population lives in public housing, as compared to 3 percent in the U.S. Not until the 1930s, following the Great Depression, did the government get involved in the provision of affordable housing. There are two diametrically opposed points of view on the topic of government assistance. Some contend that government subsidies are not only inefficient but unnecessary, while others argue that the only way to make housing affordable is through government investment.

Those that are for a nonparticipatory government ground their argument in the long history of government failure to provide shelter through numerous federal programs. Research focusing on the delivery of affordable housing without government assistance has suggested that the private sector becomes a driving force. Government participation is limited to setting policies that allow the market to work freely and reduce the hurdles to development, be it through lowering acceptable standards or increasing the number of credit alternatives to finance shelter. Those who contend that housing must be made available as well as affordable through government participation – either by means of partial subsidies or complete financing – are of the opinion that a free market marginalizes people who do not fit the "standard" mold, since they have no foundation from which to start.

More often than not, affordable housing means that which can be occupied by people in the lower levels of income strata. A federal rule establishes that to be eligible for public housing a household's income must be less than 80 percent of the median income in the area. When discussing affordable housing one cannot overlook the problem of poverty and, radiating from it, all of the causal issues traditionally associated with low-income families such as unemployment and inadequate education. Most federal government policies are redistributional in nature, striving to combat poverty. Nonetheless, 15.3 percent of occupied households are below the poverty line; over 15 million households, according to the 2005 American Housing Survey. Some examples of redistribution mechanisms are cash assistance; education and training programs; and payment in kind, namely food stamps, medical care, and housing assistance.

The government may also have a role in how financial markets facilitate housing acquisition. There are three government entities created to provide liquidity to mortgage markets: the Government National Mortgage Association (Ginnie Mae), the Federal Home Loan Mortgage Corporation (Freddie Mac), and the Federal National Mortgage Association (Fannie Mae). Ginnie Mae is a government corporation, established by the Housing Act of 1968, to make mortgage funds available to moderate-income families using government-guaranteed mortgage-backed securities. Freddie Mac and Fannie Mae are government-sponsored secondary mortgage market enterprises with the authority to regulate and ensuring improved access to affordable housing to low- and moderate-income families.

While providing affordable housing to needy populations is but one of the issues that need to be directly addressed by federal policies, it is of such fundamental importance and critical consequence that striving for constant improvement is crucial. A policy based on systematic methods to establish need has the highest potential to become a successful policy.

## The role of nonprofits in housing

Following the disastrous failures of housing projects of the 1960s, nonprofit organizations started getting involved in housing assistance programs. Even those nonprofits, whose primary mission in the 1970s and 1980s was to foster economic development, started to develop much needed decent and affordable housing.

This movement gave rise to Community Development Corporations (CDCs) which were introduced by the Housing and Community Development Act of 1970. In time, housing became the CDCs' largest single program area and the housing programs developed by these corporations became an alternative to public housing. In fact, they have been so successful that today they have a stronger record in providing units for the poor than in the development of commercial property or business enterprises. Local CDCs, heavily financed by the federal government, own, develop, and manage subsidized housing. They follow a consistent model whose key points are: emphasis on security; keeping the size of each development manageably small; mixing the poor with the working class; screening prospective tenants; and expelling those who commit crimes or break the rules. However, they usually build houses that simply comply with minimum aesthetic and durability standards, which can sometimes reduce neighborhood strength.

## Housing affordability

Affordability is a crucial element of housing delivery, whether the government or nonprofit sector is involved in giving families access to housing, or whether families, regardless of income, are left to the market's devices. What does affordable mean? The classic government definition of "affordable" is that the expenditures in rent and utilities consume no more than 30 percent of a household's pre-tax income. Although this percentage has been used as a rule-of-thumb by consumers as well as by financial institutions to estimate "how much" housing one can afford, there is no scientific explanation for the "30 percent" figure.

Housing affordability can be achieved by different means. It can be made more affordable through innovative design, less stringent building and land development standards, lower financing costs, or improvements in other areas such as education and employment. These strategies will indirectly result in making shelter accessible to people by way of increasing their skills, marketability, and consequently, their income levels.

A number of organizations compute indices to measure the ability of "typical" families to afford a home. Some of these indices are:

• Affordability Index, calculated by the National Association of Realtors (NAR), measures the ability of a household earning the median family income to qualify for a conventional loan covering 80 percent of the median existing single-family home price in its area.

• Housing Opportunity Index, calculated quarterly by the National Association of Home Builders (NAHB), computes the median family income and the percentage of all homes sold during the quarter which a family earning the median income could afford.

The NAR Affordability Index shows how affordability has varied over the past 30 years (see Table 17.4). An Affordability Index of 100 indicates that the typical (median) family can afford to purchase the median-priced home; the low indices in the 1980s indicate that families could not afford a home. The Affordability Index has been declining for over 10 years but still indicates that families are able to afford purchasing a house.

The NAHB Housing Opportunity Index shows the increase in the national median price and consequent declining opportunity index over the past three years (see Table 17.5). An HOI of 50 percent, for instance, indicates that only 50 percent of homes sold in a given area were affordable to the typical (median) family.

Affordable housing is a crucial element in any comprehensive plan or improvement program, be it local or regional. In an attempt to promote a better quality of life for the entire population, regardless of income level, community planning must address the issue of housing and relate it to all other basic components of a plan such as health care, education, and

■ **Table 17.4** Housing Affordability Index

| Year | Median price of existing single-family home | Median family income | Composite affordability index |
|------|---------------------------------------------|----------------------|-------------------------------|
| 1975 | 35,300 | 13,719 | 123.5 |
| 1980 | 62,200 | 21,023 | 79.9 |
| 1985 | 75,500 | 27,735 | 94.8 |
| 1990 | 97,300 | 35,353 | 113.7 |
| 1995 | 117,000 | 40,612 | 132.4 |
| 2000 | 147,300 | 50,733 | 121.9 |
| 2005 | 219,000 | 55,823 | 111.8 |

Source: National Association of Realtors 2007.

**Table 17.5** Housing Opportunity Index

| Year | 2004 | | 2005 | | 2006 | |
|------|------|------|------|------|------|------|
| Quarter | HOI (%) | National median price | HOI (%) | National median price | HOI (%) | National median price |
| I | 61.2 | $187,000 | 50.1 | $225,000 | 41.3 | $250,000 |
| II | 55.6 | $204,000 | 45.9 | $241,000 | 40.6 | $250,000 |
| III | 50.4 | $225,000 | 43.2 | $253,000 | 40.4 | $248,000 |
| IV | 52.0 | $219,000 | 41.0 | $254,000 | 41.6 | $247,000 |

Source: NAHB 2007.

employment. Programs based on a comprehensive examination of current conditions and realistic projections have a better chance to succeed than those dictated by policies whose foremost interest is creating a final product, whether it suits the target population or not.

## Delivery of affordable housing

Different approaches have been adopted in the development of affordable housing programs. Two of the most traditional approaches are the supply-side and demand-side approach. In basic terms, supply-side solutions are public housing and the incorporation of options with the private providers, while demand-side solutions consist of improved financing or other subsidies and incentives such as negative income tax payments or rent certificates.

- *Supply-side approach:* the main characteristic of supply-side assistance programs is that the government builds or subsidizes new housing to be occupied by those who meet the established criteria. Supply-side policies generally increase housing consumption.
- *Demand-side approach:* the main characteristic of demand-side assistance programs is that existing housing stock is occupied by families who need shelter and receive government assistance. The government hands out coupons to be used as income supplements by those who meet the

established criteria to rent existing housing. This approach has been successfully used by nonprofit organizations.

Government programs established by supply-side policies, namely new public housing, are generally more expensive than those devised by demand-side policies, or cash payments. Demand-side policies tend to increase housing prices. The danger of surplus demand is that increasing housing affordability for all income levels lays the groundwork for all home prices to increase. Equilibrium can be eventually achieved because home builders and developers would recognize advantageous opportunities in building more, thus increasing the supply and subsequently driving prices down.

### Housing assistance programs

The U.S. Department of Housing and Urban Development (HUD), created as a cabinet-level agency in 1965, guides development in urban areas, implements national housing policies, allocates funding, and manages federal housing assistance programs. HUD's mission is to provide shelter and increase homeownership, ensuring a decent quality of life through economic and community development. Federal housing programs are often identified by statutory title or section numbers from the legislative measures that created them (see Box 17.3). That is why several

## BOX 17.3 SUMMARY OF SELECTED LEGISLATION AND RESPECTIVE PROGRAMS

| *Statute* | *Purpose* | *Programs* |
|---|---|---|
| National Housing Act of 1934 | Created the Federal Housing Administration (FHA) and the Federal Savings and Loan Insurance Corporation (FSLIC) | Title I, Title XI, Section 203(b), (h), (k); Section 204(g); Section 207; Section 213; Section 220; Section 221(d)(2), (d)(3), (d)(4); Section 223(e), (f); Section 231; Section 232; Section 234; Section 242; Section 245(a); Section 251; Section 255 |
| U.S. Housing Act of 1937 | Created the public housing program, established and funded the Public Housing Administration (PHA) | Section 8, Section 23, Section 24 (HOPE VI), Section 32, Section 34 |
| Housing Act of 1959 | Created programs to support the elderly and the disabled | Section 202, Section 811 |
| Fair Housing Act of 1968 | Prohibited discrimination in housing (also known as Civil Rights Act of 1968) | Title VIII |
| Housing and Urban Development Act of 1968 | Established rental and homeownership programs and partitioned Fannie Mae into Fannie Mae and Ginnie Mae | Section 3, Section 106, Section 235, Section 236 |
| Housing and Community Development Act of 1970 | Created the Federal Experimental Housing Allowance Program and Community Development Corporation | Title V (PD&R, PATH), Section 512 |
| Housing and Community Development Act of 1974 | Created CDBG grants | Title I (CDBG), Section 106, Section 107, Section 108, Section 109 |
| National Affordable Housing Act of 1990 | Created programs to empower the most needy | Title II (HOME), HOPWA, Shelter Plus Care, Section 808, Section 811, HOPE |
| Housing and Community Development Act of 1992 | Established Youthbuild, Low-Income Housing Preservation and Homeownership, and Energy Efficiency Mortgage programs | Section 184; Section 513; Section 542(b), (c) |
| Quality Housing and Work Responsibility Act of 1998 | Public housing reform | Title V |
| American Homeownership and Economic Opportunity Act of 2000 | Implemented changes to Section 8 and reduced regulatory barriers, among other provisions | Title III, Section 103 |
| American Dream Downpayment Act of 2003 | Authorized downpayment assistance to low-income first-time homebuyers | HOPE VI (reauthorization) |

housing assistance programs are known as Title I or Section 8, for instance.

Before HUD was created, the first federal program that attempted to improve the quality of life for those living in urban areas ended up hastening the decline of these areas. This was the Urban Renewal program, instated by the Truman Administration in 1949. The second round of urban programs began during the Kennedy Administration. They consisted of small, foundation-financed efforts that eventually became the base for the Community Action program called the "War on Poverty." This program failed mostly due to political reasons. Nonetheless, it created a paradigm for antipoverty programs such as the Community Development Block Grants during the Ford Administration and the Urban Development Action Grants during the Carter Administration. In the 1980s, the dominant idea was Enterprise Zones, renamed Empowerment Zones in the 1990s.

Two of the most significant pieces of housing legislation were the Housing and Urban Development Act of 1968 and the Housing and Community Development Act of 1974. The former established programs to make homeownership achievable for low-income families (Section 235) and increased the low-income rental housing stock (Section 236). More than 400,000 housing units were subsidized under Sections 235 and 236 in 1970. The latter provided Community Development Block Grants (CDBGs), which stipulated that cities and states could decide how they wished to spend funds that were allocated as block grants instead of categorical grant and loan programs. CDBG funds continue to help improve community facilities and services, revitalize neighborhoods, advance economic development, and many other activities.

Funds from formula grant programs, such as CDBG, can only be allocated to state and local governments that have a five-year Consolidated Plan and annual action plan. The Consolidated Plan must conduct needs assessments of low-income housing, homeless persons, and special needs populations, as well as analyze the local housing market. Other formula grant programs include the Emergency Shelter Grant (ESG), the HOME Investment Partnerships Program (HOME), and Housing Options for People With AIDS (HOPWA). Amounts for formula grant programs are determined by HUD. The formulas used include measures of need such as the extent of poverty; housing age and overcrowded conditions; and population growth rates in relation to other metropolitan areas. According to HUD, each of these formula grants has special characteristics and goals:

- *CDBG:* Must "benefit low- and moderate-income persons, aid in the prevention or elimination of slums and blight, or meet certain urgent community development needs." Activities may include property acquisition and rehabilitation or clearance, improvement of public facilities and infrastructure, provision of public services, homeownership assistance, assistance to for-profit businesses for economic development activities, and Section 8 assisted housing. The CDBG program started in 1974 and today is the largest and most continuously run of HUD's formula grant programs.
- *ESG:* Awarded to cities and states to "increase the number and improve the quality of emergency and transitional shelters for the homeless." Funds may be used to rehabilitate and operate facilities and to provide social services.
- *HOME:* Granted to states and other jurisdictions to implement "local housing strategies designed to increase homeownership and affordable housing opportunities for low- and very low-income Americans." Both owners and renters may use HOME funds for new housing construction or housing rehabilitation. HOME funds may not be used for public housing and jurisdictions must match them.
- *HOPWA:* Established by the AIDS Housing Opportunity Act, it is the only federal housing program providing rental housing assistance for low-income persons living with HIV/AIDS. States, local governments, and nonprofit organizations may compete for grants, which may be used to provide emergency and transitional housing, shared housing arrangements, community residences, and single-occupancy dwellings.

Other federal housing assistance programs include:

- *FHAP – Fair Housing Assistance Program:* Assists state and local agencies to enforce fair housing laws and coordinate them across jurisdictions, maintaining consistency with Title VIII of the Civil Rights Act of 1968 as amended by the Fair Housing Act of 1988.
- *FHIP – Fair Housing Initiatives Program:* Established by the Housing and Community Development Act of 1987, provides funding to prevent and eliminate discriminatory housing practices. Through several initiatives, state or local governments and private entities may participate in the implementation of programs.
- *LIHTC – Low-Income Housing Tax Credit:* Created by the Tax Reform Act of 1986, this has become the best tool for creating affordable housing in the US today. State and local LIHTC-allocating agencies issue an offset against income taxes (tax credit) for the acquisition, rehabilitation, or new construction of rental housing targeted to lower income households. According to HUD's national LIHTC database, 13,648 projects and 985,560 units were placed in service between 1995 and 2004, most of them targeting families.
- *RC/EZ/EC – Renewal Communities/Empowerment Zones/Enterprise Communities:* Enables public–private partnerships to revitalize communities and attract additional investment for economic and community development.
- *SHOP – Self-Help Homeownership Opportunity Program:* Makes grants to nonprofit organizations to provide and facilitate self-help housing. Home buyers contribute "sweat equity" and are assisted by volunteers; community participation is a requirement of the Program.

## Public housing

The public housing program dates back to the Housing Act of 1937. It was created "to provide decent and safe rental housing for eligible low-income families, the elderly, and persons with dis-abilities." Not all federally assisted housing is public housing. This program allows local jurisdictions to establish housing authorities that issue tax-free bonds to build, own, and operate low-cost housing. Public housing ranges in type from single-family houses to high-rise apartments. While in the US less than 2 percent of the population lives in public housing, in other countries of the industrialized world as much as 20 percent of the population lives in government-subsidized housing.

Public housing is limited to low-income families and individuals; eligibility is based on annual gross income, age and disability, and citizenship or immigration status. Low-income limits are set at 80 percent and very low-income limits at 50 percent of the median income for the area in which the applicant lives. The housing agency is responsible for checking the applicants' references to make sure they will be good tenants. Demand is often higher than HUD is able to meet, so housing agencies place prospective tenants on a waiting list; preference is given to families with the greatest needs.

As of 2007, there were approximately 1.3 million households occupying public housing, with 40 percent occupied by elderly persons, 43 percent by households with children, and 12 percent by disabled persons. The majority of families with children, 56 percent, are single-parent households and 88 percent of the elderly live alone. Almost half of public housing households are black, 48 percent, and 10 percent are Hispanic. Education levels are low; 58 percent of heads of households occupying public housing have not finished high school.

Some of the public housing programs include:

- *Capital Fund* – Funds are used by housing authorities to modernize public housing developments.
- *Demolition/disposition* – Funds are used to help eliminate old, run-down public housing.
- *Family self-sufficiency* – Incentives for communities to develop local strategies to help assisted families obtain employment; the intention is to enable the community to achieve economic independence and self-sufficiency.

- *Homeownership* – Sale of public housing to eligible residents or resident organizations; the intention is to encourage homeownership.
- *HOPE VI* – Revitalization of Severely Distressed Public Housing – also known as Urban Revitalization Demonstration (URD) program, was established in 1993 and encourages public housing agencies to partner with the private sector to create mixed-finance and mixed-income affordable housing. HOPE VI funds may be used for major reconstruction and rehabilitation, provision of replacement housing, planning and technical assistance, and supportive services. HOPE originally stood for Homeownership Opportunity for People Everywhere. HOPE I and III are inactive programs.
- *Housing Choice Voucher Program* – Established in 1983, housing vouchers became a permanent program through the Housing and Community Development Act of 1988. This program allows the sale of public housing to resident management corporations so that low-income families can have access to affordable privately owned rental housing. The program targets low- and very low-income families and is run by a public housing agency.
- *MTW* – Moving to Work – Similar to Family self-sufficiency, MTW allows housing authorities to give incentives to families to "become economically self-sufficient, achieve programmatic efficiencies, reduce costs, and increase housing choice for low-income households."
- *RHIIP* – Rental Housing Integrity Improvement Project – Targets HUD's high-risk rental housing subsidy programs.
- *ROSS* – Resident Opportunities and Self-sufficiency – Similar to Family self-sufficiency and MTW, provides funds for supportive services including empowerment and economic self-sufficiency activities for residents.
- *Section 8* – Approved by the 1937 Housing Act, this is still the legal authority under which several housing assistance programs are run. It subsidizes new and existing low-income rental housing with rent certificates or housing vouchers, increasing low-income tenants' choice of housing.

- *Section 32* – public housing homeownership – Established by the 1937 Housing Act, this allows public housing agencies to sell public housing units to low-income families. Current occupants have preference and proceeds are used to meet other low-income housing needs.

In terms of typology, public housing usually comes in three standard types: high-rise buildings, garden apartments, and single-family structures (freestanding or attached). Most units are efficiencies and one-bedroom apartments (48 percent); a quarter of the units have two bedrooms; and a third have three or more bedrooms.

## Housing needs assessment

Housing needs are the first determinant in studying the housing market. Through a methodology of needs assessment, the number of housing units needed to shelter the population of any given area can be established. A vital step in meeting the housing needs of any community is quantifying that need. A number of methodologies have been developed with the purpose of estimating housing needs. Some are developed exclusively for estimating the needs of homeless people, others of low-income populations, and yet others of the population considered to be the average household. Whatever the methods applied, having a replicable formula to calculate needed housing for any group in society suggests that problems can be tackled including homelessness, unaffordability, and the declines in both quality of life and middle-class purchasing power.

HUD considers that "worst case housing needs" pertain to families or individuals who earn less than 50 percent of an area's median income and either spend 50 percent or more on housing, or live in substandard units. HUD's 2005 Affordable Housing Needs Report to Congress found that the need for decent, affordable rental housing increased a great deal between 2003 and 2005. During this period there was a 16 percent increase in the number of households with worst case housing needs, the largest increase since the report was first submitted to Congress in 1991. This increase resulted in 5.5 percent of American households, a total of six million, in need of adequate housing. The most affected groups were those with extremely low incomes (those making no more than 30 percent of an area's median income) and those with very low incomes (those making no more than 50 percent of an area's median income). Among the households in need, 22 percent were elderly and 39 percent were families with children, the latter being the group with the largest increase in the period. Non-Hispanic whites were the largest ethnic group with worst case needs, amounting to 52 percent of households.

## The case for homeownership

Homeownership rates have increased steadily but slowly over the past 25 years, from 64 percent in the early 1980s to 69 percent today. There have been diverging responses for and against the idea of government encouragement of homeownership through intervention on market conditions such as incomes, housing costs, and mortgage financing availability. Active government involvement to equalize housing availability could only be justified if there were concrete proof that the positive effects of homeownership outweigh the initial investment. This affirmative assumes that the role of government is to advance and improve the quality of life of society as a whole.

The most basic thrust of federal housing policy during the past half century has been the encouragement and facilitation of widespread homeownership. The major laws encouraging homeownership were the Tariff Act of 1913 (the original income tax law), the National Housing Act of 1934, and the Housing and Urban Development Act of 1968. Public figures involved in national housing legislation have asserted that society gains from homeownership in three ways:

1   Homeowners maintain their dwellings in superior condition, potentially saving society resources by extending the life of the housing stock. Homeowners also contribute to neighborhood stability.
2   Homeowners save at a higher rate than renters, thus permitting a higher rate of national investment and greater economic growth.
3   Owner occupants are more active in the local community and, in this sense, are better citizens than renters. Wider homeownership lends political stability to a nation because of the greater economic stake owner occupants have in the existing economic and political systems.

The increase in the proportion of family income required to buy a new house is alarming. Some studies have concluded that there is little justification for government-sponsored promotion of homeownership because it is an achievable goal for a majority of households. This assertion may be interpreted differently by those who would like to be homeowners earlier in life since, according to data collected by the Census Bureau, homeownership is indeed an achievable goal but only later in life. It may be better not to encourage homeownership if the benefits to society are small relative to their cost.

## Conclusion: the future of housing

Housing will continue to be a challenging subject. While some argue for better designed communities which provide quality of life to residents, others argue for quantity through reduced costs and defend housing everyone adequately. Design inadequacies, as well as lack of replacement reserves, operating funds, and specialized property management, may cause subsidized developments to fail. The most successful housing programs to date in the US have been those that applied a consistent methodology to address problems comprehensively, evaluate needs, and develop realistic plans of action. Housing plays an important part in community development, yet its production is not all that is necessary to achieve the goal of community development.

Housing trends stem from demographic ones. Changes in population, family structure, and lifestyles will cause changes in housing preferences and typology. Some of the characteristics that will have an impact on demographic trends will be immigration, aging and life expectancy, racial and ethnic issues, and changes in the meaning and structure of the "nuclear" family, particularly family size and household formation rates. Demographic trends are also impacted by economic conditions.

Recent trends have led toward the reduction and possible elimination of regulatory barriers. Barriers can be any regulations or policies that hinder development of affordable housing. The American Homeownership and Economic Opportunity Act of 2000 created a regulatory barrier clearinghouse. It offers proposed solutions to identified barriers; and strategies or policies that mitigate the negative effect of barriers. Reform strategies are needed to eliminate obstacles to building or maintaining affordable housing.

Success stories in housing may make the public more willing to endorse government efforts, especially in areas that foster an overall improvement in the quality of life, such as education, childcare, health, public safety, and job training, if such means are perceived as necessary to provide adequate and affordable housing. The stigma attached to public housing must be eliminated and its image, equated with crime and seedy living conditions, needs to be changed.

In order to guarantee continued success in providing affordable housing to American families, particularly to enable those who need additional assistance, programs need to support market mechanisms that allow and enhance community development. Low-income housing can be hindered by well-intentioned historic, environmental, and public review regulations that make developments more expensive.

Increasing access to affordable housing and homeownership, supporting community development, and eliminating discrimination should remain priority goals for government, nonprofit, and private entities concerned with housing and community development issues. Promoting integrated approaches that provide adequate housing in viable communities can advance the expansion of economic opportunities for low- and moderate-income families. Creating a suitable living environment for all should be paramount.

---

### CASE STUDY: COMMUNITY DEVELOPMENT CORPORATIONS' ROLE IN HOUSING, AN EXAMPLE FROM MIAMI, FLORIDA

Nonprofit organizations play an important role in the provision of affordable housing and in community planning and development. Initially dedicated to economic development, these organizations got involved in affordable housing development in the 1960s to fill the gap in housing provision left by the government. Community Development Corporations (CDCs) were part of this movement; housing became their main area of intervention and they became the main alternative for provision of public housing. Most financing of CDC activities comes from federal programs, but they also use state and local funding for some of their projects. CDCs develop, revitalize, own, and manage subsidized housing.

CDCs are formed by residents, small business owners, faith-based congregations, and other local stakeholders to revitalize a low and/or moderate income community. There are about 4600 CDCs in the US. Their sizes and levels of organization and investment run the gamut; however, they typically provide affordable housing, a variety of social services, jobs, and community development through reinvestment projects. CDCs are locally based and the majority operate in urban areas. About two-thirds serve a single neighborhood or multiple neighborhoods within a single city. CDCs have access to public sector development incentives and subsidies, so retail development has been one of their expanding areas.

The Fifth National Community Development Census, completed in 2005 by the National Congress for Community Economic Development, reports that CDCs have developed over a million units of affordable housing, created more than 750,000 private sector jobs, and provided $1.5 billion dollars in loans nationwide (see Table 17.6). Sixty-five percent of the people served by CDCs are low-income or very low-income and 22 percent are poor. This means that 87 percent of the people served make less than 80 percent of their areas' median income. This census also reports that 77 percent of the CDC housing units in urban areas are rental properties. In terms of special needs populations, most CDCs serve the disabled, elderly, persons with substance abuse problems, persons living with AIDs, and formerly homeless persons. In addition, they serve child care centers, schools, health care centers, arts and cultural centers, community or recreation centers, and senior centers in their developments.

■ **Table 17.6** CDCs production nationwide, 1988 to 2005

| Profile | 2005 | 1998 | 1988 |
| --- | --- | --- | --- |
| Number of CDCs | 4,600 | 3,600 | 1,500–2,000 |
| Housing production | 1,252,000 units | 650,000 units | 125,000 units |
| Commercial/industrial space production | 126 million square feet | 65 million square feet | 16 million square feet |
| Number of jobs created | 774,000 | 247,000 | 26,000 |

Source: National Congress for Community Economic Development 2005.

The Miami Beach, Florida Community Development Corporation (MBCDC) provides a good case study to illustrate the activities of a locally based CDC. This CDC was founded in 1981 and has made affordable housing available to the local workforce through its various initiatives. The CDC works in partnership with local organizations, following a balanced and comprehensive community development program based on revitalization, historic preservation, economic development, and community building. One of its main goals is to provide safe, decent affordable housing to low- and moderate-income residents of Miami Beach. The MBCDC has assisted almost 300 residents to become first-time homebuyers. It has also developed about 200 affordable rental units which it operates for people with special needs, the elderly, low-income families, and workforce housing. Rehabilitating affordable housing is an important component in economic development and improving quality of life for residents; however, it is only one of the activities of the MBCDC. For instance, the CDC's interest in historic preservation and urban design has led to the revitalization of the Art Deco District in South Beach, which has reemerged as a world-class tourist destination.

MBCDC began its housing development programs with the restoration of the fire-damaged Kelwyn Apartments as rental housing for low-income families. It then acquired and rehabilitated Madison Apartments, which include two- and three-bedroom units; The Jefferson; and Crespi Park Apartments in North Beach. To provide affordable housing for people living with AIDS, MBCDC has partnered with the People with AIDS Coalition to acquire and rehabilitate Shelbourne House and The Fernwood. Today, MBCDC owns nine apartment buildings and several scattered site units that provide affordable housing in Miami Beach for low-income individuals, families, the elderly, and people with special needs (see Table 17.7).

In addition to affordable housing projects, the MBCDC provides grants and loan funds from city, county, and the state to assist merchants, property owners, and arts organizations in the rehabilitation of commercial properties on major corridors such as Washington Avenue, Lincoln Road, Ocean Drive, Collins Avenue, and Espanola Way, as well as throughout the Art Deco District. It also provides such grants and funds to rehabilitate tourist-oriented businesses including numerous hotels. The MBCDC transformed the Colony Theater from a vacant movie house into a performing arts center and provided construction management services to restore the Coral Rock House as the Hispanic Community Center. In turn, these projects have attracted private investment to the area, creating new jobs and opportunities. In addition, the MBDC projects have spurred the development of new organizations and associations.

**Table 17.7** Miami Beach CDC properties

| Project | Year renovated | Number of bedrooms | Low-income units | Special needs units |
|---|---|---|---|---|
| Kelwyn Apartments | | | X | |
| Madison Apartments | 1999 | 1, 2, & 3 | 17 | |
| Crespi Park | 2002 | | 10 | 6 elderly |
| Shelbourne House | 1998 | Studios & 1 | | 24 living with AIDS |
| The Fernwood | 2002 | 1 & 2 | | 18 living with AIDS |
| The Jefferson | 1999 | Studios & 1 | | 27 elderly |
| Scattered Sites | 1994 | 1, 2, & 3 | 4 | |
| Knightsbridge | 2000 | 1 | 9 | 9 elderly |
| Aimee I | 2003 | 1 | | 9 elderly |
| Aimee II | 2004 | 1 | | 16 elderly |
| 1551 Pennsylvania Avenue | 2003 | 1 | 15 | 5 homeless |

Source: Miami Beach Community Development Corporation 2007.

MBCDC also offers several programs to assist with other needs, from financial support, mortgage loans and contract documentation, to safety, evacuation, and hurricane preparedness. Their mission is:

to enhance the quality of life of Miami Beach and achieve neighborhood revitalization through a comprehensive community development program that pursues and balances historic preservation and urban design, economic vitality and increasing job opportunity, and support for a diverse, eclectic and successful neighborhood social fabric.

## Keywords, concepts, and definitions

Most of these definitions are necessary to clarify terms used in relation to housing, particularly by government agencies such as the Census Bureau and the Department of Housing and Urban Development:

- *Households:* The people who occupy a housing unit. A household is the human component of housing. Oftentimes, households are used as economic units, particularly in relation to indicators.

- *Householder:* A person living in the household who owns or rents the residence. The householder may or may not be the head of a family. A family householder lives with one or more family members, related by birth, marriage, or adoption. A nonfamily householder lives alone or with non-relatives such as foster children or domestic partners.
- *Family:* A group of two or more people who occupy the same place of residence and who are related by birth, marriage, or adoption.
- *Housing units:* The structures in which people live.

There are several housing typologies; thus a housing unit may be a house, an apartment, a mobile home or trailer, a group of rooms, or a single room occupied as separate living quarters.

## Review questions

1  What are the dimensions of housing affordability?
2  How has the decades-long US promotion of homeownership made an impact?
3  What is a housing needs assessment and what roles does it play in community development?

## Bibliography and additional resources

Dreier, P. (1997) "The New Politics of Housing," *Journal of the American Planning Association*, 63 (1): 5–27.

HUD USER. US Department of Housing and Urban Development. Available online at www.huduser.org.

Johnson, J. et al. (2003) *Issue Papers on Demographic Trends Important to Housing*, Policy Development & Research, US Department of Housing and Urban Development. Available online at http://www.huduser.org/Publications/PDF/demographic_trends.pdf (accessed April 24, 2007).

Masnick, G.S. (2002) "The New Demographics of Housing," *Housing Policy Debate*, 13 (2): 275–321.

Miami Beach Community Development Corporation. Available online at http://www.miamibeachcdc.org (accessed November 3, 2007).

National Association of Home Builders. Available online at www.nahb.org.

National Association of Realtors. Available online at www.realtor.org.

National Congress for Community Economic Development. Available online at http://www.ncced.org/.

National Congress for Community Economic Development (2005) *Reaching New Heights: Trends and Achievements of Community-Based Development Organizations*. Available online at http://www.ncced.org/documents/NCCEDCensus2005FINALReport.pdf (accessed November 3, 2007).

National Low Income Housing Coalition. Available online at http://www.nlihc.org.

NeighborWorks America: Available online at http://www.nw.org.

Riis, J. (1890) *How the Other Half Lives*. New York: Charles Scribner's Sons.

South Florida Community Development Coalition. Available online at http://www.floridacdc.org.

Struyk, R.J. (1977) *Should Government Encourage Homeownership?* Washington, DC: The Urban Institute Press.

U.S. Census Bureau, Current Housing Reports, Series H150/05 (2006) *American Housing Survey for the United States: 2005*. Washington, DC: US Government Printing Office.

U.S. Census Bureau. Available online at www.census.gov (accessed May 2–7, 2007).

U.S. Department of Housing and Urban Development. Available online at www.hud.gov (accessed April 18–28, 2007).

# 18 Neighborhood planning for community development and renewal

## Kenneth M. Reardon

## Introduction

Interest in participatory approaches to neighborhood planning has skyrocketed among city residents, professional planners, elected officials, and scholars during the past two decades. This renewed interest in resident-led planning is the result of a number of powerful economic, social, and political trends affecting our nation's major metropolitan regions. The increasingly uneven pattern of development characterizing many of our metropolitan areas has led to a disturbing expansion in the number of economically distressed neighborhoods where the quality of life is often shockingly low. The failure of Urban Renewal, and other centrally conceived revitalization strategies to address the critical social problems confronting these neighborhoods, has undermined confidence in and support for top-down urban regeneration efforts. Without the concurrent movement of funds to support these efforts, the responsibility for basic municipal services has shifted from federal and state government to local communities. This has forced many villages, towns, and cities to transfer responsibility for these programs to local nonprofit agencies and community-based organizations. The total quality management movement, which stresses the importance of continued improvements to help an increasingly diverse citizenry, has encouraged municipal planning directors and city managers to emphasize more participatory approaches to governance. Community development and planning professionals have also been forced to adopt more collaborative forms of practice due to pressure from cultural identity groups, including African-Americans and Latinos, for a greater voice in public policy decisions affecting their communities. Finally, since funders of urban revitalization projects more frequently mandate the active participation of local stakeholder groups at every stage of the planning, design, and development process, the movement toward participatory neighborhood planning is further reinforced (Peterman 2000).

## The rising tide

Evidence of the growing popularity of participatory neighborhood planning is widespread. There has been an explosion in the number of community-based organizations, particularly community development corporations (CDCs), that are involved in resident-led planning, design, and development (Brophy and Shabecoff 2001). The number and variety of municipal government planning departments that have established active neighborhood planning units is also impressive. Cities as diverse as New York, Washington, DC, Savannah, Austin, Chicago, Portland, Los Angeles, and San Diego have created specialized units to help neighborhood residents design and implement revitalization strategies toward improving their own quality of life. Within the past ten years, the Annie E. Casey Foundation, the nation's largest family foundation, has teamed up with the American Planning Association to sponsor two national symposiums to establish principles of good practice and alternative models for collaborative neighborhood planning. In recent years, growing numbers of colleges and universities have established ambitious community/university partnership programs and centers to support collaborative approaches to community problem solving, planning, and development. Those involved in these efforts, along with scholars who have documented and evaluated this work, have contributed to the development of other important practice-oriented publications such as *City Limits, Shelterforce, The Neighborhood Works, and Progressive Planning* as well as many other academic journals and books including: *Journal of the Community Development Society*; *the Michigan Journal of Community Service Learning*; Healey's *Collaborative Planning: Shaping Places in Fragmented Societies*; Rohe and Gates's *Planning with Neighborhoods*; Jones's *Neighborhood Planning: A Guide for Citizens and Planners*; William Peterman's *Neighborhood Planning and Community-based Development*; Forester's *The Deliberative Practitioner*; and Rubin's *Renewing Hope within Neighborhoods of Despair*.

Another indication of the growing importance of these endeavors is the number of regional and national foundations that have established ongoing programs to support resident-led neighborhood planning. Among these foundations are: The Ford Foundation, the Rockefeller Brothers Foundation, the Annie E. Casey Foundation, and the Wachovia Foundation. A final indication of the increasing importance of this work is the rising number of graduate students concentrating on and/or specializing in affordable housing, economic development, and community development, in order to prepare to work as neighborhood planners.

## Emerging principles

While various professionals take different approaches to the practice of participatory neighborhood planning, set of principles of good practice has emerged in recent years. Most practitioners are committed to

a model of participatory neighborhood planning that seeks to:

- Improve the overall quality of life enjoyed by poor and working-class families by adopting a *place-based approach* to community development, one that emphasizes the importance of a healthy environment characterized by high-quality public services, living wage jobs, affordable housing, and supportive local institutions.

- Involve a broad cross-section of local residents, businesspeople, institutional leaders, and elected officials as decision makers at every step of a *resident-led* research, planning, design, and development process.

- Connect previously uninvolved and often ignored residents, including youth, seniors, immigrants, ethnic/racial/religious minorities, and others, with participatory neighborhood planning processes. The involvement of previously marginalized residents would *produce plans* to address the most critical issues confronting the community, *expand political support* for such plans, and thereby increase their chances of adoption and implementation.

- Recognize the need to complement traditional "bricks and mortar" approaches to community revitalization by implementing *comprehensive revitalization* strategies that invest in high-quality educational, health care, public safety, job training, and business development programs, in order to rebuild the social capital base of economically challenged communities.

- *Rather than a deficit-based* approach to community development, pursue an *asset-based* approach to plan and initiate increasingly ambitious programs, one that builds upon the significant knowledge, skills, resources, and commitment that every established community possesses, regardless of its economic status.

- Identify and cultivate local residents' organizing, research, planning, and development knowledge and skills to enhance the *organizational capacity* of local community-based development organizations.

- Recruit *strategic public and private investment partners* willing to commit long-term resources to support projects designed to restore the health and vitality of the community.
- Engage local residents in *municipal, regional, state, national, and international public interest campaigns* that promote both redistributive economic and community development policies and participatory decision making.
- Embrace a *reflective approach to professional practice* that encourages regular reframing of the planning problems being addressed, in order to maximize the positive outcomes of the neighborhood planning process.

## A short history of participatory neighborhood planning

Participatory neighborhood planning has a long but often overlooked place in American history. Patrick J. Geddes, the Scottish botanist turned planner who made seminal contributions to early planning theory, methods, and practice, successfully encouraged residents of several of Edinburgh's poorest neighborhoods to undertake a series of slum improvement campaigns in the 1890s with the help of his students. Geddes's revitalization efforts, which he referred to as "conservative surgery," featured the systematic collection and analysis of data describing local conditions within their regional context. They also included active resident participation to formulate innovative yet workable solutions to local problems, and strongly mobilized local citizens and sympathetic supporters to implement the key elements of his strategies on their own. Through his conservative surgery approach to community development, local residents transformed trash-filled lots into vest-pocket parks, restored dilapidated housing through the establishment of worker cooperatives, and enhanced the appearance and functionality of local public markets. Geddes sought to encourage local residents and officials to undertake increasingly challenging urban regeneration projects and to document their efforts as they organized exhibitions, adult education courses, summer school programs, and extension activities at his Outlook Tower. This building was viewed by many as the first sociological laboratory dedicated to nurturing civics – the science of community building and town planning (Welter 2002).

In the years immediately following the first national planning conference held in Washington, DC in 1909, Charles Mulford Robinson wrote a series of influential newspaper articles. These urged residents of the nation's rapidly expanding industrial cities to initiate a wide range of physical improvement schemes at the neighborhood and municipal levels of government to restore environmental health and enhance the aesthetic appearance of their neighborhoods (Robinson 1899). Robinson's campaign, which came to be known as the "City Beautiful Movement," was enthusiastically embraced by local garden clubs, women's organizations, religious institutions, and business associations. These and other organizations came together in hundreds of American cities to encourage local officials to improve basic sanitation services, establish building construction and maintenance codes, implement urban design schemes, install public art, and expand public playgrounds and parks to protect the health and enhance the quality of life enjoyed by local residents. Among Robinson's many publications was his *Third Ward Catechism* (McKelvey 1948), which contained his philosophy of neighborliness and civic improvement through cooperative action.

In the 1920s, an interdisciplinary team of design professionals and social scientists created the Regional Plan for New York and its Environs. This was sponsored by the business-led Regional Plan Association of New York and focused considerable attention on the design and construction of healthy residential areas in cities such as New York that were struggling to cope with intense urbanization pressures. Clarence Perry offered his planned unit development (PUD) concept based upon lessons learned from Forest Hills, Queens, the highly successful residential community conceived as a model working-class development by the Russell Sage

Foundation (Heiman 1988). Perry believed that a significant portion of the New York region's future growth could be accommodated in well-designed residential communities of 10,000 to 15,000. He envisioned their organization around common open spaces and community school centers that would serve as focal points for civic and social life. In Perry's scheme, commercial services would be restricted to the periphery of medium-density residential communities that offered a mix of rental and homeownership housing. Traffic would be minimized so as to encourage pedestrian activity by laying out most of the proposed streets as culs-de-sac that would follow natural topography. The scheme would also limit the number of streets that continued through the length and width of the planned developments. In 1928 Clarence Stein and Henry Wright designed Radburn, a planned community near Paterson, New Jersey that was organized around pedestrian-oriented neighborhoods reflecting many of Perry's ideas (Stein 1957).

While a number of exciting neighborhood planning initiatives were advanced within the New Deal programs of the Roosevelt Administration, these were soon curtailed due to the manpower and supply needs of the nation's war production effort. As growing numbers of industrial workers crowded into northern cities to participate in wartime manufacturing and nation-wide rationing prioritized the needs of the troops over those of the state-side civilian population, the housing, community facilities, and infrastructure of the worker housing districts of these cities suffered. Following the war, returning servicemen and women quickly came to understand the impact which years of intense use and deferred maintenance had inflicted on their former urban neighborhoods. Eager to secure employment with the growing number of firms seeking lower cost greenfield locations in rapidly expanding suburban communities, young workers soon found it possible to purchase new homes in those communities with assistance from GI benefits. The movement of firms and people from older central city neighborhoods soon resulted in falling property values and rising vacancy rates. These conditions made it increasingly

difficult to finance the improvement of residential and commercial buildings in the inner city. As economic, social, and physical conditions in these neighborhoods deteriorated, local business leaders, organized through the Urban Land Institute, successfully lobbied for the passage of the Taft, Ellender, and Wagner Housing Act of 1949.

Between 1949 and 1973 more than 2000 local communities sought federal funding to conduct large-scale clearance, infrastructure improvement, and new residential, commercial, and civic development under the Federal Urban Renewal Program. Local communities initiated the Urban Renewal Program in their municipalities by establishing Local Renewal Agencies (LRAs) charged with identifying "blighted areas" of the city where occupancy rates and property values were falling. A detailed plan would then be prepared, often with the assistance of outside consultants, to demolish seriously deteriorating structures within the target area, install state-of-the-art infrastructure, construct needed public buildings and facilities, reduce the selling prices of the remaining vacant land to stimulate private investment, and provide public insurance to those lenders financing new development within these redevelopment areas.

To support this dramatic expansion in the role of the federal government in urban real estate markets, advocates of the Taft, Ellender, and Wagner Housing Act of 1949 based their arguments on the need to improve the quality and availability of affordable housing for poor and working-class families. However, the impact of the federal Urban Renewal Program on these families was devastating. More than 600,000 housing units, most of which were affordable to those living on modest incomes, were destroyed through the Urban Renewal Program. Of the 100,000 units of new housing constructed through this program, only 12,000 were affordable to those of modest means. During the program's first five years, displaced renters, including the poor, elderly, and infirm, received no relocation assistance. In fact, the majority of the families displaced by the activities of Local Renewal Agencies moved into substandard housing for which they paid higher rents. Finally, six out of every ten families displaced by the Urban Renewal Program were African-American or Latino. While the program produced millions of square feet of attractive new retail, commercial office, and performing arts space, millions of poor and working-class families of color suffered as a result of the program's clearance activities (Anderson 1966).

The widespread pain and suffering that the federal Urban Renewal Program inflicted on hundreds of poor and working-class communities of color throughout the United States produced widespread protests. Mel King, a community educator and settlement house worker in Boston's South End, organized low-income residents to occupy a parking lot where their former homes had been demolished by the Boston Redevelopment Authority. The protest continued until community leaders were placed on their citizens' advisory board. The success of Boston's "Tent City" occupation encouraged other communities to stand up to the federally funded bulldozers targeting their neighborhoods (King 1981). In Lower Manhattan, the residents of the Cooper Square neighborhood, led by advocacy planner Walter Thabit, successfully opposed a clearance-oriented renewal plan proposed by Robert Moses. For the past 40 years, Cooper Square residents and leaders have doggedly pursued the implementation of their own preservation-oriented approach to community revitalization (Thabit 1961). With the support of progressive planner Chester Hartman, residents of the Yerba Buena community of San Francisco waged an inspired but sadly unsuccessful campaign to save this long-term immigrant enclave from federally funded clearance (Hartman 1974). Along with the growing influence of the Civil Rights Movement, these oppositional planning efforts led to a significant shift in federal policies toward revitalizing economically distressed urban neighborhoods. The passage of the Economic Opportunity Act of 1964 funded the creation of nonprofit organizing, planning, and development organizations. These groups used local resources to leverage outside funds for the implementation of critical educational, job training, small business

development, health care, and legal services to improve the quality of life for the urban poor. The Act used the term "maximum feasible participation" to describe the poor's central role in setting future economic and community development priorities for the local community action agencies (Moynihan 1970).

The emphasis on community-based and resident-led planning and development, in both the Equal Opportunity Act of 1964 and the subsequent Demonstration Cities and Metropolitan Development Act of 1966, prompted leaders of Cleveland's Hough Area and Brooklyn, New York's Bedford-Stuyvesant neighborhood to create multi-purpose economic development organizations in the mid-1960s. The organizations hoped to attract private investment in the revitalization plans of local residents. In the late 1970s and 1980s, as the Carter, Reagan, and Bush Administrations sought to balance the federal budget by reducing spending on a wide range of urban development programs, the number of community development corporations exploded. Seeking to promote sustainable development in our nation's urban and rural communities, some CDCs addressed affordable housing, job training, small business assistance, youth development, and public safety. By 2005, the National Congress for Community Economic Development (NCCED) estimated that the number of professionally staffed community development corporations exceeded 4600 (NCCED 2006: 4). Many of these organizations were engaged in various forms of participatory neighborhood planning. In low-income communities, the efforts of local CDCs and other local nonprofits were frequently encouraged by municipal planning agencies that collaborated with them to initiate Comprehensive Community Revitalization Initiatives (CCRIs). The financial support, leadership training, and technical assistance required by these efforts were often provided by the national network of financial intermediaries that arose in the early 1970s including the Local Initiatives Support Corporation, Enterprise Partnership Communities, the Neighborhood Reinvestment Corporation, and Seedco. In many communities, the efforts of these institutions to advance the goals of participatory neighborhood planning were reinforced through the work of architects, landscape architects, and urban planners working for Community Design Centers sponsored by local professional associations or nearby universities.

## Types of neighborhood plans

Today, most medium- and large-scale cities have several planners, often organized into a specialized unit within their planning departments. They are responsible for assisting local residents and leaders to design comprehensive strategies that enhance the unique quality of life available in their residential neighborhoods. These planners tend to work with local residents to pursue one of the following six types of neighborhood planning strategies based upon the environmental, economic, and social conditions confronting their communities:

- *Growth management strategies* – These plans are developed by neighborhoods seeking to encourage growth while protecting long-term residents, businesses, and institutions from displacement.
- *Preservation strategies* – These plans are developed by local residents eager to protect historically, culturally, and aesthetically significant places and structures. These sites are central to the community's identity from but threatened by loss through neglect, insensitive reuse, or demolition.
- *Stabilization strategies* – These plans are formulated by those committed to reducing and/or eliminating the out-migration of people, businesses, institutions, and capital from a community, which, if unabated, may undermine its long-term viability.
- *Revitalization strategies* – Through a comprehensive revitalization program featuring investments aimed at improving the community's physical fabric and rebuilding its social capital base, these plans are designed by residents seeking to restore

the former vitality of an area in severe decline from long-term disinvestment.
- *Post-disaster recovery strategies* – These plans are pursued by local leaders eager to rebuild neighborhoods following devastation from a significant natural disaster such as an earthquake, tornado, hurricane, flood, or fire.
- *Master plan strategies* – These plans relate to the design and construction of new, often mixed-use communities at former urban "brownfield" sites or ex-urban "greenfield" locations.

In a dynamic, rapidly changing region, residents of a given neighborhood may undertake several of the above-mentioned types of neighborhood planning as environmental, demographic, economic, and social conditions affecting their communities, cities, and regions change.

## Steps in the neighborhood planning process

While there are clearly different types of neighborhood plans, most tend to be produced through a process that reflects the following seven steps: steering committee formation; basic data collection and analysis; visioning and goal setting; action planning; plan presentation, review, and adoption; implementation; and monitoring, evaluation, and modification. The following section provides a brief description of the major activities and deliverables produced at each of these steps in the neighborhood planning process (Jones 1990).

*Steering committee formation*

Following the decision by local residents, community leaders, and municipal officials to prepare a neighborhood plan, steps must be taken to identify the major stakeholder groups in the community. One-on-one meetings should then take place with representatives of these groups to inform them on the pending neighborhood planning process; elicit their views regarding the overall goals for the process; and invite each stakeholder group to identify one or more individuals to serve on the steering committee. This body serves a number of critical functions in the planning process such as legitimizing the overall effort; serving as spokespersons and defending the initiative from both internal and external challenges; designing a process uniquely suited to the civic, social, and cultural history of the community; and mobilizing others, especially those who have previously been uninvolved in community affairs to become active participants in the effort.

Working with the steering committee, neighborhood planners draft a preliminary scope of services and a schedule for the process. When this work has been completed, an ambitious media campaign is designed and implemented to inform all those living, working, and serving in the target area as to the goals, objectives, activities, and opportunities for participation in the upcoming neighborhood planning process. Among the typical elements of such a media campaign are ads in local newspapers; articles in community newspapers and publications; appearances by steering committee members on local radio stations and cable television networks; storefront posters, sidewalk tables, church bulletin notices, and pulpit announcements; flyers sent home with local schoolchildren; and a kickoff press conference featuring the participation of the steering committee, council persons, and the Mayor. The use of a project logo, common graphic layout, and motto, such as "Northside Turning the Corner" or "Southside Blooming," helps establish the project in the consciousness of local residents, leaders, and officials (Bowes 2001).

### Data collection and analysis

Patrick Geddes encouraged citizens and planners to "survey before plan." Following this advice, neighborhood planners tend to collect and analyze a wide range of data to determine the environmental, economic, and social assets and challenges confronting the community in which they are working. The typical neighborhood plan is based upon the collection and analysis of the following types of data:

- A detailed social history of the community highlighting its past successes in overcoming local social divisions to solve critical community problems.
- A systematic review of past public and private studies, reports, and plans for the area.
- A longitudinal analysis of population, education, employment, income, poverty, and housing trends based upon the U.S. Census.
- A study of the land features, water resources, and other natural assets of the community as well as its recent land use patterns and zoning history.
- An investigation of current building conditions, site maintenance levels, infrastructure, and community facilities quality.
- A survey of residents' perceptions of existing conditions, desired development directions, and needed improvement projects.
- A parallel survey of the opinions of local "movers and shakers" regarding existing conditions, future development directions, and specific improvement projects.

Under optimal conditions, residents would participate in the development of the instruments used to gather these data, actively engage in the collection of this information in the field, and collaborate with professional planners in interpreting the meaning and implications of such analyses for local planning and policy making. Usually the results of each of these data-collection efforts are carefully analyzed. Following the completion of all of the above-mentioned data-collection activities, residents are invited to revisit their data analysis using the Stanford Research Initiative's "SWOT analysis" technique (SRI 2007). When the major findings from each dataset have been categorized according to the Harvard system — that is, whether they represent a current strength (s), a current weakness (w), a future opportunity (o), or a future threat (t) — local residents are asked to organize these observations into major themes. Described by Bernie Jones in his volume on neighborhood planning, the "PARK System" provides a useful approach to thinking about a preliminary planning response to existing conditions. Residents and planners can work together using this system to determine which of the following planning treatments or "interventions" each important neighborhood feature should receive.

- *Preserve (P)* – Take steps to save a valued neighborhood characteristic.
- *Add (A)* – Pursue opportunities to add or expand a specific neighborhood feature.
- *Remove (R)* – Undertake actions to remove a particularly offensive local feature.
- *Keep out (K)* – Initiate policies, programs, and actions to keep a particularly noxious threat from undermining the local quality of life.

## Visioning and goal setting

Several times during the steering committee formation and data collection and analysis phases of the neighborhood planning process, representatives of local stakeholder groups and neighborhood-based institutions as well as individual citizens are asked to describe their version of an "ideal" or "improved" neighborhood. Over time, a series of alternative future development options emerge for community residents to consider. Neighborhood planners typically work with members of the steering committee to prepare a draft community profile, SWOT analysis, and alternative future development scenarios report for consideration as well. These materials are often distributed via a newsletter and/or website a week before a community-wide meeting to review these documents. During what is often referred to as a Neighborhood Summit, local residents review and comment on the historical, archival, and research documents; environmental conditions; building/site/infrastructure analysis; resident interviews; and other official interview data collected by the neighborhood planning team. They also examine and revise the SWOT analysis that has been jointly prepared by the neighborhood planners and steering committee.

Having done so, they review, and occasionally amend and/or expand, the alternative future development scenarios that have emerged during the neighborhood planning process. The afternoon of the Summit typically focuses on the pros and cons of each alternative development scenario and selection of a preferred path forward. Having established what they will work toward (the end stage), those attending the Summit proceed to craft an overall development goal to guide their work. They finish their Summit activities by identifying the major development objectives that will enable them to achieve this goal. Among the most common development objectives featured in such plans are those that seek to improve public safety; restore the urban environment; expand employment and entrepreneurial opportunities; provide access to quality and affordable housing; serve the special educational, health, and transportation needs of youth and senior citizens; ensure the provision of a full range of basic municipal services; and encourage ongoing citizen mobilization for economic and community development. Before residents, businesspeople, private sector funders, institutional leaders, and elected officials leave the Neighborhood Summit, they are challenged to join an action planning team. The team will be charged to develop a set of immediate-term (year one), short-term (years two and three), and long-term (years four and five) collaborative projects to achieve a specific objective.

## Action planning

For several months after the Summit, a significant number of participants in the neighborhood planning process meet in small groups to formulate developmentally based action plans in order to achieve the redevelopment objectives articulated at the Summit. With the assistance of neighborhood planners, these groups often begin their work by brainstorming the longest possible lists of plausible economic and community development proposals to help the neighborhood gradually accomplish its development objectives.

Following what is often referred to as a "blue sky" session, in which any and every proposal is considered, the Action Planning Team often invites representatives of sympathetic public and private funding agencies to provide initial input regarding proposals. They rate them as very likely, somewhat likely, and unlikely to receive political and financial support from local economic and community development policy makers and regional funders. These economic and community development professionals are also encouraged to add their own suggestions to the community's list of "would-be" projects. With input from these friendly development professionals, members of each action team make a preliminary attempt at prioritizing their project list using the following template for guidance.

For Year 1, they are encouraged to identify one high-priority project that could be implemented by the neighborhood's existing volunteer base with little or no outside funding or technical assistance. They are further encouraged to identify two high-priority projects for Years 2 and 3. These could be implemented with what they hope would be a slightly expanded volunteer base; $50,000 or less in local and outside funds; and a small amount of high-quality technical assistance. Finally, they are encouraged to identify three high-priority projects that could be implemented in Years 4 and 5 of the neighborhood implementation planning process with a significantly expanded volunteer base; $200,000 to $300,000 or less in local and outside funds; and a significant amount of high-quality technical assistance. Table 18.1 is a preliminary planning tool that may be used to assist neighborhood action teams in identifying a tentative list of projects for the broader community to consider.

Once each Action Planning Team has prepared an initial list of development projects for their assigned area, these are reviewed by other process participants and finally adopted. When this has occurred, members of each action team and their neighborhood planners consult the economic and community development literature to identify both principles of good practice and models of program excellence. Armed with this knowledge, they develop a specific action plan that includes the following information for each proposed development project:

- name
- detailed description
- rationale
- major implementation steps
- lead organization/agency
- supporting institutions
- approximate costs
- potential funding sources
- location requirements
- comparable model programs
- design requirements/guidelines
- possible sources of technical assistance

*Plan presentation, review, and adoption* – Once the action plans have been written and approved by all those participating in the neighborhood planning process, a complete version of the draft neighborhood plan is prepared under supervision of the steering committee. Copies of the plan are made available for review at the town or city hall, public libraries,

■ **Table 18.1** Proposed action plan: neighborhood housing improvement plan

| Time frame | Immediate term | Short term | Long term |
|---|---|---|---|
| Year | (Year 1) | (Years 2–3) | (Years 4–5) |
| Number of projects | One Project | Two projects | Three projects |
| Volunteer base requirements | Current | Slightly expanded | Significantly larger |
| Approximate costs | $0 – in-kind | $50,000 each | $200k to $300k each |
| Needed technical assistance | None | Modest/short term | Significant/ongoing |
| Projects | 1 | 2 | 4 |
| Projects | | 3 | 5 |
| Projects | | | 6 |

community centers, schools, and senior citizen centers. In addition, a PDF version of the plan is typically posted on either the local municipal and/or dedicated neighborhood planning website. If possible, prior to a final neighborhood meeting held to publicly discuss, review, and vote on the plan, an executive summary thereof in newsprint form is also prepared and distributed to each household, business, religious institution, and community nonprofit organization. Following a formal vote by neighborhood residents expressing their approval of the plan, the document is usually forwarded to the neighborhood's city council representative. It is accompanied by a resolution asking the local government to formally adopt the plan through an ordinance, thereby amending it to the community's existing master and/or comprehensive plan. In reviewing and approving neighborhood plans, the specific process that local, county, and state governments follow is somewhat different in each state. In general, neighborhood plans are officially approved through the following steps:

- Local residents are publicly informed that a particular neighborhood plan is being considered for approval.
- The city planning commission is given the opportunity to review the plan; hold one or more public hearings to elicit resident input; and vote on the document.
- The city council is then given an opportunity to review the plan, go over the planning commission's resolution in favor of its adoption, and hold one or more public hearings to elicit resident input. Before voting on the plan, the council generally asks municipal planners to complete both an environmental impact statement and a historic preservation report to assess the plan's likely impact on the local environment and its historic and cultural base. If these reports show little or no negative ecological or historical impacts, the city council is likely to approve the document.
- Depending on the form of local government, the Mayor may have an opportunity to complete an independent review of the document. If he or she has the right to veto the plan, the document may be returned to the city council for revision.
- Once the neighborhood plan has been approved by the local city council and Mayor, adjacent municipalities, and/or the county where the plan was generated may have the opportunity to review and comment on the likely impact of the plan on the metropolitan region.
- Assuming a positive review by the surrounding municipalities and/or the county where the plan was generated, the document is then sent to the secretary of state's and/or attorney general's office to determine if it meets basic state standards.
- If state officials believe the plan meets state standards, it is then returned to the city clerk, who files the document as an official amendment to the master plan.

The legal authority of officially approved neighborhood plans differs by state. In some states (e.g. New York), local governments are bound to consider neighborhood and master plans when making significant physical planning decisions. However, they are not bound to follow the policies of officially approved neighborhood and master plans.

*Plan implementation* – Following the adoption of the plan by municipal officials, local residents who have participated in the process often mobilize their neighbors, local businesspeople, institutional leaders, and elected officials to carry out the immediate-term projects that often do not require either extensive technical assistance or significant levels of outside funding. As success is achieved on these more modest self-help projects, efforts are made to identify nonprofit organizations, one of which may be willing to serve as the lead agency for the more ambitious development projects contained in the plan. Such projects require significant organizing, research, planning, and fundraising activities and, in communities where networks of highly effective community-based organizations remain, it is preferable that agencies with deep programmatic expertise in a policy area undertake related development projects. While this may be preferable, a special effort must also be made to coordinate each agency's development projects in order to capture their synergistic benefits.

In communities where the disinvestment process has proceeded unabated for a long time, there may either be a single or no existing nonprofit agency capable of carrying out the more ambitious development projects contained in the neighborhood plan. In the former situation, efforts may be needed to assist community-based organizations in carrying out the neighborhood's more ambitious development agenda. In the latter case, local institutions may need to collaborate and establish a community development corporation to carry out the proposed development agenda. Whether implementation occurs through a single community-based development organization or a number of different nonprofits, a varied funding base should be developed to implement the neighborhood plan. Such a multipronged development strategy will reduce disruption in the plan's implementation by changing politics, policies, or personnel within a single funding agency.

*Monitoring, evaluation, and revision of neighborhood plans* – Like the U.S. Constitution, neighborhood plans are living documents designed to provide local civic leaders and elected officials with general policy and program development guidance. Economic and political conditions, as well as local, regional, state, and federal policy contexts, change rapidly and require participants in the neighborhood planning process to evaluate the impact of their activities on a regular basis. In many such processes, members of the steering committee are brought together on a quarterly basis to evaluate the extent to which the neighborhood action plan is being effectively carried out and having the desired impact. In light of changing conditions, modifications are routinely made in the neighborhood action plan to better achieve the neighborhood plan's overall development goals and objectives. For example, a new federal grant program may become available, which may enable the neighborhood to expand one of its programmed activities or to move it ahead of the neighborhood's overall neighborhood planning timetable.

Likewise, the decision of a nearby CDC to undertake one of the most challenging projects contained in the plan may enable the neighborhood to reevaluate its role in the project from that of developer to monitoring agency.

## The global movement toward participatory neighborhood planning

The historic argument is that broad-based participation can only be achieved by incurring significant delays and additional costs during the implementation phase of a project. However, this is contradicted by the growing list of very successful economic and community development projects that have been carried out using highly participatory planning methods. Increasing numbers of planners and designers, along with the elected officials with whom they work, have come to appreciate the important contribution that active participation by residents, business owners, and institutional leaders can make. At each step of the planning and design process, these collaborations both improve the quality of specific development proposals and broaden their political base of support. Due to the growing body of participatory neighborhood planning projects being undertaken by community development corporations; municipal, county, and regional planning agencies; and private planning and design firms in the U.S., planners and designers in other parts of the world have been inspired to undertake similar efforts. Such work is being strongly encouraged by the European Union and the United Nations Research Institute for Social Development. Planners and planning scholars have an important new venue in which to share their participatory planning and design ideas, methods, and practices. For instance, those who belong to one of the seven disciplinary associations organized on a regional basis throughout the world by the Global Planning Educators Interest Group now have regular meetings and publish comparative research in planning.

## Conclusion: challenges on the horizon

Looking toward the future, there are numerous challenges confronting those committed to various forms of participatory neighborhood planning. First, we need to further refine the core curriculum in professional planning programs by providing students with a stronger grounding in urban ethnography, participatory action research methods, and community organizing theory and methods. Second, we must expand opportunities for neighborhood activists, institutional leaders, practicing planners, and elected officials to acquire basic training in the theory, methods, and practice of participatory neighborhood planning. Third, to explore various policy initiatives that could accelerate the implementation of important local initiatives, we should assist those who become involved in participatory neighborhood planning efforts to come together on city-wide, regional, state, and national levels. Finally, we need to encourage planning scholars to undertake more systematic evaluations of various participatory neighborhood planning models to provide a more robust empirical basis upon which to pursue the practice.

---

### CASE STUDY: PROMOTING CITIZEN PARTICIPATION IN NEIGHBORHOOD PLANNING

Overcoming resident skepticism regarding the possibilities for change is often the single most important challenge confronting those involved in participatory neighborhood planning. Citizens living in economically distressed communities have witnessed local employers moving away, unemployment and poverty rates soaring, local retail stores closing, credit availability declining, municipal services deteriorating, and long-term residents departing. They have also observed several generations of elected officials and appointed administrators promising, as did Herbert Hoover, that prosperity was "just around the corner." This widening gap between existing conditions and promised improvement has left many residents of low-income communities highly skeptical of government-sponsored or -endorsed community renewal efforts (Schorr 1997). The following section describes how planners and planning students, working in a once-vibrant resort community in the Catskills Mountains region of New York State, succeeded in involving a large and representative cross-section of residents of the Village and Town of Liberty, New York to create a comprehensive economic development strategy for their community.

In 2003, representatives of a local family foundation asked Cornell University's Department of City and Regional Planning to assist residents, business owners, and officials from the Village and Town of Liberty in formulating a comprehensive community economic development plan. While the community had worked hard to create a sense of optimism and momentum regarding its future, this work had fallen on the shoulders of a very small number of volunteers. While developing the plan, planners realized they must use the process itself to identify and recruit new volunteers to assist in implementing the final draft. Therefore, the planners pursued a highly participatory "bottom-up, bottom-sideways" approach. To encourage previously uninvolved individuals to participate in this process, they used the following methods.

#### One-on-one meetings

Working with the foundation staff, and elected and appointed officials from the Village and Town of Liberty, the planners identified more than 40 civic leaders who represented the community's major stakeholder groups. The planners contacted each of these individuals to elicit their perceptions regarding conditions in the community, the need for a comprehensive economic development plan, their proposed goals and objectives for such a plan, and their willingness to participate in the process. Special effort was made to reach out to area youth and to members of the Latino, Hassidic, and Muslim communities that had recently migrated to the community. Despite their numbers, these groups had not participated previously in local community planning activities.

## Steering committee

Following a series of more than 40 one-on-one interviews, 25 local leaders representing a broad cross-section of the community were invited to serve on the steering committee for the Liberty Economic Action Plan (LEAP). The primary functions of the steering committee were to assist the planners in formulating a basic research design for the project; encourage local residents, business leaders, and elected officials to participate in the process; serve as spokespeople for the effort; and defend the undertaking from outside criticism and attack.

## Community media campaign

Members of the steering committee subsequently worked with project planners to devise an ambitious local media campaign that would inform local residents about the project and encourage them to actively contribute to the effort. The local media campaign featured a kickoff press conference; a weekly news update which appeared in the community's weekly newspaper; storefront posters; weekend informational tables and sandwich boards set up along Main Street; regular updates presented at Liberty Volunteer Fire Department meetings; bulletin and pulpit announcements at area churches, synagogues, and mosques; and notices posted on a community bulletin board in the heart of the Village's downtown.

## Social history project

One of the planners' first activities was a series of oral history interviews with residents and officials who had helped lead successful community development projects in the past. The stories of these community renewal efforts were collected to remind Village and Town residents of their history of collaboration to overcome critical economic, social, and political problems confronting their community. These interviews were also helpful in identifying effective leaders who might be recruited to participate in the newly initiated comprehensive community renewal program. A "Short and Glorious History of Liberty's Community Improvement Legacy," which recalled the efforts to restore and preserve the town's original public buildings, among other tales, was distributed throughout the community.

## Door-to-door campaign

During the week following the project's kickoff press conference, 25 students and six steering committee members spent three days visiting every house, business, and institution in the Village and the Town. They informed occupants about the economic development planning process that was underway, elicited their ideas regarding the Village and Town's future, and invited them to an initial community planning meeting to be held at the Liberty Volunteer Fire Department headquarters. During the course of a single weekend, more than 1200 informational brochures were distributed describing the goals, objectives, and desired outcomes of the economic development planning process.

## Community mapping

Approximately 60 local residents came to the first community meeting. Following a brief introduction to the goals and objectives of the planning process, residents were asked to join six-person teams for the purpose of sharing their knowledge of the community. Each team was seated at a round table on which there was a very large base map of the Village and Town and a set of colored markers. They were invited to use their purple markers to identify their community's most significant subareas (i.e., communities of interest or neighborhoods); green markers to locate important community resources or assets; red markers to identify areas of concern; and orange markers to identify untapped resources to advance their economic development efforts.

Within minutes, the noise level in the room rose as residents introduced themselves to each other and began to share their vast knowledge of the community with those at their tables. After approximately 30 minutes, a spokesperson from each table was asked to share the highlights of his or her team's mapping with the entire assembly. Attendees were impressed, not only by the collective community knowledge of the residents and business owners, but by the significant number of both current and untapped assets and resources the community had at its disposal.

## Camera exercise

At the end of the first community meeting, residents were asked to assist the steering committee in documenting the most important attributes of the community. Each meeting attendee, regardless of age, was given a simple disposable camera with 27 exposures with which to identify nine each of Liberty's most important assets, serious problems, and underappreciated, untapped resources. Along with the camera, each volunteer was given a simple log to provide captions for each photo. During the week following the preliminary meeting, more than 40 Liberty residents were seen taking photos of the community's many natural areas, historic buildings, residential neighborhoods, and community facilities. As they did so, other residents asked what they were doing. Following a brief explanation, they would invite these individuals to join them in the planning process by attending the next meeting.

## Shoe box planning

Those attending the second community planning meeting were again seated at round tables in groups of six, where they were given 50 of the photos to sort into four shoe boxes. Following the SRI's strategic planning tool, affectionately referred to as the "SWOT" analysis, residents were asked to place photos depicting positive community traits in the box marked "S" for current strengths; negative community attributes in the box marked "W" for current weaknesses; potential opportunities in the box marked "O" for future opportunities; and possible threats in the box marked "T" for future threats. Following their initial sorting of these images, the teams worked together to group photos within each box according to overarching themes. For example, images of historic structures in the "current strengths" category might be organized under the theme, "strong building stock," while photos of illegal dumpsites from the "current weaknesses" category might be placed under the theme, "Environmental Degradation." This meeting ended with the teams working together to integrate their preliminary assessments of existing conditions and future development possibilities.

## Spike Lee exercise

As adults collaborated to develop a consensus regarding Liberty and its future, middle-school students were asked to share their perspectives on the community and its future through an art activity referred to as Spike Lee's "The Good, the Bad, and It's Gotta Change Now, Baby" program. On a Friday afternoon, teams of university students visited the Liberty middle school where they worked with 50 12- to 14-year-olds on a mural project. Following a mid-afternoon snack of pizza and soda, each middle schooler was given a 30 {x} 40-inch piece of newsprint in order to share their sense of the "Best Liberty had to offer" young people, the "most serious problem affecting Liberty youth," and the one improvement project they would like to see take place if they were "Mayor for a day." Within minutes, the room quietened down as the participating youth sketched their images of the city and proposals for improving the quality of urban life. As the students did so, university volunteers took their photographs and collected basic biographical information so that they could create the kinds of informational plaques that are traditionally attached to art displays in museums.

Several weeks after this activity, the students' work was displayed as part of an interactive community-building exhibit organized by the steering committee and aided by the student planners at the Liberty Historical Museum. Each student's work was neatly matted, framed, and accompanied by a small biographical plaque with their name, age, career aspiration, family profile, and picture. More than 200 Liberty residents visited this installation to gain a better understanding of how their community's young people view their Village and Town. Many of those attending this event had not participated in the community's past planning activities but admitted that they were drawn into the process by their children's engagement.

## Community-building exhibition

A third community planning meeting was organized to involve residents in a final review and analysis of the conditions and projections data generated by the planning students. However, it was preceded by a week-long interactive planning exhibit at the Liberty Museum's Main Street facility, sponsored by the steering committee and assisted by the Museum's Board of Directors. The students presented the data organized by theme through a series of interactive installations designed to elicit additional resident input for the planning process.

Among the interactive installations was a board where local residents could place pins to identify locations within the community to which they regularly drove. They were then asked to connect the location of their home, workplace, child's daycare center, or older children's school with rubber bands.

Over the course of the week, the community's most heavily traveled areas became apparent, enabling residents to think about this question posed by the planners: "Are there ways we could simplify your life by re-imagining the location of certain community facilities (i.e., daycare centers, schools, retail stores, the "Y," etc.)?" Another exhibition gave people the opportunity to mount a fake pulpit, wear a Burger King crown, and as "King for a Day," pronounce the single most important improvement they would make in Liberty if they were given the chance. Each resident's photo was taken as "King" and their proposals were documented in the exhibit. Local residents were both amused by the photos and impressed by the dozens of innovative ideas proposed by their neighbors.

## Envision Liberty Week

One of the most enjoyable and productive activities that the planning organizers undertook was "Envision Liberty Week," which occurred two-thirds of the way through the planning process. It was advertised by a banner spread across Main Street featuring an enormous *faux* pair of Armani sunglasses accompanied by the slogan, "Envision Liberty: Making a Great Community Even Better, October 15–22." The event began with the week-long planning exhibition at the Liberty Museum, already described. On Friday evening, local volunteers and student planners cleaned, relit, and reopened the Village's Art Deco era theater.

More than 80 residents came to an event entitled "Liberty Today: A Community on the Move" in which the student planners discussed with residents the planning and policy-making implications of their activities including creating or amending archives; doing building surveys; conducting oral histories with residents and officials (movers and shakers), and holding focus groups. The evening ended with both a dialogue on the strengths and weaknesses of four alternative development scenarios and a vote on their preferred option. The following morning, more than 100 residents appeared for the second part of the program entitled "Guided Visualization: Crafting Liberty's Future."

As people gathered at the theater, they were organized into 15-person groups and assigned a facilitator, recorder, and image maker (a.k.a. artist). These groups were then asked to follow their facilitator to one of the many vacant storefronts on Main Street, which had been cleaned out, equipped with chairs, and supplied with ample amounts of newsprint, markers, tape, and a CD player. A total of eight groups spent the morning engaged in a rather unusual "Envisioning Activity" in which they were asked to get as comfortable as they could in their chairs and close their eyes while listening to the jazz music of Miles Davis. Each facilitator asked the participants to imagine that they were home alone (no kids, parents, spouse, or pets), sitting in their favorite room, resting in their favorite chair, drinking their favorite New York State wine, and falling into a very deep and restful sleep.

They were then asked to imagine that the years begin to fly by, just as they did for another Upstate New Yorker – Rip Van Winkle. They passed through 2004, 2005, 2006, 2007 … and soon it was 2018. They were asked to imagine that Liberty, New York had, through local resident and official effort, become everything they every wanted it to be! They were then asked to imagine that they were stepping out of the front door of their home and, with their virtual video camera, recording the most exciting aspects of the "New and Improved" Liberty.

Next, they were asked to take a few minutes to document the most powerful, transformed, and uplifting aspect of the New Liberty community. At the count of three, participants were asked to wake up and share their most powerful image of the New Liberty community with the group. As they shared their visions of the New Liberty, the artist who had accompanied the group to their storefront created a large image of the described scene in what we referred to as an "idea bubble." Following the reports, each participant was given a set of five green dots and a skull-and-crossbones sticker. Finally, they were invited to tour the images that now covered the walls of the room and, using the green dots or skull-and-crossbones stickers, to approve the visions they found most compelling or to disapprove projects they would fight the hardest to oppose.

After this exercise, the groups reconvened at the Liberty Museum where they briefly shared a quick summary of all their ideas while taking a few extra minutes to describe their most popular proposal. Following their reports, all the participants were given a second set of green dots and skull-and-crossbones stickers to share their views on the entire set of proposals developed by the group. Envision Liberty Week ended with the identification of five action areas which residents and officials viewed as critical to the community's future economic development. Among these were community organization, small business assistance, quality affordable housing, workforce training and development, youth leadership, and urban design. Before the meeting was adjourned, each participant was asked to both identify which of these community development areas they were most interested in and to set a date to meet with those sharing these interests.

## Action teams

In the months following Envision Liberty Week, to assist Liberty in making progress toward its overall economic development goals, more than 100 Liberty residents met to identify a set of immediate, short, and long-term development projects that could be undertaken by the community in their specific program areas. While the five action planning teams worked in somewhat different ways, they all tended to follow a similar process. Each team began work by inviting their members to "brainstorm" the longest list of possible development projects within their individual program area without consideration of feasibility. Having done so, each team identified two to three local funders with considerable development experience in their area to share their "quick and dirty" evaluation of: (1) their sense of the long-term value of each potential project, and (2) their assessment of the feasibility of funding such a project in the current local and regional context. Following these two steps, each group was asked to produce a list of six to eight high-priority projects that could be undertaken during the next five years. When this list was completed, two or three residents, along with a student planner, were assigned to research the basic literature related to this project as well as model projects successfully undertaken by communities of Liberty's size. For several weeks, these small teams worked on three to five briefing papers summarizing their research on "best practices" and "model projects." Then, prior to determining the final list of projects to be pursued in each program area and creating phases for each one, the briefing papers were sent to each member of the action team.

## Final community forum

When this work was completed, another community-wide meeting was held in which residents and officials from the Village and Town had the opportunity to discuss and adopt a final list of project proposals. Community participation was bolstered as the meeting was held at the corporate headquarters of the sponsoring foundation and, especially, because it was attended by that institution's founder, a highly respected national business figure and the President of Cornell University.

In the year following the completion of the Liberty Economic Development Strategy, the Village and Town had worked with their allies to accomplish many things. With the help of the local foundation, the two municipalities came together to establish the Liberty Community Development Corporation, to mobilize public and private resources, and thereby to implement its very ambitious economic and community development agenda. With funds of over $100,000 it had raised, and in cooperation with the Cornell Extension Service, this group implemented an after-school business internship program for "at-risk" high school students. Youth participating in the process worked together to secure land for the construction of a skateboard park; complete a design for the project; and raise funds to build the first improvements needed to establish that facility.

The Village, with the assistance of the Historic Preservation students and alumni of Cornell University, succeeded in cleaning and stabilizing a Mondrian-inspired commercial building that had stood vacant for many years at the corner of a prominent downtown intersection. This resulted in purchase of the building by an outside developer and adapted as an antique furniture and vintage clothing store. Finally, again with the assistance of Cornell planning students and faculty, the Village and Town completed plans for downtown traffic and transportation improvement and revitalization of the Village's major public park. The collaboration also produced a new subdivision proposal for the nearby hamlet of Swan Lake, based upon the behavioral guidelines contained on the Torah, to meet the needs of this rapidly growing Hassidic community.

## Keywords

Participatory neighborhood planning, community-based development organizations, resident-led planning and design, empowerment planning, asset-based community development, participatory action research.

## Review questions

1 How does an asset-based approach to community development impact neighbourhood planning versus a deficit-based approach?

2 What is a SWOT analysis and how is it used in neighborhood planning?

3 Describe a process to encourage citizen participation in neighborhood planning and development decision making.

## Bibliography and additional resources

Afshar, F. and Pizzoli, K. (2001) "Editors' Introduction," *Journal of Planning Education and Research*, 20(3): 277–280.

Anderson, M. (1966) "The Federal Bulldozer," in J.Q. Wilson (ed.) *Urban Renewal: The Record and the Controversy*, Cambridge, MA: MIT Press, pp. 491–508.

Bowes, J. (2001) *A Guide to Neighborhood Planning*, Ithaca, NY: City of Ithaca Department of Planning.

Brophy, P. and Shabecoff, A. (2001) *A Guide to Careers in Community Development*, Washington, DC: Island Press.

Hartman, C. (1974) *Yerba Buena: Land Grab and Community Resistance in San Francisco*, San Francisco, CA: Glide Publications.

Heiman, M. (1988) *The Quiet Evolution: Power, Planning, and Profits in New York State*, New York: Praeger Publishers, pp. 30–97.

Jones, B. (1990) *Neighborhood Planning: A Guide to Citizens and Planners*, Chicago, IL: Planners Press, pp. 1–38.

King, M. (1981) *Chains of Change: Struggles for Black Community Development*, Boston, MA: South End Press, pp. 111–118.

McKelvey, B. (1948) "A Rochester Bookshelf," *Rochester History*, 10(4): 7–13.

Moynihan, D.P. (1970) *Maximum Feasible Misunderstanding*, New York: Free Press, pp. 75–101.

National Congress for Community Economic Development (NCCED) (2006) *Reaching New Heights: Trends and Achievements of Community-based Development Organizations, 5th National Community Development Census*, Washington, DC: NCCED.

Peterman, W. (2000) *Neighborhood Planning and Community-based Development: The Potential and Limits of Grassroots Action*, Thousand Oaks, CA: Sage, pp. 1–32.

Robinson, C.M. (1899) "Improvement in City Life: Aesthetic Progress," *Atlantic Monthly*, 83 (June): 771–185.

Schorr, L.B. (1997) *Common Purpose: Strengthening Families and Neighborhoods to Rebuild America*, New York: Anchor Books, p. i–xxviii.

SRI (2007) *Timeline of SRI International Innovations*, Palo Alto, CA: SRI International.

Stein, C. (1957) *Towards New Towns for America*, Cambridge, MA: MIT Press, pp. 37–64.

Thabit, W. (1961) *An Alternative Plan for Cooper Square*, New York: Cooper Square Community Development Committee and Businessmen's Association.

Welter, V. (2002) *Biopolis: Patrick Geddes and The City of Life*, Cambridge, MA: MIT Press.

# 19 Measuring progress
## Community indicators, best practices, and benchmarking

## Rhonda Phillips and Robert H. Pittman

Measurement and evaluation of community development progress is not only challenging, it is vital. Communities must be able to demonstrate the value and outcomes of their activities in order to be accountable to citizens, to secure funding, and to assess the efficacy of their programs. Community indicators may be used to evaluate the progress of communities and community development organizations. Communities face many needs and opportunities, and must allocate limited funds and human resources as efficiently as possible to successfully achieve their goals across these areas. Best practices and benchmarking are valuable tools in community decisions on development program structure, operations, and follow-up modifications.

## Introduction

What is evaluation? Simply put, it is a way to figure out the importance, value, or impact of something. There are numerous ways to "figure it out" and numerous "things" that may need to be evaluated. So that it is clear what is being evaluated and which approaches will be utilized, evaluation is typically conducted in a methodical manner with a defined process or approach. Thus, evaluation may be defined as a systematic determination of the value or quality of a process, program, policy, strategy, system, and/or product or service including a focus on personnel (Davidson 2005).

Getting what a community wants in the future requires evaluation. Past performances can be reviewed to estimate future outcomes but, more importantly, evaluation should be included in the continuous cycle of program and policy development and implementation. Evaluation is not a one-time effort; it should be ongoing and periodic. Evaluation helps communities to develop, evolve, and improve in a constantly changing environment. Every time something new is tried – be it a policy, strategy, program, process, or system – its value must be considered (Davidson 2005). In community development, evaluation is particularly critical because citizens' quality of life is affected by such policies, programs, strategies, and so on. If the impact and outcomes have not been soundly evaluated, may it be said that one approach is better than another or has a more positive influence?

While all communities are unique, many share common problems and issues. These may be addressed by previously developed strategies and solutions. To avoid wasting resources and "reinventing the wheel" when confronting an issue, a community should first conduct research into such proven best practice solutions. Benchmarking, or measuring one or more aspects of a community or program against its counterparts, is also a useful way to measure progress and provides additional perspectives on community indicators. This chapter discusses how indicators, best practices, and benchmarking can be used to assess community development progress.

## Community indicators

Given the importance of evaluation, contrasted with the complexity and barriers to conducting it, what should a community or community development organization do? Among evaluation techniques in the field of community/economic development planning, the use of community indicators is reemerging. These indicators were first used over 100 years ago but their new application is more beneficial and useful.

When used as a *system*, indicators hold much promise as an evaluation tool. What makes community indicators any different from other measures of community development such as job growth or changes in per capita income? The key is developing an integrative approach, to consider the impacts of development not only in terms of *economic* but its *social* and *environmental* dimensions. A community indicators system reflects collective values, providing a more powerful evaluative tool than simply considering the economics of change and growth. When properly integrated into the early stages of comprehensive community or regional planning, community indicators hold the potential to go beyond mere activity reports because they can be used systematically, making it easier to gauge impacts and evaluate successes. Furthermore, these indicators incorporate frameworks of performance and a full spectrum of process outcomes, both of which facilitate evaluation and decision making.

Just what are community indicators? Essentially, they are pieces of information that combine to provide a picture of what is happening in a local system. They provide insight into the direction a community is taking; whether it's improving or declining, moving forward or backwards, increasing or decreasing. Combining indicators creates a measuring system to provide clear and accurate information about past trends, current realities, and future direction in order to aid decision making. Community indicators may also be thought of as a report card of community well-being. It is important to note that these systems generate much data. It is the analysis of these data that may be used in the decision making and policy/program improvement processes.

There are four common frameworks used for developing and implementing community indicators systems in the U.S.: (1) quality of life; (2) performance evaluation; (3) healthy communities; and (4) sustainability.[1] A summary of each type is presented below, and a case study is provided to illustrate the usefulness of community indicators.

## Quality of life

Quality of life is reflective of the values that exist in a community. Indicators may be used to promote a particular set of values by making clear that residents' quality of life is of vital importance. If agreement can be reached, the advantage of this type of system is its strong potential to stimulate all types of community outcomes, not the least of which is evaluating progress toward common goals. The disadvantage is that measuring quality of life is a political process. What defines "the good life" differs among individuals, groups, and institutions.

The most notable example of this quality of life framework is the "Jacksonville Community Council Inc." Many U.S. community indicator projects are based on this model, which started in 1974 and has since become a part of the Floridian city's ongoing evaluation and decision making (Swain 2002). It attempts to integrate indicators into overall planning activities while monitoring for consistency with the comprehensive plan and other plans. The system has ten indicator categories, and annual quality of life reports and indexes are released on each.

## Performance evaluation

This type of indicator is mostly managed by state or local governments as a way to gauge the outcomes achieved by their activities. It is very beneficial as an evaluative technique because it provides reports on progress and outcomes, usually annually or semi-annually. It is typically part of the annual budgeting process so that adjustments can be made for priority areas.

An example of this type of indicator system is "Oregon Benchmarks" (2007). It uses urgent and core indicators to prioritize budgetary considerations and subsequent policy and programmatic responses. It incorporates citizen participation through public forums and hearings but elected and appointed officials usually have final responsibility.

## Healthy communities

This approach is gaining popularity as it attempts to cultivate a sense of shared responsibility for community health and well-being. It focuses on indicators that reflect health care in the phases of life that often do not show on typical economic indicators concerned with working adults: prenatal, early childhood, and youth. Healthy communities also prioritizes education and other human development facets including social concerns.

An example is Hampton, Virginia's "Healthy Families Partnership" (2007). By focusing on healthy children and families, Hampton has garnered desirable community and economic development outcomes. Some of the indicators in its system have provided remarkable evidence of progress on a variety of challenging issues.

## Sustainability

Community indicator systems can provide the mechanism for monitoring progress toward balanced or sustainable development because they provide information for considering the impacts of development, not only in economic terms but in social and environmental dimensions. The concept of sustainable development includes such characteristics as broad citizen participation, ongoing assessment, and a guiding vision. Indicators are consistent with these principles. The difficulty with this approach is to fully integrate the use of indicators into overall community planning so that sustainability can be a reality instead of rhetoric.

There are several examples of this approach to developing indicators systems. Santa Monica, California's "Sustainable City Plan" (2007) is one of the most progressive of its type in the U.S. The City selected eight areas, each with specific goals and indicators to measure progress. Using their own community as an experiment, they now vigorously practice what they preach by converting the City's vehicle fleet to natural gas systems and instituting green building standards for all new or renovated buildings. Another example is Seattle, Washington. "Sustainable Seattle" (1991) is a nonprofit group that gauges the City's progress on a variety of indicators. Again, many similar efforts at local levels are modeled after this program. In response to this nonprofit initiative, the City created the Office of Sustainability and Environment to encourage integration of the indicator system with overall city functions and planning activities.

## Best practices and benchmarking

"Best practices" and "benchmarking" are terms commonly used in business and industry. They came into vogue in the 1980s when a host of books on business competitiveness were published. Since that time, these terms have been widely applied to many disciplines including community and economic development.

How are best practices and benchmarking defined? Consider these definitions from various sources.

## Best practices

The processes, practices and systems identified in public and private organizations that performed exceptionally well and are widely recognized as improving an organization's performance and efficiency in specific areas.

(The U.S. General Accounting Office)

The term best practice generally refers to the best possible way of doing something; it is commonly used in the fields of business management, software engineering, medicine and increasingly, in government.

(Wikipedia: The Free Encyclopedia)

## Benchmarking

The process of identifying, learning and adapting outstanding practices and processes from any organization, anywhere in the world, to help an organization improve its performance.

(International Network for Small and Medium (Business) Enterprises)

Measuring how well one country, business, industry, etc. is performing compared to other countries, businesses, industries and so on. The benchmark is the standard by which performance will be judged.

(European Union)

As seen in these definitions, best practices and benchmarking are concepts with broad applicability in a number of disciplines. Community and economic development organizations at all levels have adapted these tools to gauge success and track progress toward their goals.

## Applications in community development

As discussed throughout this book, community development is a broad discipline. Economic development is a part of community development and there is a broad literature on best practices and benchmarking in economic development. However, community development also entails leadership, infrastructure development, effective local government, health care, workforce development, and so on. Therefore, when applying best practices and benchmarking to community development, one must specify a particular aspect of the field. If the focus is on strategic planning and visioning as applied to both public and private sectors, there is a wide body of best practices and benchmarking literature. Likewise, if the focus is on local transportation, there is a corresponding body of literature.

An Internet search for "community development" and "best practices" yields over 200,000 websites reflecting the eclectic nature of the discipline. Some of the topics include:

- leadership development;
- venture capital;
- mapping relationships between individuals and groups in the community;
- public health;
- homelessness;
- substance abuse;
- managing real estate;
- project financing;
- housing;
- grant development.

The list goes on an on. Clearly, best practices and benchmarking in community development must be defined in terms of specific aspects of this broad field.

Utilizing case studies of particular situations in community and economic development is certainly one of the most useful ways to identify best practices and conduct benchmarking. However, by definition, case studies deal with a particular situation. One would need to gather many of them in a specific area in order to identify a trend or common best practices theme. Surveys of similar organizations can provide a broader set of best practice parameters. Both techniques can provide useful best practices and benchmark information. Case studies can provide depth of information while surveys can provide breadth.

## Why are best practices and benchmarking important?

As noted above, while every community is unique, they all face many of the same issues. How can more industry be recruited? How can transportation options for low-income workers be improved? How can a strategic plan be developed when community factions oppose each other? These are just some examples of the plethora of issues that communities may face.

Rather than try to address these issues in a vacuum, doesn't it make sense to learn from the ways that other communities have addressed them? So much can be learned from examining the innovative solutions that other communities have developed to address similar issues or problems.

Another reason why best practices and benchmarking are important in community development is that they serve as evaluation tools for programs and policies. Take economic development as an example. Many communities set economic development goals such as "Create 100 net new jobs each year for the next five years by recruiting new companies." This is an admirable goal but, given the history of the community and resources devoted to economic development, is it realistic? What if the community doesn't meet the goal? Should it dismiss the economic development staff or lower its sights?

Before taking any action, the first question that should be asked is "Why didn't we meet the goal?" The answer could be that the economy has taken a downturn during the year and the number of new company locations and expansions is much lower as a result. When economic development goals are not met, it is usually for reasons outside of the community's control. By the same token, when a goal is not met in a particular year, the mistake is often to "throw the baby out with the bathwater" by totally changing an economic development program.

Because there are reasons for not meeting community goals, whether in economic development or other areas outside of its control, program evaluation should consist of looking at the "outputs" of the program (i.e., did we meet our goals) and the "inputs" (i.e., resources, staff, structure, productivity). One way to measure inputs such as program resources inputs is to see if all the action steps and items agreed upon were executed in accordance with the strategic plan. Another way is to compare a community's programs to those of similar communities that have proved successful. Does a community allocate a comparable amount of budget and staff time to a particular program? Are the program elements similar to those of other successful communities? This exercise, according to the definitions above, is what best practices and benchmarking analysis is all about.

## Finding existing best practices/benchmarking research

As discussed above, in applying best practices and benchmarking to community development, the first step is to define the specific field of interest. As an example of best practices and benchmarking, let us consider the field of economic development. Numerous and varied organizations have done considerable research into best practices in several areas of economic development including organizational structure, marketing, business retention and expansion, and business start-up.

Below are some organizations that can be valuable sources of information on best practices and benchmarking in economic development:

- National and international government organizations such as the U.S. Economic Development Administration, the Department of Commerce, the Small Business Administration, the Department of Housing and Urban Development, and the Department of Labor.
- National and international nongovernmental organizations (NGOs) such as the United Nations, the World Bank, and the Organization for Economic Cooperation and Development (OECD).
- National and international professional associations such as the International Economic Development Council, the Community Development Council, the Council of State and Community Economic Development Agencies, and the National Association of State Development Agencies.

- Regional professional associations such as the Southern Economic Development Council, the Mid-American Economic Development Council, and the Northeast Economic Developers Association. In addition, most states have economic development professional associations that can be good sources of best practices and benchmarking information.
- State departments of economic development and related state agencies such as labor, community development, or transportation.
- Utilities. Many utilities have active economic and community development programs and sponsor or conduct best practices and benchmarking studies.
- Universities. Many universities have economic development, business institutes, or related departments that pursue best practices and benchmarking or related research.
- Consultants. While studies by private consultants are usually proprietary to the client, economic and community development consultants are often willing to share information with other communities.

A second way to obtain studies and data on best practices and benchmarking is via the Internet. An Internet search for "community development," "economic development," and "best practices" yields hundreds of thousands of web pages. Here are a few useful sites:

- U.S. National Agricultural Library, USDA, Rural Information Center, Rural Resource Center: www.nal.usda.gov/ric/ruralres.
- National Governors Association: www.nga.org.
- Federal Reserve Bank of Chicago, Consumer and Economic Development Research Information Center, LesLe (Lessons Learned): www.chicagofed. org/community_development/lesle/index.cfm.

In addition to these general sites, an Internet search also turns up thousands of references to case studies and/or best practices analyses by communities who have dealt with specific economic development areas.

Since such studies are usually in the public domain for their communities' benefit, many are available to download. Alternatively, one may call the relevant community contact who is usually willing to share and discuss the study.

A third way to identify information on best practices and benchmarking is to search research and newspaper databases. For example, a search of the *ProQuest* database, which includes journals and newspapers, yielded the following on economic development best practices and benchmarking:

- "The New Transit Town: Best Practices in Transit-Oriented Development," D. Brand, *Choice*, September 2004.
- "Tucson AZ Panel Evaluates Best Practices in Economic Development," Tim Steller, *Knight Ridder Tribune Business News*, June 13, 2004.
- "European Cities Join Up to Share Best Practices," Ben Walker, *Regeneration and Renewal*, March 19, 2004.
- "Creating Regional Wealth in the Innovation Economy: Models, Perspectives and Best Practices," J.M. Nowakowski, *Choice*, June 2003.
- "TQM Benchmarking For Economic Development Programs," Eric Canada, *Economic Development Review*, summer 1993.
- "Economic Development Best Practices: The Best and the Worst," James. H. Renzas, *Economic Development Review*, fall 1994.
- "Best Practices Study Takes a Look at NW Ohio Development Effort," *Toledo Business Journal*, July 1, 1996.
- "Benchmarking University–Industry Technology Transfer," Louis G. Tornatzky, *Journal of Technology Transfer*, June 2001.

## Conducting a best practices/benchmarking study

It can sometimes be difficult to find best practices and benchmarking studies, or information dealing with a community's specific topic of interest, that are based on communities of similar size and circumstances. For example, a community might want to know how regional economic development marketing programs have been set up in other areas of similar population. What kind of budget and staff requirements have been involved? How has the regional marketing entity be organized? The community might also want to benchmark themselves against closely comparable areas in terms of employment and industry base.

Fortunately, it is not very difficult to develop a survey form and solicit other communities or regions to participate in a best practices/benchmarking study. Communities are usually happy to participate in such studies, especially if they receive the results in order to benefit from the time and effort they invested.

Best practices/benchmarking studies can be anonymous or not. The decision would be influenced by the questions in the survey and the preferences of the participating community. In the case of economic development information, much of that is usually public (e.g., marketing budget, staff size) because public monies are often involved.

Here are the steps in conducting a best practices/benchmarking study for a community:

- Identify the topic. The more specific and definable the better.
- Identify the communities or areas with which you want to be compared. Usually four to six communities are sufficient for a best practices comparison. As mentioned above, often the comparison communities are of similar size and situation. However, some prefer to benchmark themselves against exemplary communities they hope to emulate.
- Call the appropriate representative(s) from the comparison communities and explain the process. Determine whether the participants prefer to remain anonymous or not. Decide whether you will share the results with the participants in return for their cooperation.
- Develop the survey form; an example is given in Box 19.1. Decide whether the survey is going to

be administered in written form – either hard copy or Internet – by telephone, or in person. If possible, travel to the comparison communities to administer the survey in person in order to get the best results. Among these are the many opportunities for elaboration and "off-the-record" conversations.
- Administer the best practices/benchmarking survey and collate the results. Compare your community against the others on all questions and analyze how yours is different or alike. Based on the survey results, develop program recommendations for the future.

## Conclusion

Evaluation is critical, as communities and community development organizations cannot rely solely on intrinsic perceptions of success. Evaluations must be integrated into the continuous cycle of organizational management and operations. There are numerous approaches to evaluation. One of these, community indicator systems, holds much promise as a means to integrate ongoing evaluation into overall development efforts.

Best practices and benchmarking are useful tools to address common issues and problems and benefit from collective wisdom and experience. These tools may also be used to evaluate the structure and effectiveness of ongoing programs. Community development is an extremely broad field so, in order to be manageable and meaningful, a best practice topic must be as narrowly defined as possible.

Best practices/benchmarking studies and data may be obtained from Internet searches which yield relevant organizations, associations, and literature. Probably the best way to obtain specific, applicable best practices results is to conduct a study, picking participant communities that provide "apples to apples" comparisons. Best practices/benchmarking studies can be a relatively low-cost way to ensure that a community's programs are on track to produce the desired results.

## BOX 19.1 SAMPLE BEST PRACTICES SURVEY

### Survey questions on regional economic development marketing programs

1 Name of regional marketing organization.
2 Names of counties and municipalities covered by the marketing organization.
3 Total population of area served by marketing organization.
4 How many years has the regional marketing organization been in existence?.
5 Please indicate if the organization engages in the following types of marketing activities and how effective you believe they are:

| Activity | Effectiveness | | | |
| --- | --- | --- | --- | --- |
| | Very effective | Somewhat effective | Not effective | Don't do |
| National advertising | | | | |
| Local/regional advertising | | | | |
| Direct mail | | | | |
| Email | | | | |
| Trade show attendance | | | | |
| Consultant visits | | | | |
| Prospect visits | | | | |
| Networking | | | | |
| Telemarketing | | | | |
| Other (please list below) | | | | |

6 How many full-time staff members are there? Part-time?
7 Would you please provide the titles of each staff person and a brief description of their primary job responsibilities.
8 If possible, can you share with us what the total annual budget is for the regional marketing organization? Can you provide us with any breakdown of budget allocation across major areas (e.g., advertising and other marketing activities, staff, rent)?
9 Please describe the organization's funding. Is it public, private, or a mixture? How do you determine the financial contribution from each county or municipality?
10 What local in-kind contributions does the regional marketing organization receive?
11 Do you have a written marketing plan? How often do you update it?
12 Please provide any other information that you believe will help contribute to a successful regional marketing effort.

Thank you for your time and cooperation!

## CASE STUDY: COMMUNITY INDICATORS – HERNANDO COUNTY, FLORIDA

Comprehensive indicators have been developed for Hernando County, Florida as part of its desire to plan strategically for development outcomes in the community. The indicators are used as a system, considering the relationship with overall comprehensive planning and community economic development efforts by both the public and private sectors. The following is excerpted from Phillips (2005: 121–128). The case uses a set of economic, social, and environmental indicators to gauge changes and outcomes in the county.

### Economic indicators

Gauging the economic health of a community is vital. The ability of a local economy to provide job opportunities and income to residents is paramount. It also impacts the remaining community factors such as social and environmental aspects. The economic indicators for Hernando County are divided into six areas of county revenue sources: population, employment, income, poverty, housing and real estate. Table 19.1 provides examples of these indicators, the data sources used to calibrate them, and the reasons for including each.

**Table 19.1** Economic indicators

| | Indicator | Data source | Update | Reason |
|---|---|---|---|---|
| County revenue sources | Property tax (taxable real estate value) | County property appraiser | Yearly | Indicates the value of homes in the area, thus the economic prosperity of residents. |
| Population | Incorporated and unincorporated | Florida Office of Economic and Demographic Research: www.state.fl.us/edr/population/citypop2000.pdf. Also available at: http://fcn.state.fl.us/lcir/databank/popdata.html | Yearly | Shows where the majority and minority of people live and may determine the allocation of services. |
| Employment | Total employment and number of jobs | www.webcoast.com/dhernand.htm. Current information available at bureau of labor statistics: http://stats.bls.gov/news.release/cpi.toc.htm | Yearly/quarterly | Strong indicator of the economic health of a community. |
| Income | Median, household, personal income | Florida Research Economic Database: http://fred.labormarketinfo.com/. Also BLS: http://stats.bls.gov/blshome.htm | Yearly | Indicator of the economic health of community. |
| Poverty | % of population on public assistance | U.S. Census | Yearly extrapolations | If large percentage of population is on public assistance, may indicate a lack of jobs or declining economic base. |
| Housing and real estate | Total number of housing units (low income, single family) | Local builders' association | Yearly | Indicates total strength of real estate market. |

## Social indicators

The social indicators selected include a variety of areas and reflect the issues identified during the community assessment activities. There are 72 indicators divided into six categories: education, crime, health and wellness, transportation, organizations, and government. Social indicators attempt to determine how the community is changing in a variety of contexts. These indicators answer such questions as: Is crime increasing? Is the quality of the educational system improving? Are transportation needs being met? Social indicators are vital to overall community health and well-being and, as such, provide the basis for economic development outcomes. It is difficult to consider community economic development planning goals and activities without addressing concurrently the social impacts of proposed actions. Table 19.2 provides some example of social indicators.

▧ **Table 19.2** Social indicators

|  | Indicator | Data source | Update | Reason |
|---|---|---|---|---|
| Education | School population | School board/local government | Yearly | Student population data help to determine if overcrowding is an issue. |
| Crime | Number of violent crimes reported | Police department/Census data | Yearly | Crime is a key measurement of quality of life. |
| Health and wellness | Number of hospital beds per capita or per 1000 | Local government | Yearly | A key element in the quality of life for any community is the quality and availability of health care. |
| Transportation | Time spent commuting | Survey of residents | Yearly | Measures quality of life; indicates overcrowding and inadequate infrastructure. |
| Organizations | Active volunteer organizations | Survey/Chamber of Commerce | Yearly | Provides information on how active citizens are in community affairs. |
| Government | Number of registered voters | Board of Elections | Yearly | These indicators provide a view of how involved citizens are with their government institutions and processes. |

## Environmental indicators

Environmental indicators are a vital and necessary component of an overall community evaluation system. Too often, environmental factors are pushed to the forefront after an issue reaches crisis status because a community has waited until an issue had to be addressed. A better community approach is to monitor its environmental health, resources, and status as a part of overall planning efforts. Four categories of indicators, with a total of 13 across the categories, have been selected for Hernando County's community indicators framework. These reflect "benchmarks" for gauging the impact on the environment of rates and types of growth. The four categories of water, solid waste, air, and land cover several measures within each. When combined, these categories provide a comprehensive picture of the state of the environment and use of resources in the area. Table 19.3 provides some examples of environmental indicators.

**■ Table 19.3** *Environmental indicators*

|  | Indicator | Data source | Update | Reasons |
|---|---|---|---|---|
| Water | Water use | Utility organizations | Yearly | Reveals the usage over time. |
| Solid/hazardous waste | Tons of waste | Landfill records | Yearly | Reveals the amount of solid waste and whether recycling is influential, etc. |
| Air | Average daily vehicle miles traveled | Transportation department records | Yearly | Measures road use and relates to emissions. |
| Land | Major bodies of water | Local maps | Yearly | Protects from pollution. |

## Keywords

Evaluation, performance evaluation, community indicators, community indicator, best practices, best practice, benchmarking, case studies, comparison communities.

## Review questions

1  What are community development indicators and why are they useful?
2  What are some types of community development indicators?
3  What are "best practices?"
4  What is "benchmarking?"
5  Why are best practices and benchmarking useful in community development?
6  What are the advantages to doing your own best practices/benchmarking study rather than relying on existing studies and research?

## Note

1  A full description of these frameworks was first presented in Phillips (2003).

## Bibliography and additional resources

Brand, D. (2004) "The New Transit Town: Best Practices in Transit-Oriented Development," *Choice*, 42(1): 156.

Canada, E. (1993) "TQM Benchmarking for Economic Development Programs," *Economic Development Review*, 11(3): 34–39.

Council for Community and Economic Research (Formerly "American Chamber of Commerce Research Association"). Available online at http://www.c2er.org/ (accessed December 14, 2007).

Davidson, E.J. (2005) *Evaluation Methodology Basics*, Thousand Oaks, CA: Sage.

European Union, Brussels. Available online at http://ec.europa.eu/enterprise/library/lib-competitiveness/series_competitiveness.htm (accessed December 19, 2007).

Federal Reserve Bank of Chicago, Consumer and Economic Development Research Information Center, LesLe

(Lessons Learned). Available online at www.chicagofed. org/community_development/lesle/index.cfm (accessed December 19, 2007).

Hampton, VA Healthy Families Partnership. Available online at www.hampton.va.us/healthyfamilies/ (accessed December 4, 2007).

International Network for Small and Medium (Business) Enterprises. Available online at http://www.insme.org/ t_fulltext_search.asp?target=benchmarking&x=35&y= 12 (accessed December 19, 2007).

National Governors Association. Available online at www.nga.org (accessed December 19, 2007).

National Neighborhood Indicators Partnership. Available online at www.urban.org/nnip/ (accessed December 4, 2007).

Nowakowski, J.M. (2003) "Creating Regional Wealth in the Innovation Economy: Models, Perspectives and Best Practices," *Choice*, 40(10): 1795.

Oregon Benchmarks. Available online at www.oregon.gov/ DAS/OPB/index.shtml (accessed December 4, 2007).

Phillips, R. (2003) *Community Indicators. PAS Report No. 517*, Chicago, IL: American Planning Association.

—— (ed.) (2005) *Community Indicators Measuring Systems*, London: Ashgate Publishing.

Redefining Progress. Available online at www.rprogress.org (accessed December 4, 2007).

Reese, L. and Fasenfest, D. (2004) *Critical Evaluations of Economic Development Policies*, Detroit, MI: Wayne State University Press.

Renzas, J.H. (1994) "Economic Development Best Practices: The Best and the Worst," *Economic Development Review*, 12(4): 83–88.

Santa Monica Sustainable City Program. Available online at www.santa-monica.org/environment/policy/ (accessed December 4, 2007).

Steller, T. (2004) "Tucson AZ Panel Evaluates Best Practices in Economic Development," *Knight Ridder Tribune Business News*, June 13: 1.

Sustainable Communities Network. Available online at www.sustainable.org (accessed December 4, 2007).

Sustainable Seattle. Available online at www.sustainable seattle.org (accessed December 4, 2007).

Swain, D. (2002) *Measuring Progress: Community Indicators and the Quality of Life*. Available online at http://www.jcci.org/measuringprogress.pdf (accessed December 4, 2007).

*Toledo Business Journal* (1996) "Best Practices Study Takes a Look at NW Ohio Development Effort," 12(7): 1–11.

Tornatzky, L.G. (2001) "Benchmarking University–Industry Technology Transfer," *Journal of Technology Transfer*, 26(3): 269.

U.S. General Accounting Office, Washington, DC. Available online at http://www.gao.gov/ (accessed December 19, 2007).

U.S. National Agricultural Library, U.S. Department of Agriculture (USDA), Rural Information Center, Rural Resource Center. Available online at www. nal.usda.gov/ric/ruralres (accessed December 19, 2007).

Walker, B. (2004) "European Cities Join Up to Share Best Practices," *Regeneration and Renewal*, March 19: 1.

Wikipedia, "Best Practice." Available online at http://en.wikipedia.org/wiki/Best_practice (accessed December 19, 2007).

# PART IV

# Issues impacting community development

# 20 Community development finance

## Janet R. Hamer and Jessica LeVeen Farr

Success in community and economic development projects often hinges on financing. Since many projects may not qualify for conventional financing, it is important to be aware of federal, state, and private sources of capital for community and economic development. This chapter provides an overview and brief explanation of different types of community and economic development financing. It also discusses how to put a financing deal together.

## Introduction

Economic development is a comprehensive strategy that integrates a wide array of activities that help sustain and grow a local economy, and it is a critical component of community development programs at the neighborhood, city, and state level. Creative financing tools are usually required to accomplish economic development objectives. Most community and economic development finance programs are public–private partnerships designed to fill the funding gaps not covered by the private market alone. Programs are available to help finance economic development activities such as business recruitment and retention, job creation, small business assistance, and real estate development. The purpose of this chapter is to examine the broad community development finance industry focusing primarily on the tools that are provided through this industry for economic development project financing.

## Overview of basic community development finance vocabulary

The first step toward understanding community development finance is to become familiar with some basic terms.

*Community development finance:* A lending process designed to stimulate community and/or economic development. With the use of community partners and/or lending enhancements, financial institutions are able to lend to borrowers who cannot meet conventional credit underwriting standards.

*Subsidy:* Any financial assistance granted to an individual or organization.

*Debt:* Any money, goods or services owed to someone else. Debt may take the form of mortgages, other kinds of loans, notes, or bonds.

*Equity:* The ownership interest in a project after debt and other liabilities are deducted.

*Grants:* A gift usually given by a foundation, a government agency or the philanthropic community that may take the form of money, land, or in-kind services. Grants may provide equity to a project and can reduce the amount of debt required.

*Credit enhancements:* Special arrangements and programs that mitigate the credit risk associated with the borrower or the project, thereby affecting the evaluation of a potential borrower's creditworthiness. The enhancements may include mortgage insurance, tax credits, rent supplements, interest rate subsidies, loan guarantees, favorable structure, terms, conditions and pricing of credit products, underwriting flexibility, loan to value ratios and tax abatements.

*Interest subsidy:* A grant to reduce the interest a borrower is required to pay on a loan. Subsidies may take the form of a direct cash grant to a lending institution to lower (or buy down) the bank's interest rate; a government sponsored, low-interest loan subordinated to a participating lender; or a lower than market rate loan to a qualified borrower as the result of an advance from a public entity.

*Loan guarantee:* Repayment of loans may be guaranteed through private or public sector sources. Loan guarantees are used to reduce risk of loss to a lender and are usually considered a secondary source of repayment for a portion of the debt in the event of default. A loan guarantee can also improve a project's ability to secure private financing, or qualify loans for sale on the secondary market.

## Doing a deal: who are the players in community development finance?

Public–private partnerships play a key role in the success of most economic development projects. Frequently there are multiple partners involved in a project so it is very important that their roles be defined early in the planning process. Another key step is identifying the "visionary leader" for the project, who is often the person or organization that develops the first concept of the project and acts as motivator and spokesperson. This section looks at the various players most often involved in economic development projects.

*Government:* Local, state, and federal governments will typically be involved in an economic development project in three ways: (1) providing funding, (2) approving permits, and (3) acting as landowner or developer. Local government, in particular, is

always a critical player and needs to be brought into the planning process as early as possible.

*Financial institutions:* The Community Reinvestment Act (CRA) specifically encourages banks to help meet the credit needs of the communities in which they operate including low- and moderate-income neighborhoods. The CRA creates an incentive for banks to participate in the financing of community and economic development activities. Financial institutions generally provide the market rate financing on projects, invest in loan funds, and purchase tax credits. Larger banks may involve their bank-owned community development corporation. This type of corporation, either for-profit or nonprofit, is capitalized by one or more banks for the purpose of making debt and/or equity investments in projects that promote community and economic development.

*Community Development Financial Institution (CDFI):* CDFIs are private sector, financial intermediaries with community development as their primary mission. Some are chartered as banks or credit unions and others are nonregulated nonprofit institutions that gather private capital from a range of investors for community development lending and investing. CDFIs provide lending and equity financing for projects that typically could not be financed solely by a conventional financial institution.

*Community Development Corporations (CDCs):* CDCs are community-based organizations owned and controlled by community residents engaged in community and economic development activities, the majority of which are nonprofit 501(c)(3) organizations. The role of nonprofit CDCs is critical to the success of many economic development projects. CDCs have access to funding sources not available to the private sector or government partners, and they can also serve as the critical link with the surrounding neighborhood and community leaders. CDCs may take many forms including faith-based organizations. They also differ greatly in capacity and areas of focus. For instance, larger CDCs may have the capacity and experience to act as the developer and builder of a project, but many are required to outsource these activities to a third party.

*National intermediary:* National intermediaries are organizations that mediate between community-based organizations and large-scale sources of capital. These intermediaries function at the national level aggregating capital from sources such as foundations, corporations, and the government and disbursing it to local organizations for capital projects, operating support, and predevelopment financing along with technical assistance and capacity building. Local Initiative Support Corporation (LISC), Neighborhood Reinvestment Corporation (NRC), and Enterprise are the three largest national community development intermediaries. National intermediaries may serve limited geographic areas.

*Economic Development Commissions/Chambers of Commerce:* These entities are an important link with the business community and potential investors and developers of the economic development project. They may be structured as member-supported private organizations, government-supported entities, or divisions of local government. These organizations are essential players in the recruitment of new businesses to the project and can also mobilize support for the project with local government officials and potential private investors.

*Investor:* An investor is any organization, corporation, individual, or other entity that acquires an ownership position in a project, thus assuming some risk of loss in exchange for anticipated returns. In the case of economic development projects, the investor may be any combination of entrepreneur, small business owner, developer, purchaser of tax credits, or other large corporation.

## Doing a deal: finding sources of money

Putting together the financing for an economic development project requires convening the relevant players and identifying the resources they have available. Government has always played a significant role in economic development financing, but funding from some traditional federal government programs has been declining, making state and local

government funding increasingly more important. There is a growing emphasis on public–private partnerships that bring multiple players together and use limited public resources to leverage private sector investment.

Financing for economic development activities is divided into several broad categories including low-cost loans and loan guarantees, subsidies or grants, and tax credits or abatements. This section reviews the financial resources offered by all levels of government and other players in the community development finance industry.

## Federal government resources

The federal agencies that provide financing for community and economic development are the U.S. Department of Housing and Urban Development, the U.S. Department of Agriculture, the U.S. Department of Commerce, the Economic Development Administration, and the U.S. Environmental Protection Agency. In some cases, funding is passed through state or local agencies that use the funds for a variety of community development purposes. In other instances, funding is provided directly from the federal agency to a project. Actual grantees may be local or state government, nonprofit organizations, or for-profit entities depending on the source of funds and regulations.

### U.S. Department of Housing and Urban Development (HUD)

HUD has several programs that provide grants, loans, and loan guarantees to support economic development.

- *Community Development Block Grant Program (CDBG):* CDBG is one of the most flexible federal programs intended for use by cities and counties to promote neighborhood revitalization, economic development, and improved community facilities and services principally to benefit low- and moderate-income persons and communities. Specific uses of the funds are left to the discretion of the local governments. Examples of eligible economic development activities include infrastructure improvements, small business loan programs, capitalization of revolving loan funds, purchase of land, and commercial rehabilitation activities. Local governments may provide funding for projects in the form of loans or grants. Historically, 8–15 percent of CDBG funds have been used for economic development purposes.

Through the CDBG entitlement program, HUD allocates annual CDBG grants to entitlement communities. Entitlement communities are the principal cities in large Metropolitan Statistical Areas (MSAs), other metropolitan cities with a population greater than 50,000, and urban counties with populations greater than 200,000. These communities are allowed to develop their own community development programs and priorities. States administer CDBG funds for all non-entitlement communities (i.e., those communities that are not part of the CDBG entitlement program). Non-entitlement communities are cities with populations of less than 50,000 or counties with populations under 200,000. Annually, each state develops funding priorities and criteria for selecting projects and awards grants to local governments that carry out community development activities.

- *Section 108 Loan Guarantee Program (Section 108):* Section 108 is the loan guarantee provision of the CDBG program that allows jurisdictions to borrow against their future CDBG allocations. This program provides communities with a source of financing for economic development, housing rehabilitation, public facilities, and large-scale physical development projects.
- *Brownfields Economic Development Initiative (BEDI):* HUD also provides financial support for the redevelopment of "brownfields" which are environmentally contaminated industrial and commercial sites. The BEDI provides grants on a competitive basis to entitlement communities as defined above. Non-entitlement communities are eligible as supported by their state governments. BEDIs must be used in conjunction with loans guaranteed under the Section 108 Program. Communities fund projects with the BEDI grants and the Section 108 guaranteed loan financing to clean up and redevelop brownfields.

*U.S. Department of Agriculture Rural Development (USDA)*

The USDA provides financing to support economic development in rural communities through several programs that offer grants and zero-interest loans.

- *Rural Business Enterprise Grants:* The Rural Development, Business and Cooperative Programs (BCP) make grants under the Rural Business Enterprise Grants (RBEG) Program to local governments, private nonprofit corporations, and federally recognized Indian tribal groups to finance and facilitate development of small and emerging private business enterprises located in any area other than a city or town that has a population of greater than 50,000 and the urbanized area contiguous and adjacent to such a city or town. Local government, private nonprofit corporations, and federally recognized Indian tribes receive the grant to assist a business.
- *Rural Business Opportunity Grants:* This program provides grants to pay the costs of providing economic planning for rural communities, technical assistance for rural businesses, or training for rural entrepreneurs or economic development officials.
- *Rural Economic Development Loans and Grants:* This program provides zero-interest loans to electric and telephone utilities financed by the Utilities Program, an agency of the United States Department of Agriculture, to promote sustainable rural economic development and job creation projects.

*U.S. Department of Commerce,*
*Economic Development Administration*
*(EDA), Public Works and Economic*
*Development Program*

The EDA is another source of financing for economic development projects. The Public Works and Economic Development Program provides matching grants for public works and economic development investments to help support the construction or rehabilitation of essential public infrastructure and facilities necessary to generate or retain private sector jobs and investments, attract private sector capital, and promote regional competitiveness including investments that expand and upgrade infrastructure to attract new industry, support technology-led development, redevelop brownfield sites, and provide eco-industrial development.

*Small Business Administration (SBA)*

The SBA provides a number of financial and technical assistance programs for small businesses including direct and guaranteed business loans, venture capital investments, and disaster loans. The Small Business Administration is the largest single financial backer and facilitator of technical assistance and contracting opportunities for the nation's small businesses.

*U.S. Environmental Protection Agency*
*(EPA), Brownfield Program*

The EPA provides financial and technical assistance to support the redevelopment of brownfields for economic development purposes. The Brownfield Program provides funding to state, local, and tribal governments to make low interest loans and grants to be used to assess environmental conditions of brownfields sites and fund cleanup activities.

# Federal tax credit programs

Federal tax credits have become some of the most important tools for financing economic development projects because they create an incentive for more private sector involvement, which allows greater leveraging of resources and encourages more public–private partnerships. Each federal tax credit program is structured differently, but all programs are designed to provide equity for projects by allowing the owner or developer of the project to sell or transfer the tax credits in return for up-front equity investments. Individuals or for-profit corporations purchase the tax credits for use against their federal income tax liability.

Tax credit programs help finance economic development activities ranging from business start-ups to real estate development. The major federal tax credit programs are as follows:

- *New Markets Tax Credit:* The New Markets Tax Credit (NMTC) was created in 2000 to spur private investment in low-income urban and rural communities and to provide capital to businesses and economic development projects in these communities. Individuals or corporations can invest in Community Development Entities (CDEs) that provide loans, equity, and other financing to qualified low-income businesses. Those investing in the CDE will receive a credit against their federal income taxes totaling 39 percent of the cost of the investment over a seven-year period. This program is administered by the Treasury Department's Community Development Financial Institutions (CDFI) Fund and the Internal Revenue Service (IRS).
- *Historic Preservation Tax Incentives:* Historic Preservation Tax Incentives support the rehabilitation of historic and older building and encourage the preservation and revitalization of older cities, towns, and rural communities. The tax incentive rewards private investment in rehabilitating historic properties such as offices, rental housing, and retail stores. Currently, the tax credit for

certified historic structures is equal to 20 percent of the renovation or construction costs, and a 10 percent tax credit is available for buildings built prior to 1936 that are non-residential. This program is administered by the National Park Service, the IRS, and State Historic Preservation Offices.

- *Federal Brownfields Expensing Tax Incentive:* This cost deduction provides a business incentive to clean up sites contaminated with hazardous substances and is intended to offset the costs of cleanup. This program is administered by the EPA and the IRS.

Tax credit programs bring together many of the community development finance players including government entities, businesses, investors, financial institutions, and nonprofits. While the popularity of these programs has been increasing, they are considered some of the more complex forms of financing, and smaller developers and projects still tend to look for simpler and more traditional sources of financing.

## State resources

State governments play an important role in economic development financing, particularly in rural areas. Much of the federal funding is passed down through state agencies to reach the non-urban or non-entitlement communities. In addition, many of the tools that local governments rely on to finance economic development (e.g., tax increment financing, property tax abatements) are enabled by state legislation.

Every state has tax credit programs to encourage economic and community development. These programs are frequently tied to the state's individual economic development strategy and may be tailored to meet the specific goals of that strategy. Most state tax credit programs are tied to new business investment and job creation and are not typically as broad and flexible as the federal tax credit programs. Many states have created state enterprise zone programs. Enterprise zones are economically depressed areas

targeted for revitalization through tax credits and other incentives given to companies that locate or expand their operations within the zone.

States also have the ability to offer favorable financing for economic development activities by issuing bonds, particularly Industrial Revenue Bonds (IRBs). Issued by a public authority, IRBs qualify for federal tax exemption. The proceeds of the bond issue may be used to offer lower cost financing for new industrial development.

## Municipal funding sources

Local governments also fund economic development projects and have worked to develop creative financing tools that do not generate higher tax burdens for local taxpayers. Tax Increment Financing is one of the most widely used of these tools along with tax abatements, tax credits, and providing access to lower cost financing.

### Tax Increment Financing (TIF)

TIF is a mechanism that allows the future tax benefits associated with new real estate development to pay for the present cost of the improvements. Generally, TIF is used to finance costs such as land acquisition, infrastructure improvements, utilities, parking structures, debt service, and other related development costs.

Almost every state has passed legislation that gives local governments the authority to designate TIF districts for redevelopment. These districts are typically physically or economically distressed areas where private investment is not likely to occur without some additional public subsidy. Based on the expectation that property values in the district will rise as a result of that redevelopment, the city splits the property tax revenues from the district into two streams: the first consisting of revenues based on the current assessed value; the second based on the increase in property values – the "tax increment." The tax increment is diverted away from normal property tax uses (i.e., schools, public safety) and may be used to repay the cost of new development in the TIF district.

TIF is popular with local officials because it is a flexible tool and, as a result, it is one of the most frequently used sources of funding for community economic development projects. However, because local officials have made TIF so widely available, there is some concern that developers have come to expect it for all projects. Thus, some critics contend that it no longer serves its purpose as a targeted development incentive.

In addition, the use of TIF has raised concerns because it diverts property tax revenue away from traditional uses. As a result, many states have tightened TIF legislation to control the amount of money that can be shifted from these uses to finance TIF debt. Some states have also established performance requirements or set-asides, such as the provision of affordable housing units, for projects financed with TIF.

### Property tax abatement

Property tax abatements are another tool used by local governments to promote economic development in areas designated as blighted, distressed, or in state enterprise zones. Tax abatements are defined as either tax forgiveness or tax deferral until a certain date. Most states limit the maximum time that taxes can be forgiven or deferred to 10 to 12 years. Tax abatements are usually offered in targeted areas where local communities are hoping to encourage new investment, redevelopment, or create new job opportunities by attracting new businesses.

Tax abatements are one of the more frequently used tools because property taxes are one of the items that fall under state and local government control. Unlike TIF, tax abatements do not directly divert spending from other programs, so abatements may be more politically feasible.

### Municipal bonds

Local governments can provide favorable financing to encourage economic development by using their bonding authority to issue private activity bonds including industrial revenue bonds. These bonds are backed by the revenue generated by a project and not by the government. Because these bonds are tax exempt, local government can use the bond proceeds to offer a lower cost source of project financing. Some local governments attach performance standards and accountability requirements to municipal bonds.

### Land acquisition

Local governments also help with the acquisition of land, either by providing land that the local government already owns or by acquiring property and transferring it to the private developer at a lower price. Eminent domain is one tool that local government may use to acquire land if it can show a public benefit from the project or if the property is blighted. Acquiring contiguous parcels for redevelopment can be costly and time-consuming for developers, so the ability of local government to assist with this process can be very valuable.

### Fee waivers or deferrals

Fee waivers or deferrals are other tools used by local governments. Local governments may reduce fees for building permits, water and sewer connections or impact fees[1] for projects that have a public benefit, such as redeveloping a blighted area, creating new commercial space in a distressed community, or developing affordable housing.

## Other financing resources

In addition to government programs, there are several other important tools and programs used to finance economic development projects including Revolving Loan Funds, the Main Street Preservation Program, and programs offered through the Federal Home Loan Bank System.

### Revolving Loan Funds (RLFs)

A Revolving Loan Fund is a self-replenishing pool of funds structured so that loan payments on old loans are used to make new loans. RLFs provide a lower cost and more flexible source of capital that may be used in conjunction with conventional sources to reduce the total cost of debt to the borrower and to lower the risk to participating lenders. Initial funding for RLFs often comes from a combination of public sources (federal, state, or local government) and private sources (financial institutions and philanthropic organizations). State and local government funding may come from tax set-asides, bonds, or a direct appropriation. Federal funding to capitalize loan funds can come from HUD, USDA, and the Department of Commerce programs. The source of the public funds used for capitalization can impact the eligible uses for the loan fund proceeds. Revolving loan funds are most often used for small business assistance but may also be used to fund real estate development projects, land acquisition, and construction financing.

## The Main Street Preservation Program

The Main Street Preservation Program is offered through the National Trust for Historic Preservation. In the 1970s, the National Trust developed its pioneering Main Street approach to commercial district revitalization, an innovative methodology that combines historic preservation with economic development to restore prosperity and vitality to downtowns and neighborhood business districts. It has been proven effective in communities of all sizes, from small towns to large cities. Main Street programs are locally operated by independent nonprofits or city agencies. There is typically a state-wide (or city-wide in larger cities) organization that coordinates a network of affiliated Main Street programs. The coordinating Main Street organization generally has an application process through which a community can be designated as a Main Street program and provide direct technical services, networking, and training opportunities to their affiliates.

## The Federal Home Loan Bank System

The Federal Home Loan Bank System is a government-sponsored enterprise and another important partner for economic development finance. The 12 district banks provide stable, low-cost funds to financial institutions for home mortgage, small business, rural, and agricultural loans. Through grants, subsidized rate advances, reduced rate advances, and technical expertise, the district banks assist member financial institutions in their community economic development efforts. The Community Investment Program is a lending program that provides below-market loans enabling member financial institutions to extend long-term financing for housing and economic development that benefits low- and moderate-income families and neighborhoods. This program is designed to be a catalyst for economic development because it supports projects that create and preserve jobs and help build infrastructure to support growth.

## Foundations

Privately funded and community foundations have historically been a source of grant funds for nonprofit organizations and may be utilized by nonprofit partners in an economic development project. In addition, to increase the impact of their resources, some foundations have added financial instruments known as program-related investments (PRIs) to their traditional grant portfolios.

PRIs include loans, loan guarantees, and equity investments. They may finance a variety of community economic development projects including neighborhood shopping centers, revitalization projects, small businesses, micro-enterprise funds, and nonprofit intermediaries. Like grants, PRIs have as their primary purpose the achievement of the foundation's mission and philosophy. Unlike grants, PRIs are considered financial instruments which should produce financial returns for the foundation.

## Conclusion: putting it all together

Community and economic development projects require many partners and sources of funding, so the real issue is how to bring it all together. How should these projects be structured, what role should each partner play, and how should potential risks be mitigated to successfully complete these projects?

All of the players fulfill an important role in financing an economic development project, and each player must understand their respective role and the role of the other partners. However, identifying a point person/organization to manage the project is critical to its success. This person should act as liaison between all of the different entities involved and provide oversight and management. It is the role of this point person to coordinate all project activity and define the roles of each participant. In addition, there is typically a visionary leader of the project who develops the first concept and serves as the "cheerleader" and spokesperson for the project.

It is also critical to the success of the project that an evaluation of the financial strengths and capacity of all partners be undertaken early on in the process. This is particularly important when evaluating nonprofits and community development corporations. Many of these organizations are dependent on grant funds from public and private funding sources for their long-term administrative and operating expenses and organizational stability. In addition, the missions and goals of many nonprofits are built on the strengths of their leaders (executive director and board leadership), and evaluation of their management structure and succession plan may also be necessary.

Most economic development projects require multiple financing sources, commonly known as "layered financing," which may include both traditional and non-traditional loans, investments, and grants from all levels of government and the private sector. Most financing programs are designed to bridge the gap between what the developer needs to finance a project and what the private lender is willing to lend. Gap financing can also help mitigate risk and may give a private lender or investor more confidence in the deal. A case study illustrating the financing of an economic development project (Tangerine Plaza) is included at the end of the chapter.

There are several challenges to successfully completing an economic development project, due largely to the inherent complexity and the number of different players typically involved. First, each partner has a different tolerance for risk. Government and nonprofit partners are typically more willing to take risks than the private, for-profit players such as financial institutions and investors. This risk may be mitigated through tax abatements, tax credits, loan guarantees, interest rate subsidies, and other credit enhancements.

Second, funding programs may have unique guidelines and underwriting requirements. Some programs require a higher income ratio to service the debt (debt service ratio), while others may focus on loan-to-value ratios or the strength of its guarantees. Matching the different sources of funding to ensure that all requirements are met can be challenging but is an essential step to determine the viability of a project.

Third, the partners may not be aware of the available resources that other partners can bring to the project. Because these projects typically involve multiple sources of funding, it is critical that the lead player in the project be knowledgeable of all available resources and communicate levels of commitment to all parties.

Fourth, getting all of the partners together at the right time can be challenging. Most sources of funding for community and economic development projects have different funding cycles with unique eligibility requirements (such as requiring site control and land-use approval prior to applying for funding). Coordinating the timing of the funding and approval processes is critical to the success of the project.

Fifth, it can be difficult to allocate the transaction costs for the project between the players. Who does what and when needs to be agreed upon at the beginning of the process and monitored by the lead person or organization.

Sixth, each partner may be looking for a different return on the project. The public sector partners evaluate the return on economic development projects in terms of the impact on the community including the amount of new investment, the increase in the tax base, the number of new jobs created, and the overall improvements in the quality of life. The private sector partners will be focusing on the financial return from the project. Thus, it is important to consider the "double bottom line" for these deals, and to weigh both the financial and social return to the community associated with each project.

Finally, it is important that the use of public financing and subsidies is evaluated to ensure that the incentives are used responsibly to meet the needs of both the project and the local community. The use of public subsidies is critical for many community economic development projects, but, because these resources are limited, it is important that they are used for projects that provide a measurable benefit to the public.

## CASE STUDY: TANGERINE PLAZA

Please refer to the following list for an explanation of the acronyms used throughout the following case study.

AMI:        Area Median Income
CDE:        Community Development Entity
CDFI:       Community Development Financial Institution
EDGE:       Economic Development and Growth Enhancement Program from the Federal Home Loan
            Bank of Atlanta
FHLB        Atlanta: Federal Home Loan Bank of Atlanta
NLPWF:      Neighborhood Lending Partners of West Florida
NMTC:       New Markets Tax Credits
OCS:        Office of Community Services

### Project history and overview

Tangerine Plaza was the first new significant commercial development to be done in the Midtown area of St. Petersburg, Florida in many years. The initial concept for the shopping center was developed in the year 2000 by two individuals who were committed to redeveloping this area of St. Petersburg. Midtown is a low-income (average income is 47 percent of area median income, 33 percent are below poverty level), predominantly African-American community that has been designated by the local government as a Community Redevelopment Area.

Tangerine Plaza is a 47,000-square foot neighborhood shopping center that is anchored by a Sweetbay Supermarket, the first full-service grocery store in the neighborhood. The remaining retail space at the center is occupied by smaller local retail tenants.

Complex community economic development projects like Tangerine Plaza often involve a variety of partners and multiple sources of financing. The lead developer for Tangerine Plaza was Urban Development Solutions, a nonprofit organization that had previously been involved with affordable housing projects in the Midtown area. Urban Development Solutions had purchased several parcels of land that were abandoned and condemned, with the goal of building more affordable housing. As they continued to assemble land, however, it became clear that the real need in the community was commercial development and, specifically, a full-service grocery store. To develop the retail center, Urban Development Solutions formed a for-profit limited liability company.

Urban Development Solutions approached the City of St. Petersburg with this concept, and the City agreed to assist in the assembly of the remainder of the 32 parcels needed for the development. As is typical of urban redevelopment, obstacles such as nonconforming lots, inappropriate zoning classifications, and numerous lien and title problems had to be addressed. The City agreed to buy the lots that had been purchased by Urban Development Solutions, as well as the remaining lots that were not already under the developer's control. The city then cleared all of the liens and other encumbrances against the properties, rezoned all of the property for neighborhood commercial development, and replatted the lots into one parcel. Once the land was ready for construction, the city agreed to a 99-year lease with Urban Development Solutions with an annual payment of US$5. The entire land assembly process took two and a half years.

## Project financing

A complex financing package that involved numerous partners was required for the development of Tangerine Plaza. Construction financing for the project was provided by Neighborhood Lending Partners of West Florida (NLPWF), a state-wide CDFI. As described earlier, the City of St. Petersburg provided the land (valued at $1,150,000) to be leased for 99 years at US$5 per year. Additional funding for the project came from a $700,000 federal grant from the Office of Community Services and a $10,000 donation from the principals of the developer.

During the time it took to assemble the property, construction costs escalated, almost doubling from the original estimates. In order to cover the increasing construction costs, New Markets Tax Credit (NMTC) funding was secured to provide the gap financing. The New Markets Tax Credit syndicator was LISC (Local Initiatives Support Corporation), a national intermediary, and the two investors in the transaction were large financial institutions. Finally, the project was awarded an EDGE (Economic Development and Growth Enhancement Program) Loan from the Federal Home Loan Bank of Atlanta for its permanent first mortgage financing at a below-market rate of 3.4 percent.

The New Markets Tax Credit allocation was a very important component of the project financing. Due to the higher construction costs, there was insufficient funding to pay the developer's fee, so the New Market Tax Credit proceeds helped with the construction financing and payment to the developer. Second, the developers of the center had a strong desire to lease space in the shopping center to local small business owners, so the New Market Tax Credit proceeds were also used to assist in the tenant build-outs, equipment purchases, and rent abatement for the small businesses locating in the center. The complete financing package for the project is shown in Table 20.1.

## Project impact and results

Construction of Tangerine Plaza was completed in November 2005, and, since opening, the Midtown Sweetbay store has set the record for the highest increase in sales for all Sweetbay Supermarkets. The property tax revenue on this property has increased from $6000 in the year 2000 to over $110,000 in 2006. Since this is a designated Community Redevelopment Area, a portion of the tax revenue may now be utilized to fund additional redevelopment projects in the surrounding neighborhood. In addition, Sweetbay received state tax credits for job creation and for hiring neighborhood residents.

**Table 20.1** Summary of sources and uses of funds

| Line | Sources | Construction | NMTC | Total |
|------|---------|-------------|------|-------|
| A | NLPWF EDGE loan (from FHLB Atlanta) | $3,500,000 | $300,000 | $3,200,000 |
| B | City of St. Petersburg | $1,336,500 | | $1,336,500 |
| C | NLPWF CDFI funds | $661,500 | | $661,500 |
| D | OCS grant | $700,000 | | $700,000 |
| E | Sweetbay/Kash-n-Karry | $394,960 | | $394,960 |
| F | Sweetbay/Kash-n-Karry | $374,252 | | $374,252 |
| G | City of St. Petersburg | $75,000 | | $75,000 |
| H | Private donation | $10,000 | | $10,000 |
| I | NMTC equity | | $2,784,149 | $2,784,149 |
| J | Total sources | $7,052,212 | $2,484,149 | $9,536,361 |

| Line | Uses | Construction | NMTC | Total |
|------|------|-------------|------|-------|
| 1 | Land lease (prepaid in full) | $495 | | $495 |
| 2 | Site development | $750,270 | | $750,270 |
| 3 | Building shell/local tenant improvements | $4,498,758 | | $4,498,758 |
| 4 | Anchor tenant improvements | $1,000,000 | | $1,000,000 |
| 5 | Permits and government fees | $39,466 | | $39,466 |
| 6 | Professional fees | $377,310 | | $377,310 |
| 7 | Pre-development soft costs | $133,000 | | $133,000 |
| 8 | Construction financing costs | $209,301 | | $209,401 |
| 9 | Insurance | $26,353 | | $26,353 |
| 10 | Developer overhead | $17,159 | $116,195 | $133,354 |
| 11 | Working capital | | $504,859 | $504,859 |
| 12 | Developer profit | | $429,943 | $429,943 |
| 13 | Community outreach program | | $200,000 | $200,000 |
| 14 | Tenant assistance | | $500,000 | $500,000 |
| 15 | NMTC financing costs | | $264,003 | $264,003 |
| 16 | NMTC organization and placement fees | | $456,609 | $456,609 |
| 17 | CDE/investment fund escrows | | $12,540 | $12,540 |
| 18 | Total uses | $7,052,212 | $2,484,149 | $9,536,361 |

Perhaps the most important achievement of this project is the "social equity" generated in the community, and the track record needed to continue redevelopment of the Midtown area. The success of Tangerine Plaza is widely recognized and, in 2006, the project was awarded the Florida Redevelopment Association's Roy F. Kenzie Award for "outstanding new building project" and was also honored as "Best of the Best," the top redevelopment project in the state.

## Lessons learned and best practices

There are a number of lessons to be learned from the development of Tangerine Plaza and suggested best practices for the financing of similar community economic development projects.

## Lessons learned

The Tangerine case study illustrates some of the typical challenges of redevelopment projects, particularly those that are located in a low-income or distressed areas.

1    First, it can be extremely difficult to assemble the numerous parcels of property required for the project if they are owned by different individuals or have title problems and liens.
2    Second, to secure the financing needed to develop a commercial center it is often necessary to secure the commitment of a major retail anchor (in the Tangerine Plaza case study it was the grocery store chain). Finding the right anchor tenant that is willing to commit up front can be a time-consuming process.
3    Third, if the project is located in a depressed real estate market area, it will likely yield below-market rental rates. Since rental rates are used to determine property values for purposes of obtaining construction and/or permanent financing, it can be difficult to obtain sufficient financing to develop projects in depressed real estate markets.
4    Fourth, the complexity and length of time required to put together a deal can lead to significant increases in construction costs, which may require identifying additional sources of gap financing to cover the increased cost of the project.
5    Fifth, it can be difficult to get the timing of the financing package aligned. In the case of Tangerine Plaza, the commitment for the permanent financing was not completed when the construction loan closed, as some lenders would have required.
6    Sixth, there may be issues among the financial partners over control of the collateral. In the case of Tangerine Plaza, the city and the lenders had issues over control of the leased land.

## Best practices

This case study also highlights several examples of best practices for financing community economic development projects:

1    First, a nonprofit can be a key player to involve in a project because it can help with community creditability, and often has better access to gap financing in the form of low interest loans and grants.
2    Second, if a nonprofit organization is involved, it must have the capacity and sophistication to take on the project or the financial ability to obtain this capacity through additional staff or outside consultants.
3    Third, all players involved in the process must be focused, consistent, and patient in working through the problems, which may take an extended period of time.
4    Finally, a strong and committed local government that has the ability to attract funding and is willing to do what it takes is critical to turning a project idea into a reality.

## Keywords

Community development, economic development, finance, Community Development Block Grant programs, new market tax credits, tax increment financing, Revolving Loan Funds.

## Review questions

1   Who are some of the typical "players" in a community finance deal?
2   What are some of the major sources of federal government money for community development?
3   What are some state and local resources for community development financing?
4   What are some challenges often faced when putting a community development finance deal together?

## Note

1   Impact fees are typically one-off charges imposed against new development to pay for the cost of development. Impact fees may be used to cover the costs of new roads, infrastructure, schools, or public services.

## Bibliography and resources list

Fannie Mae Foundation. Available online at www.knowledgeplex.org.

Federal Home Loan Banks. Available online at http://www.fhlbanks.com/.

Federal Reserve Board of Governors. Available online at www.federalreserve.gov.

International Economic Development Council. Available online at www.iedconline.org.

National Park Service Historic Preservation Tax Incentives. Available online at http://www.cr.nps.gov/tax.htm.

National Trust for Historic Preservation Main Street Program. Available online at http://www.mainstreet.org/.

Office of the Comptroller of the Currency. Available online at www.occ.treas.gov.

Opportunity Finance Network. Available online at www.opportunityfinance.org.

U.S. Department of Agriculture Rural Development. Available online at http://www.rurdev.usda.gov/.

U.S. Department of Commerce Economic Development Administration. Available online at http://www.eda.gov/.

U.S. Department of Housing and Urban Development. Available online at http://www.hud.gov/economicdevelopment/.

U.S. Department of the Treasury CDFI and New Markets Tax Credit. Available online at http://www.cdfifund.gov/.

U.S. Environmental Protection Agency-Brownfields Program. Available online at http://www.epa.gov/brownfields/.

U.S. Small Business Administration. Available online at http://sba.gov/.

# 21 Securing grants for community development projects

## Beverly A. Browning

There are numerous public and private sources of grant funding to assist with community development, from building infrastructure to workforce development to health care. Many communities approach grant seeking in a haphazard manner, without doing their homework first or taking the time and resources to develop a good proposal. Even if a community is most deserving of a grant, if its grant package is not up to standard, it will most likely be rejected. This chapter provides an overview of how to research, write, and win grant proposals

## Introduction

Finding grant funding opportunities for community development projects is easy if Internet and other investigative skills are honed. To get started, become familiar with some of the websites that track funding for community development projects. One of the major sources of federal funding for community and economic development is the Community Development Block Grant program.

*http://www.hud.gov/* – This website, available through the U.S. Department of Housing and Urban Development (HUD), should always be the first one to access when conducting federal-level research on community development funding. HUD awards Community Development Block Grants (CDBGs) directly to state and local governments to revitalize neighborhoods; expand affordable housing and economic opportunities; and improve community facilities and services. Recipients of the grants include more than 900 metropolitan cities and urban counties as well as 49 states which, in turn, distribute the funds to smaller communities. Funds from this entitlement program may be used for parks, infrastructure improvements, environmental cleanup, and other projects benefiting the community. Check the HUD website to find deadlines for the annual applications.

## Tracking other federal funding

*www.grants.gov* – The purpose of this website is to provide a simple, unified electronic storefront for interactions between grant applicants and the federal agencies that manage grant funds. There are 26 federal grant-making agencies and over 900 individual grant programs that award more than $350 billion each year. The grant community – including state, local, and tribal governments; academia and research institutions; and nonprofits – need only visit one website to access the annual grant funds available across the federal government.

Here are a few suggestions for navigating the grants.gov website. On the home page, click on the "Find Grant Opportunities" button. This will lead to another page where an applicant can sign up for the "Receive Grant Opportunity Email Alerts" service and receive daily grant funding announcements electronically. The page also gives the option to "Search Grant Opportunities." On the "Opportunities" page, click "Basic Search." On the "Basic Search" page, type "Community Development" in the "Keyword Search" dialogue box and click on "Search:"

Search results will appear in chronological order, with the most recent grant announcements appearing first. Review each announcement to determine which grant competition best fits a specific community development initiative. Investigative skills will come in handy when reading through dozens of grant announcements from the U.S. Departments of Agriculture, Housing and Urban Development, Commerce, Labor, and more. If a relevant announcement and its deadline has passed for this year, print out the information anyway and use it to track the same funds in future fiscal years. Knowing that funding may be available for an initiative in an upcoming year will enable an applicant to prepare for and approach financial issues proactively rather than reactively. The knowledge will also help the applicant gain a toehold on funding stabilization, sustainability, and staying power.

*.gov* – Web portals for any state may be found by entering www, a state's name or postal abbreviation, and "dot gov" (.gov) in the Web browser's address box. From the gateway, links to state governmental departments or agencies can be found. More specifically, one should search for direct links to state agencies that award grants for community development projects. Remember: federal monies trickle down to the state level through common state agencies such as agriculture, commerce, housing, and labor.

## Tracking foundation and corporate funding

Don't make the mistake of eliminating private sector funding for a community development project. Foundations and corporations are not interested in building bridges, paving roads, or erecting buildings. However, many such institutions are interested in funding smaller scale community development projects.

*http://www.lib.msu.edu/harris23/grants/2commdev.htm* – Michigan State University Libraries compile and manage this website. It provides links to foundation and corporate funding opportunities and is updated frequently, so all of the links are current. Many of the foundations and corporations fund nationally. One advantage of this website is its list of federal funding opportunities for community development projects. Still, it is advisable to cross-check between content found on <www.grants.gov> and this website.

*www.fdncenter.org* – This website belongs to the Foundation Center in New York City. It is the number one grant funding resource database for nonprofit organizations. With a few keystrokes, RFPs may be found in the database that match specific projects. Begin by clicking on the "Newsletters" tab at the top of the home page. Once on that page, enter an email address and select the "RFP Bulletin" from the list of newsletters. In a week or less, the RFP Newsletter will come to that email address. Current RFP alerts may be viewed on the Foundation Center's website. To see PND alerts from the Philanthropy News Digest, click in the box on the right side of the Newsletters Web page.

## Hitting the target

Before starting the time-consuming process of writing a grant application or proposal, particular sources must be determined as fitting or not to a project's funding needs. Go through the checklist in Table 21.1 for every funding opportunity to decide whether it is worth pursuing.

If the answer is "no" to any of the questions in Table 21.1, do not proceed from the preplanning stage to grant writing. Use this checklist regularly to gauge if the money matches the project, because grant awards are only given for "perfect fits."

■ **Table 21.1** Grant pre-screening checklist

| Guiding questions | Yes | No |
| --- | --- | --- |
| Is my organization eligible to apply? | | |
| Can we meet the grant application deadline? | | |
| Does the grant meet the funding priorities that we established in our planning process? | | |
| Is this a service or activity my agency is equipped to implement? | | |
| Will there be more than 10 grant awards? | | |
| Is the maximum grant award enough to fund at least 50% or more of the project budget? | | |
| Is there a geographic restriction? | | |
| Does our idea for a grant request match the funder's guidelines and interests? | | |

## Preparing grant proposals

A solid proposal package contains eight basic components: (1) the proposal summary; (2) an introduction to the organization; (3) the problem statement or needs assessment; (4) the project objectives; (5) the project methods or design; (6) a project evaluation; (7) ideas for future program funding; and (8) the project budget. The following sections summarize the content of the components.

## The proposal summary: outline of project goals

The proposal summary outlines the proposed project and precedes the proposal. It may be in the form of a cover letter or a separate page but, which ever form its takes, it should definitely be brief – no longer than two or three paragraphs. It is best to prepare the summary after the proposal has been fully developed; that way, it will encompass all the key points necessary to communicate the objectives of the project. This document becomes the cornerstone of the proposal, and the initial impression it gives will be critical to the success of a venture. In many cases, the summary will be the first part of the proposal package seen by funding agency officials and,

very possibly, could be the only part of the package that is carefully reviewed before it is considered any further.

Within the summary, describe a fundable project which can be supported in view of its local need. In the absence of receiving a grant award, alternatives should be pointed out. The influence of the project both during and after its implementation should be explained. The consequences of funding to the project should be highlighted.

## Introduction: presenting a credible applicant organization

Most proposals require that the applicant organization describe its past and present operations. For this section, data on an organization should be gathered from all available sources. Some details to include could be:

- Brief biographies of board and key staff members.
- The organization's goals, philosophy, track record with other grantors, and any success stories.
- Data relevant to the goals of the funding initiative and the establishment of the applicant organization's credibility.

## The statement: stating the purpose at hand

A key element of any proposal, the problem statement or needs assessment, is a clear, concise, and well-supported statement of the problem to be addressed. The best way to collect information about the problem is to conduct and document both formal and informal needs assessments for a program in the target or service area. The information provided by the assessments should be both factual and directly related to the problem in question. Areas to document are:

- The purpose for developing the proposal.
- The beneficiaries – who they are and how will they benefit.
- The social and economic costs to be affected.
- The nature of the problem including as much hard evidence as possible.
- How the applicant organization came to realize that the problem exists.
- What is currently being done about the problem.
- The alternatives available when funding has been exhausted; what will happen to the project and the implications.
- Most importantly, the specific manner through which problems might be solved including a review of the resources needed, how they will be used, and to what end.

There is a considerable body of literature that may be consulted on the exact assessment techniques to be used (see Chapter 9).

In addition, any local, regional, or state government planning office, or local university offering coursework in planning and evaluation techniques, should be able to provide excellent informational references. Types of data that may be collected during an assessment include historic, geographic, quantitative, factual, statistical, and philosophical. Besides conducting a needs assessment, locate relevant studies completed by colleges, and search for literature either online, at public or university libraries, or both. In addition, local colleges or universities

with a department or section related to the proposal topic may help conduct a needs assessment. They may also be interested in developing a student or faculty project. Include and highlight all pertinent findings in such a proposal.

## Project objectives: goals and desired outcome

Program objectives refer to specific activities in the proposal. It is necessary to identify all objectives related to the goals and methods to be employed to achieve the project's stated objectives. A well-stated objective contains measurable quantities, and refers to a problem statement and the outcome of proposed activities. The figures used should be verifiable. Remember: if the proposal is funded, the stated objectives will probably be used to evaluate program progress, so be realistic.

## Program methods and program design: a plan of action

The program design describes how the project is expected to work and solve the stated problem. Sketch out the following in this section:

1  The activities to occur along with the related resources and staff needed to operate the project (inputs).
2  A flowchart of the organizational features of the project. Describe how the parts interrelate, where personnel will be needed, and what they will be expected to do. Identify the kinds of facilities, transportation, and support services required (throughputs).
3  Explain what will be achieved through 1 and 2 above (outputs); i.e., plan for measurable results.

Project staff may be required to produce evidence of program performance through an examination of stated objectives, either during a site visit by the

**■ Table 21.2** The "Logic Model"

| Inputs | Activities (same as your goals) | Outputs (counting people and events) | Intermediate outcomes (individual benchmarks) | Long-term outcomes (systemic) |
|---|---|---|---|---|
| Resources dedicated to or consumed by the program<br>– Money<br>– Staff and staff time<br>– Volunteers and volunteer time<br>– Facilities<br>– Equipment and supplies | What the program does with the inputs to fulfill its mission<br>– Provide…<br>– Educate…<br>– Counsel…<br>– Create…<br>– Conduct… | The direct quantitative product of program activities<br>– Number of classes taught<br>– Number of sessions conducted<br>– Amount of educational materials distributed<br>– Number of hours of service delivered<br>– Number of participants served | Benefits for participants during and after program activities<br>– New knowledge<br>– Increased skills<br>– Changed attitudes or values<br>– Modified behaviour<br>– Improved condition<br>– Altered status | Changes in systems and processes after the funding is expended<br>– New approaches<br>– New services<br>– Stronger partnership, working agreements |

funder and/or grant reviews, which may involve peer review committees. It may be useful to devise a diagram of the program design and insert it in the proposal. To present a well-planned case for a solid program design, this author prefers to use a modified "Logic Model" (see the example in Table 21.2).

Throughout the narrative, justify the course of action to be taken. The most economical method should be used; one that does not compromise or sacrifice project quality. The expenses associated with performance of a project will later become negotiable items with the funder's staff. If everything in the proposal is not carefully justified in writing, the approved project may not entirely resemble the original concept after negotiation with the funding agency. Therefore, carefully consider the pressures of the proposed implementation; that is, the time and money needed to achieve each part of the plan. In addition, draw attention to the innovative and distinguishing features of a project to separate it from others under consideration.

Whenever possible, use appendices to provide details, supplementary data, references, and informa-

tion requiring in-depth analysis. Even if these items support the proposal, if included in the body of the design they could detract from its readability. Appendices provide the proposal readers with immediate access to details, if and when clarification of an idea, sequence, or conclusion is required. Timetables, work plans, schedules, activities, methodologies, legal papers, personal vitae, letters of support, and endorsements are examples of items best put in appendices.

## Evaluation: product and process analysis

The evaluation component is twofold: (1) product evaluation, and (2) process evaluation. Product evaluation addresses results that may be attributed to the project, as well as the extent to which the project has satisfied the desired objectives. Process evaluation addresses how the project was conducted, in terms of consistency with the stated plan of action and the effectiveness of the various activities within the plan.

Most funders require grantees to perform some form of program evaluation. The evaluation requirements of the proposed project should be explored carefully. Some funders may require specific evaluation techniques, such as designated data formats (an existing information collection system), or may offer financial inducements for voluntary participation in a national evaluation study. The applicant should ask specifically about these points.

Evaluations may be conducted by a staff member, an evaluation firm, or both. Remember to state the amount of time needed to evaluate the project, how the feedback will be distributed among staff and stakeholders, and a schedule for review and comment for this type of communication. Evaluation designs may start at the beginning, middle, or end of a project, but the applicant should specify a start-up time in the proposal. It is practical to submit an evaluation design that starts at the beginning of a project for two reasons:

1 Convincing evaluations require the collection of appropriate data before and during program operations.
2 If the evaluation plan cannot be prepared at the outset, then a critical review of the program design may be advisable.

Even if the evaluation plan must be revised as the project progresses, it is much easier and cheaper to modify a solid data-collection methodology than to flounder with a weak evaluation approach. The intended evaluation outcomes may be difficult to achieve if the problem is not well defined in the need or problem statement or carefully analyzed for cause and effect relationships. Sometimes a pilot study is needed to begin the identification of facts and relationships. Often a thorough literature search may be sufficient to rectify design problems. Evaluation requires both coordination and agreement among program decision makers (if known). Above all, the funder's requirements should be highlighted in the evaluation design.

## Future funding: long-term project planning

In this section, describe a plan for project continuation beyond the grant period and/or the availability of other resources necessary to implement the grant. If the program is for construction activity, discuss maintenance and future program funding. If the program includes the purchase of equipment, account for those expenditures.

## The proposal budget: planning the budget

Funding levels change yearly for federal and state agencies, as well as for foundations and corporations. It is useful to review the appropriations and total grants funded over the past several years to try to project future funding levels. However, it is safer to never expect that the income from a grant will be the sole support for the project. This consideration should be given to the overall budget requirements and, in particular, to budget line items most subject to inflationary pressures. Restraint is important in determining inflationary cost projections (avoid padding budget line items), but do attempt to anticipate possible future increases.

Some budget areas vulnerable to inflation are: utilities, building and equipment rental; salary increases; food; telephones; insurance and transportation. Budget adjustments are sometimes made after the grant award but getting approval for adjustments can be a lengthy process. So, when the project budget is created, be certain that implementation, continuation, and phase-down costs can be met. Consider costs associated with leases, evaluation systems, hard/soft match requirements, audits, development, implementation and maintenance of information and accounting systems, and other long-term financial commitments.

A well-prepared budget justifies all expenses and is consistent with the proposal narrative. When preparing the budget, keep in mind the following: (1) the salaries in the proposal should be similar to those of the applicant organization; (2) if new staff members are being hired, additional space and equipment should be considered; (3) if there is an equipment purchase, it should be the type allowed by the grantor agency; (4) if additional space is rented, the increase in insurance should be supported; (5) if an indirect cost rate applies to the proposal, the division between direct and indirect costs should not be in conflict and the aggregate budget totals should refer directly to the approved formula; and (6) if matching costs are required, the contributions to the matching fund should not be included unless otherwise specified in the application instructions.

## Nonprofits, accountability, and sustainability

When nonprofits request grant funding via the grant proposal or application, they put in writing their promise to carry out the proposed goals and objectives. Today, all types of funders from both private and public sectors require a statement on sustainability in the funding request. They need to know how the program/project will continue when the grant funding period is over. Because most nonprofit organizations that survive on grant monies promise to be accountable for both monies and actions to the grantor, this presents a very difficult situation.

## How are nonprofits held accountable as they strive to maintain sustainability?

Nonprofits are held accountable by the grantors. Grantors audit fiscal expenditures and program outputs and outcomes. More and more these days, nonprofit organizations are being asked to show that the work they do "makes a difference" and that they are achieving the mission for which they were created. When they do this, they are recognized as

accountable. The accountability information is generated during the evaluation process. All funders require that the evaluation plan or process is incorporated into the language of the proposal. This monitoring and/or measuring process may be used to evaluate individuals, programs, organizations, or entire communities. Evaluation is a formal attempt to get a picture of "how an organization is doing." It turns out that many nonprofit CEOs want more and better evaluation processes. However, few feel they have the time, money, or expertise to invest in developing really useful ones so an organization or community should emphasize the importance of the evaluation process in its accountability to the grantors.

## Supporting research on nonprofit accountability

Public expectations of nonprofits are significantly higher than for businesses and most nonprofits meet that higher standard. Now more than ever there is an opportunity to exceed public expectations and rebuild any trust that may have been lost. When nonprofits are perceived as acting in less than professional, irresponsible or ethically questionable ways, public reaction can quickly rise to feelings of betrayal and violation of trust. Distrust in one nonprofit can result in a "halt" to giving to all nonprofits.

(Maclean 2002)

The Internal Revenue Service has begun using a new tool, the "soft audit" – a series of detailed questions fired off to nonprofits filing 990s which fit specific profiles of potential abuse. Soft audit questions require very quick responses and can easily turn into full-fledged audit and enforcement procedures … One of the great ironies of the current push for nonprofit accountability is that we are quite likely preaching to the choir. The nonprofits that take the steps discussed above are almost certainly already being guided by strong ethical commitments that board and staff hold as core values regardless of whether a motion is made or a policy adopted.

Unfortunately, those within our sector who have difficulty separating their own private interest from the public benefit to be provided by the nonprofit may "go through the motions" of adopting policies and embracing standards of excellence.

(Sohl 2005)

The realities of today's economic environment have required that nonprofits, and those that raise money for them, make their case for support as strong as possible. Warm and fuzzy appeals that tug at the heart strings have met with limited success. Assuming that funding targets intuitively know the value of the good work being done is unrealistic.

(Ralser 2007)

## Conclusion

Sometimes the simplest things make the difference between a proposal's being read and approved or tossed in the trash can. Table 21.3 suggests how to make a proposal readable while emphasizing points the reviewer is more likely to latch onto. Table 21.4 explains what to do after submitting a proposal, and Table 21.5 provides common reasons why grants are rejected. Following these guidelines will help increase the likelihood that a grant will be funded.

**■ Table 21.3** Finishing touches

| | |
|---|---|
| Recheck your formatting! Is your narrative response formatted to match the funder's guidelines? | Did you follow the spacing, margin and font guidelines? |
| Underline key phrases. | Remember to paginate your final document! |
| Apply boldface to keywords. | Do you have enough white space to refresh the grant reader's vision path? |
| Italicize sentences that need emphasizing. | Did you use graphics to break up the narrative and make the layout visually appealing? |

**■ Table 21.4** Following up

| |
|---|
| Did the funder receive your grant application? |
| When will a funding decision be made? |
| If your grant is rejected, did you ask for reader and/or peer reviewer feedback in order to improve your narrative or budget before the next funder deadline? |
| Did you break your initial grant application or proposal into smaller components and find "just-in-case" funders? |

**■ Table 21.5** Why grant requests don't get funded

| |
|---|
| Grant applicant does not meet the funder's eligibility criteria. |
| Narrative sections fail to earn the maximum review points, resulting in a lower overall score. |
| Grant applicant does not pass the technical review (writer does not use the mandatory font size, exceeds the narrative page limitation, or does not include the mandatory forms, certifications, assurances, and required attachments). |

## CASE STUDY: A SUCCESSFUL COMMUNITY DEVELOPMENT GRANT-FUNDED PROJECT – BROOKSVILLE MAIN STREET COALITION

The Brooksville Main Street Coalition is located in a rural Cochise County, Arizona. In 2000, its official U.S. Census was 20,000; today in 2007, this disappearing farming community has grown to 40,000 residents – and they just keep coming. The focus of the Coalition's request was to seek funding for the restoration of the town's 1912 courthouse and jail. This 2000-square-foot facility is located on Main Street. Its boarded up and deteriorating appearance has been the target of gangs and more than one wayward drunk driver. The building has managed to hold up through major structural damage and pigeon infestation. The rehabilitation, restoration, and stabilization work on the Brooksville Courthouse and Jail will enable the town to open the building for public use once again. The Southern Arizona Literacy Council has already signed a Memorandum of Agreement for facility usage when the building is ready for occupancy. Its volunteers will lead tours for visitors and school groups.

### Where was the funding obtained for this project?

The Arizona State Parks Heritage Grant Program funded this project for $95,000.

### What were the success factors for getting the funding?

Applied by the Arizona State Parks Heritage Grant Program staff, high rating criteria in the following areas resulted in a funding award:

1  Extensive project planning over multiple years involving public input.
2  Affiliation with regional land-use planning commissions.
3  Updated master revitalization plan with stakeholder input.
4  Building Condition Assessment conducted by an architect approved by the Arizona Historic Preservation Commission.
5  Updated Central Business District Inventory.
6  Immediate plan for use of facility on completion of project's scope of work.
7  Capable and qualified project team members.
8  Matching funds (100 percent).
9  Ability to show high level of community impact.
10  Willingness to complete the National Register nomination process.
11  Demonstration of community involvement and support (dozens of letters of support, newspaper articles, Main Street Coalition meeting minutes, and resolutions and evidence of the matching funds' immediate availability).

### How was the project managed?

The grant-funded project was managed by a team consisting of a lead architect, the Main Street director, a structural engineer, and a project director.

## What were the successes and failures in carrying out the project?

The successes centered around winning a grant award from a highly competitive, state-funded grant program and finally addressing a downtown eyesore. Until 2006, no other city administration sought to restore this building. The failures centered around the administration at the coalition/town levels, the powers in control of the project. There was a lengthy delay in getting someone on the grantee side to sign the state's grant agreement. There was extensive arguing over the architect selected and the restoration plan/costs. Then the city manager was fired and the new city manager did not want to restore the building. Amidst all this small town bickering, the Main Street Coalition's director left her position. Great efforts have been taken to keep it all quiet and certainly to keep the funding agency from knowing of the town's or Coalition's convoluted management approach.

## How was the project evaluated?

This remains to be determined. If the state tires of waiting for the Coalition and town to move forward with the restoration, the grant award could be rescinded. Who would lose? The entire community. Who would win? No one!

## What are the plans for maintaining funding in the long run?

Once the restoration has been completed, the town has committed to maintaining the building's repair and day-to-day maintenance for the next ten years (a required element of the grantor).

## What lessons have been learned?

A community should not apply for any type of grant funding if it is going through a metamorphosis and that is the basis for sudden elected official departures, repeated negative media coverage, and an ongoing parade of newcomers who do not wish to be associated with the previous people's or administration's projects.

## Keywords

Community Development Block Grants, problem statement, entitlement program, project design, project evaluation, process analysis, philanthropy, sustainability.

## Review questions

1 What are two major websites to search for federal government grants? For corporate grants?
2 What assessment should be taken before deciding to write a grant proposal?
3 What are the key elements of a grant proposal?
4 What do accountability and sustainability refer to regarding grants? Why are they important?
5 What are some sources of information and research literature on nonprofit accountability?

6   What are some typical reasons why grants do not get funded?

## Bibliography and additional resources

Browning, B. (2005) *Grant Writing For Dummies*, 2nd edn, Hoboken, NJ: John Wiley & Sons.

—— (2006) *Winning Strategies for Developing Grant Proposals*, 3rd edn, Tampa, FL: Thompson Publishing Group.

Davidson, J.E. (2004) *Evaluation Methodology Basics*, Newcastle, UK: Sage.

Foundation Center. Available online at www.fdncenter.org (accessed December 11, 2007).

Grants.gov. Available online at www.grants.gov (accessed December 11, 2007).

Maclean, C.B. (2002) "Tips for Nonprofit Accountability." *PNN Online*. Available online at http://www.pnnonline.org/article.php?sid=734 (accessed December 11, 2007).

Michigan State University Libraries. *Grants for Nonprofits: Community Development*. Available online at http://www.lib.msu.edu/harris23/grants/2commdev.htm (accessed December 11, 2007).

Murray, K. and Mutz, J. (2005) *Fundraising For Dummies*, 2nd edn, Hoboken, NJ: John Wiley & Sons.

New, C.C. and Quick, J.A. (2000) *Grant Winner's Toolkit: Project Management and Evaluation*, Hoboken, NJ: John Wiley & Sons.

Ralser, T. (2007) *ROI for Nonprofits: The New Key to Sustainability*, Hoboken, NJ: John Wiley & Sons.

Rich, E.H. (1998) *National Guide to Funding for Community Development*, Washington, DC: Foundation Center.

Sohl, K. (2005) "Congress and IRS Demand More Nonprofit Accountability." *TACS News*. Available online at http://www.tacs.org/tacsnews/dirtemplate.asp?pID=153 (accessed December 11, 2007).

U.S. Department of Agriculture. Available online at http://www.usda.gov/ (accessed December 11, 2007).

U.S. Department of Commerce. Available online at http://www.commerce.gov/ (accessed December 11, 2007).

U.S. Department of Housing and Urban Development. Available online at http://www.hud.gov/ (accessed December 11, 2007).

U.S. Department of Labor. Available online at http://www.dol.gov/ (accessed December 11, 2007).

# 22 The global economy and community development

## David R. Kolzow and Robert H. Pittman

The term "global economy" has become a common, almost trite term in the lexicon of modern business and economics. Over the past few decades, improvements in transportation and telecommunications have made it possible to conduct business on a truly global scale. Indeed, for thousands of companies, a global presence is a competitive imperative.

It is apparent that the global economy has affected larger cities around the world through trade and growing expatriate populations. All too often, when industries important to local economies downsize, close or move offshore, smaller communities are impacted as well. Today, communities large and small must be knowledgeable about global economic trends in order to formulate effective community and economic development strategies.

The chapter begins with an overview of the increasing interconnectedness of the world economy, touching on the theory of international trade and discussing trends in foreign direct investment. Next, the chapter defines and discusses two terms often seen in the media: outsourcing and offshoring. The chapter concludes with a discussion of the impact of globalization on community and economic development. New "second-wave" and "third-wave" strategies that states and communities are using to cope with and adapt to the global economy are discussed as well.

## Introduction

It would be difficult to find a community just about anywhere in the world that isn't significantly affected by international activity today. Often, the larger urban centers get much of the attention with respect to international issues. However, rural communities can no longer afford to ignore "globalization" as they plan for and work toward their economic futures. The products which rural customers buy are increasingly made in other countries; the reservation or technology centers they contact are often in India or the Philippines; the markets for products manufactured in their communities may be sold overseas; and their new manufacturing jobs may be coming from foreign firms locating in their community.

The world's communities are now in a global marketplace where, in both manufacturing and services, competitive companies now have to play if they hope to sustain profitability. On the one hand, this competition reduces the ability to raise prices but, on the other, it can stimulate innovation and higher productivity. A company in an industrialized nation now has to compete with other companies in China, India, Singapore, or other parts of the developing world, so the attraction of businesses to rural communities in the more developed part of the world has to be something other than low wages.

The momentum for globalization is only going to increase. The world is being brought closer together through the expanded exchange of goods and services, information, knowledge, technologies, and culture. However, over the past few decades, the pace of this global integration has become much faster and more dramatic owing to unprecedented advancements in technology, communications, science, and transportation.

## Why should people care about globalization?

Globalization has triggered a lot of discussion over the past few years. Some of this discussion is about how domestic industry suffers as other countries dramatically expand their exporting and open up their economies to investment. The criticism is that inequalities in the current global trading system benefit certain developing countries at the expense of the more developed ones. On the other side of the coin, countries such as China and India that have opened up to the world economy have experienced high levels of growth in gross domestic product (GDP). According to the International Monetary Fund (IMF), this is significantly reducing their levels of poverty and is creating new and growing markets for goods and services.

Unfortunately, a number of developing countries have not benefited from the gains of globalization, particularly in Africa. Exports from this continent continue to focus on a narrow range of primary commodities that experience considerable fluctuation in price and demand. Most of the African nations have seen relatively little value-added manufacturing or processing of their raw material commodities.

Generally, however, both developing and developed national economies are becoming steadily more interdependent as the international flows of trade, investment, and financial capital increase. The reduction of important obstacles to globalization, such as the high costs of tariffs and the complexity of trade regulations from country to country, is opening up new opportunities internationally. Around the world, people are buying more imported goods; a growing number of firms now operate as multinational companies; and companies and investors are continuing to grow their investments in the developing nations to take advantage of lower operating costs and rapidly expanding markets.

## International trade

Why is it important that trade flows relatively freely across the world? Companies trade because doing so allows them to concentrate on what they do most competitively, which generally results in increased productivity and sales revenues. This trade allows people to enjoy a higher standard of living because imports have a lower price tag. It also allows a country to exchange what it produces with what others produce, which provides a mutual benefit and leads to more efficient national economies. Free trade fosters this balancing of economies, although it might be said that completely free trade does not fully exist. More than 200 years ago, Adam Smith said it well in his *Wealth of Nations*:

> It is the maxim of every prudent master of a family never to make at home what it will cost him more to make than to buy. If a foreign country can supply us with a commodity cheaper than we ourselves can make it, better buy it of them.
>
> (Smith 1904)

International trade and investment play a much greater role in the economic life of the developed countries than in past decades. With cheaper goods available across the globe, demand grows as more people can afford to buy them. Growing international markets increase the standard of living globally, which leads to increased demand for the technologically advanced and innovative products and services of domestic companies.

It has been frequently stated that freer trade creates jobs. The idea is that, as other countries lower their trade barriers to U.S. goods, exports to these countries can be increased. In turn, those exports increase the overall sales and profits of domestic companies, which can then go out and hire more workers. For example, many domestic "growth" companies in the U.S. are those that continually increase their level of exporting to a global marketplace (Richardson and Lewis 2001).

For most countries, free trade does not appear to

create jobs across all economic sectors. It leads to more jobs in some sectors and fewer in others. For example, lower wage assembly-type jobs, such as apparel, textiles, shoes, and consumer electronic goods, tend to relocate to developing countries, typically lowering the cost of those products. Companies that invest in advanced technologies to raise their productivity are able to be more globally competitive. The resulting higher wages in these capital-intensive industries raise the standard of living but they aren't as labor-intensive as assembly-type and processing manufacturing plants. Even in "cheap labor" locations, the rapid spread of information technology is promoting investments in new capital equipment that increases productivity and lowers production costs.

Software designers do not sew their own shirts due to the comparative advantages found in different countries across the globe. It is also the reason why most families in the developed countries do not grow much of what they eat. They have better things to do with their time. It is the same with foreign as with domestic trade. Milton and Rose Friedman (1997), noted economists, wrote years ago: "We eat bananas from Central America, wear Italian shoes, drive German automobiles, and enjoy programs we see on our Japanese TV sets. Our gain from foreign trade is what we import. Exports are the price we pay to get imports."

In exchange for imports, countries gain income from selling other countries' competitively produced domestic goods. In the U.S., this includes such things as aerospace products, processed chickens, movies, or advanced medical devices. Other countries export electronic consumer products, apparel, and furniture to the U.S. Generally, people who send imports to the developed countries spend the resulting income in those countries. They can then buy goods they don't produce competitively or make investments in such things as real estate or auto plants. Countries benefit from the lower prices that imports provide, and the money saved may be either used to buy things made domestically or saved or invested.

Prices of goods are not cheaper globally simply because of lower wages being paid in developing countries. Prices of globally competitive products from developed countries have been holding steadier because global competition also makes it difficult to raise prices, which frequently results in the reduction of profits. Productivity increases have often been sufficient to allow for wage increases and create profits in more innovative companies. However, many firms in this competitive price situation have to cut costs to be profitable, which often results in employee downsizing.

However, it should be pointed out that importing lower cost products to a country doesn't just represent revenues for foreign firms. For example, overseas subsidiaries of American companies operating in foreign countries account for almost half of U.S. merchandise imports, and that keeps rising. Some of this is being driven by lower production costs overseas and the need to serve overseas markets from local operations. Actually, the bulk of international trade now takes place within industries as countries tend to specialize in varieties of particular goods. Multinational companies have begun to divide the production process into multiple steps at different locations to take advantage of location-specific advantages in each step (e.g., low labor costs in the production of labor-intensive parts, or the availability of software engineers in India).

The global economy is becoming dominated by large world-class corporations that emphasize competition for increased world market share. Many smaller firms may find it hard to compete on this level, given the expertise and financing needed. Smaller firms in rural communities in particular are often at an information and access disadvantage when it comes to finding help in exporting.

At the U.S. federal level, nine agencies are involved in export promotion. Two-thirds of these programs are dedicated, at least in part, to helping small and medium-sized companies. These include the Small Business Administration, the Export-Import Bank, the U.S. Foreign Commercial Service, and the Foreign Agricultural Service. These agencies provide market information, guaranteed loans, and credit insurance; identify business contacts and opportunities; and offer foreign advocacy on behalf of U.S. firms.

The exporting activities of U.S. firms, especially smaller ones, often benefit from state assistance as well. Many of the states have created international offices to help with exports and foreign investment, discovering new market opportunities, setting up the exporting process, financing export credits, determining what trade shows to attend, and so on. Many of the European and developed Asian countries maintain significant exporting offices around the world.

Exporting often starts with a few inquiries from future foreign customers. This leads to a greater interest in exports and increased participation in trade shows and other marketing events, where a company starts to find distributors and other useful contacts in a foreign market. Over time, exports grow to the point that a company can consider a sales office or other subsidiary in a foreign country. Finally, it might move part of its production there to take advantage of certain efficiencies if that market can sustain the business in question.

## Foreign direct investment

In the U.S. in particular, many rural communities have benefited over the past several decades from new manufacturing plants owned by foreign firms. This is known as foreign direct investment (FDI) and involves control of businesses or property across national borders. FDI includes corporate activities, such as businesses building plants or subsidiaries in foreign countries, and buying controlling stakes or shares in foreign companies. It does not include short-term capital flows such as portfolio investments. FDI allows firms to locate different stages of production wherever they are best suited: assembly where wages are lower, marketing where consumers are close at hand, and research and development where workers are well educated.

The United States, the world's largest economy, has historically experienced greater FDI activity than the other major industrialized economies in terms of

dollars invested. On a cumulative basis, the British remain the largest foreign direct investors in the U.S. economy, followed by French, Dutch, and Japanese investors. Although the bulk of FDI flows are among developed countries, the share of developing countries in world FDI activity is growing. Since the early 1990s, FDI has been the largest component of financial flows to developing countries, accounting for 25 percent of FDI activity globally in 2004 (United Nations Conference on Trade and Development 2006).

Most governments of lesser developed countries are very interested in attracting FDI. China and India rival one another and are aggressively challenging the United States as the world's most favored destination for foreign direct investment. China is now by far the largest recipient of FDI (UNCTAD). However, the developing countries are also seeing money flow out of their borders, even though they might prefer its investment at home. As with firms in rich countries, Third World multinationals invest abroad largely because they think they can put their money to better use there. Just as China's low wage workers attract many multinational firms from the developed world, China's own multinationals are exploring the advantage of even cheaper labor elsewhere.

Foreign firms continue to invest in the U.S. because of their interest in penetrating and expanding in its market. As a result, ten years from now a much larger amount of U.S. productive capacity will likely be directly or indirectly in foreign hands, and more multinational companies will exist. The highest level of employment in foreign-owned establishments in the U.S. is in chemicals and allied products; computers and electronic equipment; and transportation equipment (U.S. Bureau of Economic Analysis 2007).

The manufacturing plants built by these foreign firms can be a major boost to local economies in the U.S. These facilities tend to be newer and larger than existing domestic companies, pay wages on average that are higher than other local firms, and experience higher productivity and output per worker. To offset the risks that come with foreign investment, as well as to generate higher profits for economic operations management far from the parent firm, foreign investors generally tend to develop larger facilities and higher capital-intensive plants than the average U.S. firm.

Foreign direct investment is not a one-way street, especially in the U.S. American corporations are also continuing to invest abroad and its multinationals are now beginning to employ large numbers of foreigners relative to their American workforces. These American-owned firms are beginning to rely on foreign facilities to perform many of their most technologically complex activities and are to export from their foreign facilities. This includes bringing products back to the U.S.

Many people are of the opinion that U.S. manufacturing firms primarily invest in low-cost countries to take advantage of substantially cheaper wages. However, it is important to note that the highly developed countries of the world account for most of the total U.S. FDI. This reflects the continued involvement of U.S. firms in these more developed economies with their strong markets and highly skilled workers.

FDI has also been rapidly expanding in the services sector. Service functions that can be computerized, pulled out from the operation of a firm without major disruption, and exported back via telecommunication links from cheaper locations, are being identified. In many cases, these are being located offshore by multinational companies, either as parts of their own internationally integrated operations or for delivery (as "contract service providers") to other firms. In Asia, for example, the share of FDI activity in services continues to increase while that in manufacturing continues to fall (UNCTAD 2006).

While many services still need to be produced where their customers are located, others are being closely analyzed to determine their offshore potential. For example, information technologies and back-office work in areas, such as accounting, billing, software development, design, testing, and customer care, are being increasingly relocated abroad by multinational companies to developing countries. These types of services generally require higher skill levels than manufacturing or natural resource exploitation in developing countries.

## The great global job shift and outsourcing

*Business Week* (2003) highlighted the impact of global outsourcing over the past several decades on the quality and quantity of jobs in both developed and developing countries. The first wave of global outsourcing began in the 1960s and 1970s with the exodus of production jobs in shoes, clothing, cheap consumer electronics, and toys. Subsequently, routine service work, such as credit card receipt processing, airline reservations, and the writing of basic software codes began to move offshore. Today, computerization of work processes and the expanded use of the Internet and high-speed private data networks have enabled a wide range of knowledge work to become more "footloose" in its location (Gereffi and Sturgeon 2004).

The global outsourcing process combines two quite different activities: outsourcing and offshoring. Outsourcing is something which most businesses do, as they frequently need to make the decision to "make" or to "buy" specific inputs and services. While companies regularly decide whether they wish to produce goods and services in-house or buy them from outside vendors, the trend in recent years has shifted in the direction of buying them. Many have outsourced a wide range of services, such as accounts receivable, insurance, and logistics, to specialized firms.

Offshoring refers to the decision to move the supplying of goods and services from domestic to overseas locations. These activities may be carried out in facilities owned as a whole or in part by the parent firm or by overseas suppliers. The geographic shift of industries is certainly not a new phenomenon. In the early twentieth century in the United States, many industries that were established in New England, such as textile, apparel, footwear, and furniture, began to move to the South in search of available natural resources and cheaper labor. They frequently located in right-to-work states that made it difficult to establish labor unions.

The same forces behind the impetus to shift production to low-cost regions within the United States eventually led U.S. manufacturers to cross national borders to places such as Mexico, Japan, and Singapore, and eventually to most of East Asia. The reduction of tariffs has further fostered this outsourcing.

Global outsourcing continues to grow rapidly. This is because, despite all the criticism and skepticism, global outsourcing can provide huge benefits. Many firms have been able to cut costs, raise their productivity, improve the management of projects, and increase profits. Global outsourcing may be seen as a natural evolution from the domestic outsourcing practices in which many firms have engaged. In fact, global outsourcing has become a necessary business approach in order that companies maintain their profitability and market share (increasingly in a global marketplace) by using multiple offshore locations.

Another advantage of global outsourcing is that it takes advantage of the supply of specific talent wherever it is located. Jobs move to where costs are lower and new markets can be accessed. Countries like India, China, the Philippines, Mexico, Russia, South Africa, and parts of Eastern Europe, are loaded with college graduates who speak Western languages and who can handle outsourced information technology work. India is particularly well positioned in this regard.

In advanced economies such as that of the U.S., the rise of global outsourcing has triggered a lot of debate about job and production capacity loss that undermines national economies. On the other hand, many have dismissed these concerns. They argue instead that global outsourcing should be embraced as a means for more mature economies to shift out of low-value activities and traditional industries. This could free up capital and human resources for higher value activities and the development of newer industries and cutting-edge products. Clearly, such assurances are of little comfort to workers in the more developed countries who have been displaced from their jobs due to direct competition with firms and workers with low wages and increasingly high skills.

Although a good deal of publicity has swirled around global outsourcing's negative impact on developed economies like the U.S., a large and important portion of economic activities remains located in advanced economies, at least so far. This is true even as these activities have become tightly linked to activities located elsewhere in the world. Firms are now less likely to simply make products and export them; they increasingly participate in highly complex operational arrangements that transcend national borders, and that involve a growing array of partners, customers, and suppliers.

What began as simple assembly work in the 1960s and 1970s has rapidly spread up and down the production chain into a wide range of goods and services. There are few consumer products sold by retailers in developed countries today that aren't made entirely, or to a significant extent, at offshore factories in developing countries. Even products that require precision manufacturing, such as hard disk drives and many kinds of semiconductors, are becoming high-tech commodities made in capital-intensive facilities in Southeast Asia and elsewhere. Certain kinds of software programming and hardware design can now be done more cheaply in places like India, Taiwan, South Korea, and the Philippines than in the United States, Europe, or Japan. A growing array of knowledge-intensive business services is now beginning to move offshore as well. This includes engineering; design; accounting; legal and medical advice; financial analysis; and business consulting. This trend in a number of developing countries is leading their governments to pin their hopes on global outsourcing as a key driver for economic development.

It would appear that when offshore investment is primarily a result of pursuing new markets, it tends to have a minimal impact on the loss of jobs domestically. When activities are moved offshore to cut costs or to find new talent and the market remains at home, jobs are almost certainly displaced domestically.

When a company announces that it will move some or all of its production overseas, it is usually reported widely in the news media. However, several studies have shown that most of the manufacturing job losses in the U.S. have been due to productivity increases and not offshoring, outsourcing, and other structural changes. When worker productivity increases, fewer workers are required to produce the same level of output. Productivity can be increased in a variety of ways including upgrading production equipment and improving worker training. While productivity increases and offshoring act to reduce the level of U.S. manufacturing jobs, growing population and consumer demand act to increase employment.

Ward (2006) analyzed the causes of U.S. manufacturing employment decline from 17.7 million in 1990 to 14.4 million in 2004 (a net loss of 3.3 million jobs). According to his estimates, GDP growth should have added 5.7 million jobs in manufacturing over this period, but this was more than offset by estimated job losses of 7.5 million due to productivity increases. Ward estimated that 1.5 million jobs were lost from 1990 to 2004 due to "structural and competitive shifts" including offshoring. In other words, productivity increases appear to have caused the loss of many more jobs than offshoring.

## International corporate activity

Today, globalization has become a permanent and irreversible part of the life of corporations. Much of this is due to the integration of the latest information technology into corporate strategies and operations. Globalization in the 1990s differs from that of the past because it is being driven by the first truly global companies, which are major multinational operations. As the name implies, a multinational corporation is a business concern with operations in more than one country. These operations outside the company's home country may be linked to the parent by merger, operated as subsidiaries, or have considerable autonomy.

The global economy is now dominated by large world-class corporations which compete for increased world market share. The top multinational corporations are headquartered in the United States, Western Europe, and Japan. These major companies have the capacity to shape global trade and international financial flows. According to estimates by the United Nations Conference on Trade and Development (UNCTAD), the universe of multinational companies now includes some 77,000 parent companies with over 770,000 foreign affiliates. In 2005, these foreign affiliates generated an estimated $4.5 trillion in value added, employed some 62 million workers, and exported goods and services worth more than $4 trillion (UNCTAD 2006). These corporations are capital- and knowledge-intensive, and focus on where they can operate any part of their business most productively and profitably. However, despite this opportunity, they continue to locate more than 75 percent of their production, employment, and capital spending in their home countries.

Increasingly and rapidly, international corporate activity is being driven by the Internet platform. Companies are becoming "e-businesses" that build their whole operation around the Internet. These electronic business methods enable companies to link their internal and external data-processing systems more efficiently and flexibly, to work more closely with suppliers and partners, and to better satisfy the needs and expectations of their customers. E-business involves business processes that cover every aspect of the operation including electronic purchasing, supply-chain management, processing orders electronically, handling customer service, monitoring the assembly of goods, and interaction with business partners. The World Wide Web now permits this to take place on a global scale. In addition, the Internet has opened up foreign markets that for decades have been costly and difficult and enter. A simple website can attract visitors from all over the world.

The early dominance of the Internet by U.S. firms is quickly fading, however. Now, it is estimated that more than half of the world's online users don't live in the United States; they speak and write different languages, use different currencies, abide by different legal systems, and have very different preferences and tastes than Americans.

## Offshore manufacturing

Contrary to popular opinion, the majority of the investment in developed countries abroad by manufacturing firms has not gone to countries with low labor rates. For example, based on data from the U.S. Department of Commerce, Europe is by far the most popular region for U.S. foreign investment. Many companies in the developed nations are first looking to build market share in other industrialized countries by accessing customers, technology, and skills, as opposed to choosing a country due to the availability of cheap labor.

However, wage rates are as much as 90 percent lower in the underdeveloped nations. An American production worker may be three times more expensive than a Taiwanese worker and cost six times more than a Brazilian or Mexican competitor. Recent research has demonstrated, however, that low wages in developing countries go hand in hand with lower productivity. In part, this reflects the simpler machinery used because labor is cheap relative to capital. However, inferior infrastructure and education in poor countries also play an important part, providing a significant handicap even with identical technology. In a global economy, countries that let their educational standards slip risk being left behind.

Global competition is hastening the shift of the developed countries to the production of relatively high-value goods. They will be less competitive in world markets for goods produced with a high input of labor and a small input of capital and technology. Manufacturers will import or "outsource" components that can best be produced by unskilled labor.

It is not only cheap labor that is attracting production facilities. Many of the developing countries are creating high-tech industries through technology transfers from the industrial world. Multinational corporations are building state-of-the-art manufacturing facilities in developing countries. Nations such as India, China, and Mexico are turning out engineers and technicians at a faster rate and in greater numbers than is the U.S., making the availability of this talent an attraction for further investment.

## Community and economic development strategies in the global economy

With footloose firms seeking the best locations all over the world in the global economy, the competition to attract them has greatly intensified. Furthermore, many communities in the developed world that lose their traditional low-skilled industries to less developed countries may not be suitable locations for newer knowledge-based manufacturing and service industries. They may lack a skilled labor force, modern infrastructure, and other location factors important to new industries.

### Second- and third-wave strategies

Global competition has encouraged many countries, states, and communities to move their economic development strategies away from an emphasis on recruiting traditional manufacturing industries toward a more strategic approach focused on community development – making the community a more competitive location for new high-growth industries. Bradshaw and Blakely (1999) refer to the traditional recruiting approach, or smokestack chasing, as the first wave of industrial attraction efforts whereby states and communities competed against each other for company locations, often with large incentive packages. As returns from this approach diminished, states and communities created the second wave which, according to Bradshaw and Blakely, focused on growing existing businesses already in the area and developing local entrepreneurs. New business retention and expansion policies were developed (see Chapter 14) which included company surveys and visits to identify and correct problems with the local business climate; research services to help businesses expand their geographical markets; and other programs designed to increase the likelihood that businesses would stay and grow where they were currently located. Second-wave strategies to encourage new business start-ups included incubators where new businesses could share common facilities and pay less rent, loan funds to finance new businesses, and entrepreneur training programs.

According to Bradshaw and Blakely, the third wave of economic development focuses on broad strategies to make an area more attractive to modern manufacturing and service industries and increase its global competitiveness. Third-wave economic development strategies include:

- *Community development.* Comprehensive efforts — including improving local education, workforce development, infrastructure modernization, and other actions — to make a community attractive to technology companies and knowledge workers. Third-wave strategies recognize the link between community and economic development.

- *Regionalism.* Often communities in a region compete against each other for the same new companies. In this game, elected officials and citizens often believe that the community that gets the project "wins" and all other communities around them "lose." In reality, when a company moves into a community, surrounding communities, counties, and sometimes even states benefit. Workers typically commute from miles away, providing residents of different jurisdictions with new employment opportunities and increasing incomes and tax revenues where these workers live. In addition, companies that buy and sell goods or services to the new firm may locate in an adjacent community, again benefiting the region as a whole. Olberding (2002) has documented the benefits of and increase in regional economic development partnerships. Working together regionally, local governments can develop infrastructure and industrial sites, improve business retention and expansion and new business start-up programs, and enjoy larger regional marketing budgets.

- *Public–private partnerships.* New jobs and higher incomes from economic development benefit many private sector businesses in a community and region. Local banks, professional service firms (e.g., legal, accounting), retail shops, local restaurants, and manufacturing firms supplying other local firms with intermediate products all benefit as the demand for their products and services increases. In many communities, states, and regions, private sector businesses have begun to team with local governments and to support their economic development efforts. Often private sector representatives will serve on economic development committees and boards of directors. Ongoing private sector representation can help provide continuity to offset the needless changes in economic development policy often triggered by the changing of public administrations. Furthermore, it is common for private sector partners to help fund the economic development effort.

- *Industry clusters.* Certain kinds of businesses can benefit from close geographical proximity, creating a competitive advantage. Businesses within the same industry that export their products or services to other regions may benefit from a shared labor pool; an increased number of executives with industry experience in the area; and the sharing of new technology and business techniques through the local "grapevine" (see Chapter 11). Businesses can also benefit when firms that supply them with intermediate products or services locate in the area. This can save transportation costs and create a more efficient supply chain. By identifying, researching, and facilitating industry clusters appropriate to their area, states and localities can develop a significant comparative advantage in attracting businesses that are part of the clusters. One of the best examples of industry clusters is Silicon Valley in California. Because of all the advantages listed above and more, computer and related industries have come to dominate the area. Austin, Texas, and Boston, Massachusetts are similar examples of technology industry clusters.

As Bradshaw and Blakely point out, third-wave strategies can complement first- and second-wave approaches. All of the third-wave strategies listed above can help improve the "traditional" economic development activities of recruiting new businesses, retaining and expanding existing businesses, and facilitating new business start-ups. Third-wave strategies can also make communities more competitive in recruiting foreign firms. As noted above, foreign direct investment among developed countries is quite high, affording many opportunities for communities and states to recruit foreign firms. In the U.S., one does not have to look far to see automobile assembly plants for Toyota and Daimler-Benz or production facilities for electronics companies such as Philips and Sony.

## Advanced manufacturing

While total employment in manufacturing in the U.S. is declining as noted above, certain industries are adding high-skill manufacturing jobs requiring well-trained productive workers. Because of their superior education systems, infrastructure, research facilities, and so on, developed countries commonly have an advantage over less developed countries in many high-skill jobs. The number of high-skilled positions paying over US$24 an hour actually increased by 36 percent in the U.S. between 1983 and 2002. Third-wave strategies can be particularly effective in preserving and creating high-skill manufacturing jobs.

## Domestic outsourcing

When companies decide to outsource parts of their operations, it does not necessarily have to be to another country. The process of sending work from high-cost to low-cost areas within a country is often referred to as "farmshoring" or "nearshoring." Communities and regions within a country can compete for these domestically outsourced operations. As an example, Technologent, a Sun Microsystems supplier headquartered in Rancho Santa Margarita, California, established a technical support and sales call center in Ainsworth, Nebraska. The company decided to outsource this activity to a U.S. location to avoid language and cultural barriers in Asian countries (Wright 2005).

## Conclusion

It is virtually impossible to ignore the reality of the global economy and its impact on a particular community, no matter where it is located. With respect to responding appropriately to globalization, the decisions made are not easy ones, because the issues are so complex and much disagreement exists on what is happening and why. The challenge is to research these issues carefully and think about their implications for the local economy. Recognizing that full understanding is unlikely, the greater the level of understanding, the better the policy decisions and action agendas will be formulated. This chapter is only a stepping-stone into the arena of the international economy.

## CASE STUDY: SHRINKING TO PROSPERITY

Growth and development go hand in hand with prosperity in most communities. Communities grow in area and density as new industrial, commercial, and residential development occurs. However, many communities lose population for various reasons including a decline in their industrial base. As a matter of fact, it has been estimated that during the 1990s more than a quarter of the world's largest cities declined in population (Aeppel 2007). When population declines, tax revenues generally fall and it becomes more difficult and expensive to maintain existing infrastructure that was designed to support a larger population. Buildings and houses are boarded up, often due to the inability to pay property taxes, and sometimes whole neighborhoods are abandoned.

Most communities respond to this negative state of affairs by attempting to maintain services and infrastructure at current levels. The usual reaction is to try to grow the industrial base and population to where they were before – so that they match the city's existing infrastructure. In some cases this strategy is successful, but it may not be possible in many cities, especially older industrial ones. Trying to maintain infrastructure and services at previous levels with shrinking budgets is difficult and inefficient.

Youngstown, Ohio, historically a major center of steel production, faced this problem as its mills shut down in the 1970s and 1980s, due primarily to foreign competition. Population dropped by 60 percent, leaving properties abandoned and neighborhoods blighted. Rather than try to grow out of the problem, city officials faced up to the fact that the steel mills weren't coming back and it was unlikely that the lost jobs would be replaced by new industries. Instead, the city decided to adjust to its new reduced population base through "creative shrinkage" (Aeppel 2007).

In 1999, city officials decided to take a new approach, so they asked the city planner, Jay Williams, to develop a master plan more suited to a smaller population. The plan advocated a radical approach: rather than fight to keep all areas of the city from losing population and declining, allow some neighborhoods to continue to depopulate and redevelop them as urban green spaces. Reducing the city's geographic area by de-annexing certain areas was not viewed as an option, as changing jurisdictional boundaries would raise political and legal issues. City planners divided Youngstown into 127 neighborhoods and labeled them as stable, transitional, or weak. The city is now developing a plan for each neighborhood, with city resources going primarily to stable and selected transitional neighborhoods.

The city has told residents of weak neighborhoods that it will no longer invest resources and try to stabilize them. Instead, residents have been informed that eventually neighborhood streets will be dug up, blighted buildings torn down, and streetlights removed to create green spaces for parks and community gardens.

The "Clean and Green" master plan was widely embraced by Youngstown residents. The city practiced good community development principles, conducting dozens of neighborhood meetings to explain the plan and gather input from citizens. The city planner, Jay Williams, was subsequently elected Mayor and is continuing the process of public participation and remaking Youngstown one neighborhood at a time. The city has won numerous awards for its planning process including the National Planning Excellence Award for Public Outreach from the American Planning Association. "Creative shrinkage" will help Youngstown become a more desirable place to live and work. Undoubtedly, at some point in the future, Youngstown's population will rebound, and once again growth and development will accompany prosperity.

## BOX 22.1: IMMIGRATION AND COMMUNITY DEVELOPMENT

As the world "shrinks" through advances in telecommunications and transportation technologies, international migration among countries is on the increase. The movement of people across borders has many benefits (e.g., cultural cross-pollination and more efficient global labor markets) but it also presents many challenges including the assimilation of immigrants into their destination countries. This section contains a brief overview of some of the implications of immigration for the field of community development. As global migration grows, the tools and techniques of community development will become even more important in assimilating immigrants and moving communities forward.

### Global migration trends

A recent study by the United Nations (2006) reported these facts concerning global migration:

- International migrants numbered 141 million in 2005; 115 million lived in developed countries.
- Immigration is growing fastest in developing countries.
- Three-quarters of all migrants lived in just 28 countries in 2005, with one in every five migrants living in the U.S.
- Migrants constitute at least 20 percent of the population in 41 countries.

Migrants are not just subsistence workers seeking low-skill jobs. The World Bank found that 36 percent of migrants to the 20 richest countries in 2000 had a college education, a sharp rise from a decade earlier (Economist 2008a).

According to demographic projections by the Pew Hispanic Center, immigration will change the face of the United States over the next few decades (Spencer 2008). By 2030, foreign-born residents will constitute 15 percent of the population, surpassing the last peak in 1910. Almost 82 percent of the population growth in the country through 2050 will be accounted for by new immigrants and their U.S.-born descendants. Nearly one in five people in the country in 2050 will be an immigrant, compared with one in eight in 2005. In other words, the U.S. as a whole in 2050 will be demographically similar to California and Florida today. Seeking economic opportunity, immigrants are settling all across America. The proportion of Mexican-born people living in states other than the four Mexican border states plus Illinois increased from 10 percent in 1990 to 25 percent in 2000 (Economist 2008b). The change in workforce composition will be even more dramatic. It is predicted that the combination of aging baby boomers, a lower indigenous birth rate, and a higher immigrant birth rate will increase the percentage of Hispanics in the workforce to 31 percent in 2050, compared to 14 percent in 2008 (Spencer 2008).

While illegal immigration is a hot political topic and receives extensive media coverage, it is estimated that two out of every three immigrants to the U.S. are legal immigrants, many sponsored by relatives. From 2002 to 2006, approximately one million legal immigrants entered the U.S. every year, compared to half a million illegal immigrants. It is difficult to estimate the corresponding numbers for the European Union because immigrants can move freely among member countries.

## Impacts of global migration

Global migration will continue to profoundly affect the social, political, and economic systems of both origin and destination countries. Many economists argue that just like international capital flows and foreign direct investment, the movement of labor across different markets enhances global economic growth. The World Bank recently estimated that the economic gains from international migration are greater than the gains that would accrue from significantly liberalizing international merchandise trade flows, and the United Nations estimates that the negative impact on wage rates and unemployment rates in destination countries from immigration is minimal (United Nations 2006).

Immigration, legal and illegal, often puts enormous economic and social strains on destination countries and communities. The challenge of educating immigrants is well documented, as illustrated by the community of Baron, Wisconsin (see case study below). Immigrants also present unique challenges in the provision of a host of other community services including health care, criminal justice, and housing. There are many stereotypes regarding immigrants and their impacts on countries, states, and communities. While the challenges are real, studies are debunking many of the myths associated with immigration. For example, a recent study showed that immigrants do not contribute to a higher crime rate in Britain (Economist 2008c). Many studies on the impacts of immigration on countries and communities have been conducted, but a review of this growing literature is beyond the scope of this chapter.

### *Immigrants and education in Barron, Wisconsin*

Barron, located in the west-central portion of Wisconsin, has a population of approximately 3000. As of 2003, a local turkey-processing facility employed approximately 400 non-English-speaking immigrants, including Hispanics and Somalis, up tenfold over six years. Reactions from indigenous residents to this sudden influx of immigrants included:

"They're living in America: why can't they live like us?"
"If they live here, they better speak English. If you want to become part of our society, learn our language!"
"A gang of them hangs out in the park. It's scary to let our kids play there."

A group of community leaders decided to address this cultural divide by organizing a Diversity Council to foster lines of communication and mutual respect. They established an International Center where minorities could attend English classes and receive help with housing, health care, legal issues and other community needs. The founding of the Center was facilitated by a $150,000 grant from the federal Office of Refuge Resettlement, since the Somalis were classified as political refugees.

Community meetings were held to organize the Diversity Council. At first they were poorly attended, but volunteers continued to publicize their activities and contributions from local businesses were obtained. Cultural awareness and English as a Second Language (ESL) programs were started in local schools. A Diversity Day was held in the Barron Middle School where students learned dances from the immigrants' countries and listened to stories about growing up in other countries. While conflicts and misunderstandings still exist, the efforts of volunteer community developers to increase appreciation for diversity and assimilate the immigrants into local society have increased the quality of life for all residents of Barron.
Source: Principal Leadership Magazine, April 2003

The impact of immigration often depends on how successfully immigrants assimilate into the society of their destination country. Some scholars argue that today's immigrants, many of whom arrive with knowledge of American culture and speaking English, are "pre-assimilated" into American society. As evidence, they point to homeownership, one of the key measures of assimilation. About 68 percent of immigrants who arrived in the U.S. in the 1970s are homeowners – equal to the rate of natives (Spencer 2008). However, other scholars worry that many immigrants from Third World countries arrive in developed countries poorly schooled from cultures that do not place much emphasis on education, and therefore assimilation will be difficult.

## Immigration, social capital, and the future

One of the more interesting implications for the growth of immigration in the field of community development is the potential impact on social capital, defined in Chapter 1 as the resources and social relationships among persons and organizations that facilitate cooperation and collaboration in communities. Research by Harvard political scientist Robert Putnam (author of *Bowling Alone* which chronicles the decline of social connectivity in post-war America) has shown that in diverse communities, citizens display lower levels of confidence in local government and media, are less likely to be involved in local volunteer groups, and less likely to vote and have lower levels of expressed happiness (Giddens 2007). This seminal research spawned a contentious debate on whether Putnam's study measured social capital appropriately and completed the data analysis correctly. Repeated statistical analysis produced the same result.

Some scholars have explained Putnam's result by noting that it takes time for immigrants to assimilate into communities and neighborhoods and that in future years social capital will likely grow in heterogeneous communities. As the accompanying example of Cerritos, California shows, contrary to Putnam's study, diversity does not necessarily mean low social capital. Immigration presents a special challenge to community development, and the discipline will play an increasingly important role in helping countries and communities cope with the growing challenges presented by global migration.

## Diversity and success in Cerritos, California

Cerritos, a city of 55,000 people south of Los Angeles in Orange County, is by all measures a very successful and prosperous community. It boasts outstanding schools, modern infrastructure, a booming economy with high sales tax collections, lush parks, libraries, and a state-of-the-art performing arts center. The City has been well managed through a series of long-serving city managers who work closely with city council members. In a 2002 poll, 96 percent of the residents said they were satisfied with the provision of public services.

Cerritos has managed to achieve this success while becoming an international melting-pot. In 1980, whites comprised more than half of the population. By 2000, however, they comprised only 21.4 percent of the population. Other ethnic groups in Cerritos in 2000 included Latino (10.4 percent), black (6.7 percent), Chinese (15 percent), Filipino (11.7 percent), Indian (5.6 percent), Korean (17.4 percent), and mixed-race/other (11.8 percent). These diverse populations have not congregated together in isolated groups. The 2000 Census showed that the population in Cerritos is more residentially intermixed than in all but 16 other American cities. Apparently, the residents of Cerritos have learned to work together for the good of the community and have developed a high degree of social capital despite being from many different backgrounds. Perhaps they even bowl together.

Source: *The Economist*, August 16, 2007

## Keywords

Foreign direct investment, offshoring, outsourcing, second-wave strategies, third-wave strategies, regionalism, public–private partnerships, industry clusters, advanced manufacturing, domestic outsourcing.

## Review questions

1 What is comparative advantage and how is it related to international trade?
2 What is foreign direct investment? In which countries does most of the world's FDI occur?
3 What is difference between outsourcing and offshoring? Why do they occur?
4 What are some factors driving corporations to globalize their businesses?
5 What are second-wave economic development strategies?
6 What are third-wave economic development strategies? What are some key elements of third-wave strategies?
7 What is domestic outsourcing and why is it an opportunity for communities?

## Bibliography and additional resources

Aeppel (2007) "Shrink to Fit: As Population Declines, Youngstown Thinks Small," *Wall Street Journal*, May 3: A1.

Bradshaw, T.K. and Blakely, E.J. (1999) "What are 'Third Wave' State Economic Development Efforts? From Incentives to Industrial Policy," *Economic Development Quarterly*, 13(3): 229–244.

*Business Week* (2003) "A Global Corporate Migration," February 3: 118.

Economist (2008a) "Open Up," *Economist* Magazine, Vol. 386, January 5, p. 3.

Economist (2008b) "United States: The Newest Frontier," *Economist* Magazine, Vol. 387, February 23, p. 60.

Economist (2008c) "Britain: Not Guilty," *Economist* Magazine, Vol. 387, April 19, p. 41.

Friedman, M. and Friedman, R. (1997) "The Case for Free Trade," *Hoover Digest*, 1(4). Available online at http://www.hoover.org/publications/digest/3550727.html (accessed December 29, 2007).

Gereffi, G. and Sturgeon, T.J. (2004) "Globalization, Employment, and Economic Development: A Briefing Paper," *Sloan Workshop Series in Industry Studies*, Rockport, MA. Available online at http://www.soc.duke.edu/sloan_2004/Docs/Workshop_Summary.pdf (accessed December 29, 2007).

Howenstine, N. and Zeile, W.J. (1994) "Characteristics of Foreign-owned U.S. Manufacturing Establishments," *Survey of Current Business*. U.S. Bureau of Economic Analysis. Available online at www.bea.gov/scb/pdf/internat/fdinvest/1994/0194iid.pdf (accessed December 11, 2007).

Jackson, J.K. (2005) *CRS Report for Congress. Foreign Direct Investment in the United States: An Economic Analysis*, Washington, DC: Congressional Research Service.

Kotkin, J. (2007) "The Myth of Deindustrialization," *The Wall Street Journal*, August 6: A13.

Olberding, J.C. (2002) "Diving Into Third Waves of Regional Governance and Economic Development Strategies," *Economic Development Quarterly*, 16(3): 251–272.

Richardson, J.D. and Lewis, H. (2001) *Why Global Commitment Really Matters!* Washington, DC: Peterson Institute for International Economics.

Smith, A. (1904) *The Wealth of Nations*, 5th edn (Book IV, Chapter II), London: Methuen & Co.

Spencer, M. (2008) "The Increasingly Changing Face of America," *McClatchy-Tribune Business News*, February 12, Washington, DC.

United Nations Conference on Trade and Development. (2006) *Trade and Development Report*. Available online at http://www.unctad.org/Templates/WebFlyer.asp?intItemID=3921&lang=1 (accessed December 11, 2007).

United Nations General Assembly (2006) *International Migration and Development*. Available online at http://www.un.org/esa/population/migration/hld/Text/Report%20of%20the%20SG%28June%2006%29_English.pdf (accessed April 29, 2008).

Ward, W.A. (2006) "Manufacturing Jobs," *Economic Development Journal*, 5(1): 7–15.

Wright, R. (2005) "Outsourcing in America – There's No Place Like Home," *VARbusiness*, 21(15): 20.

# 23 Sustainability in community development

## Stephen M. Wheeler

The aim of this chapter is to provide some basic background on the concept of sustainability and how it may apply to both the practice and content of community development. It starts with a brief overview of the history and theory of this term, then examines its implications for a number of areas within the context of community development. There is substantial agreement in the international literature on many of these implications; however, there is no single ideal of "the sustainable community," nor any examples of such places. Rather, there are many strategies that can potentially improve the long-term health and welfare of communities by working with local history, culture, economy, and ecology. Every existing community has some features that others can learn from as well as many challenges to be addressed. For any given place, the task for professionals is to develop creative strategies and processes that will work within the local context and with its constituencies to improve long-term human and ecological welfare.

## History of the sustainability concept

The reasons why sustainability has become a leading theme worldwide are well known. Concerns such as climate change, resource depletion, pollution, loss of species and ecosystems, poverty, inequality, traffic congestion, inadequate housing, and loss of community and social capital are ubiquitous. These problems are interrelated; for example, global warming emissions are caused in part by inefficient transportation systems and land-use patterns, poorly designed and energy-intensive housing, and economic systems that do not internalize the costs of resource depletion and pollution.

As far as anyone has been able to tell, the term "sustainable development" was used for the first time in two books that appeared in 1972: *The Limits to Growth*, written by a team of MIT researchers led by Donella Meadows (Meadows et al. 1972), and *Blue-*

*print for Survival*, written by the staff of *The Ecologist* magazine (Goldsmith et al. 1972). The Meadows report in particular was significant in that it used newly available computer technology to develop a "systems dynamics" model predicting future levels of global resources, consumption, pollution, and population. Every scenario that the team fed into the model showed the global human system crashing mid-way through the twenty-first century, and so the researchers concluded that human civilization was approaching the limits to growth on a small planet. This prediction was highly controversial. But revisiting the model in 2002, with three additional decades of actual data, the team concluded that its initial projections had been relatively accurate and that humanity has entered into a period of "overshoot" in which it is well beyond the planet's ability to sustain human society (Meadows et al. 2004).

Other events in the 1970s also helped catalyze concern about the sustainability of human development patterns. The first United Nations Conference on Environment and Development, held in Stockholm in September, 1972, brought together researchers and policy makers from around the world to explore humanity's future on the planet. The energy crises of 1973 and 1979 raised global concerns about resource depletion and brought these concerns home to millions of Americans at the gas pump. Public attention to the need for sustainable development received further boosts in the early 1990s as a result of United Nations conferences, such as the "Earth Summit" held in Rio de Janeiro in 1992, and in the early 2000s as knowledge spread about the threat of global warming. Although for many years "sustainability" was dismissed as a faddish or overly idealistic term, by the early twenty-first century it had become well established as a priority in many communities worldwide.

## Perspectives

Several perspectives on sustainable development emerged early on that have characterized debates ever since (Wheeler 2004). One of these viewpoints is that of global environmentalism, which has focused on resource depletion, pollution, and species and habitat loss (Brown 1981; Blowers 1993). Some, such as the so-called "deep ecologists," have even argued that other species should be given the same rights as humans and that human population overall is too large and should be substantially reduced, presumably through wise family planning in the long run (Devall and Sessions 1985). Counter to these environmental perspectives – in fact directly opposing the limits-to-growth viewpoint – has been the approach known as technological optimism which holds that human ingenuity and technology will be able to conquer environmental problems. Although clearly this does happen sometimes, technology has not yet addressed many of the concerns described above.

A somewhat different set of perspectives, also originating in the 1970s, focus on the role of economics in addressing environmental and social prob-

lems. Economists within the newly emerging disciplines of environmental economics and ecological economics set to work to better incorporate environmental factors into economic models (Repetto 1985; Pearce et al. 1989; Costanza 1991). Some began to question on a much more fundamental level the desirability of endless economic growth on a planet with finite resources. Herman Daly, in particular, advocated a "steady-state society" with qualitative rather than quantitative economic growth (Daly 1973, 1980, 1996). Some recent economic thinkers have also advocated new forms of capitalism that better incorporate environmental and social concerns (Hawken et al. 1999; Barnes 2006).

A third main set of perspectives is that of social justice advocates, many of them in the Third World. These critics point out global inequities that have led the United States, for example, with about 4 percent of the world's population to consume some 25 percent of its resources (Goldsmith et al. 1992; Barlow and Clarke 2001; Shiva 2005). Such critics have argued that sustainable development first needs to address global disparities and that wealthier countries need to substantially reduce their consumption. Such a viewpoint has been met with varying opinions.

Finally, spiritually and ethically oriented observers have argued that the global crises facing humanity are due to misplaced values, a cognitive perspective that does not adequately recognize interdependency, and/or the lack of an ethical perspective that takes the needs of other societies and the planet into account (Daly and Cobb 1989; Goldsmith 1993; Capra 1996). These writers often build on precedents such as the "land ethic" of Aldo Leopold (1949) to argue that a new relationship between humans and the Earth and between humans and each other is necessary.

These different perspectives on sustainable development have led to different arguments, analyses, and proposals ever since. For example, economists tend to assume that market mechanisms, such as emissions trading systems or steps to set the proper prices on natural resources and pollution, will be able to address sustainability problems. Many environmentalists, on the other hand, argue for strong regulation by the public sector and public investment in areas such as alternative energy and land conservation. Equity activists call for radical rethinking of global capitalism and tend to be highly critical of institutions such as the World Bank and the World Trade Organization. Meanwhile, ethically or spiritually oriented thinkers seek leadership and education toward a different set of societal values and, in some cases, seek guidance within organized religious traditions. Elements of all of these approaches seem useful at different times, and an awareness of all of these perspectives is important to form an understanding for pragmatic application of sustainability ideas within communities.

## Sustainability definitions and themes

Despite the extraordinary influence of the sustainable development concept, no perfect definition of the term has emerged. The most widely used formulation is that issued by the United Nations Commission on Environment and Development (the "Brundtland Commission") in 1987, which defines sustainable development as "development that meets the needs of the present without compromising the ability of future generations to meet their own needs" (World Commission 1987). However, this definition is problematic, since it raises the difficult-to-define concept of "needs" and is anthropocentric, discussing the needs of humans rather than those of ecosystems or the planet as a whole.

Other definitions have similar problems. For example, relying on the notion of the "carrying capacity" (the inherent ability of a community or region to support human life and maintain environmental well-being) is difficult, since this is hard to determine for human communities. Relying on concepts such as maintaining natural and social capital is problematic, since these entities are difficult to measure and would require a complex economic calculus. One preference is simply to define sustainable development as development that improves the

long-term welfare of human and ecological communities and then move on to a more specific discussion of particular strategies.

## Approaches

Sustainable development tends to require certain approaches on the part of community development leaders and professionals. One obvious starting point is to emphasize the long-term future. Rather than thinking about the next economic quarter, the next election cycle, or even the next 10 or 20 years (as is common in local planning documents), it becomes important to think what current development trends mean if continued for 50, 100, or 200 years. Often short-term trends that seem acceptable become disastrous when viewed in the longer term. One essential starting place is getting the public and decision makers to understand the long-term implications of current trends in addition to their near-term impacts.

Another main approach within sustainability planning is to emphasize interconnections between community development issues. Land use, transportation, housing, economic development, environmental protection, and social equity are all related. Historically, one main problem in planning has been that these issues have been treated in isolation; for example, highways have been planned without considering the sprawling land-use patterns that they will stimulate, and suburban malls and big box stores have been encouraged without realizing that they may lead to disinvestment and poverty in traditional downtowns. Viewing any given topic from a broad and holistic perspective, each individual decision can be better tied into sustainable community development as a whole. Likewise, developing an understanding of how actions at different scales interrelate is important as well. Building, site, neighborhood, city, region, state, national, and global scales fit together; actions at each scale must consider and reinforce actions at other scales. Recent movements such as the New Urbanism have emphasized a similar coordination of action at different scales (Congress for the New Urbanism 1999).

Another theme within sustainable community development is attention to place. Local history, culture, climate, resources, architecture, building materials, businesses, and ecosystems provide a rich and valuable context for local sustainability efforts. Working with these resources is also a way to build community pride and identity. For example, restoring a stream or river front can create an attractive new amenity for a community; identifying and restoring historic buildings can help give character to a neighborhood.

Tied to an emphasis on place is an acknowledgment of limits. Any given place can only handle so much change before it becomes something different (which is of course sometimes desirable). There are limits to the number of people or the amount of traffic that can be accommodated easily in any given community without undermining those place-based attributes that community members value. Likewise, there are limits to the quantities of resources that our society as a whole can use without damaging either local or global ecosystems. "Growth" itself must be reconsidered within a sustainable development paradigm, following Daly's notion, moving from quantitative expansion of goods consumed to qualitative improvement in community welfare. An organization named Redefining Progress has in fact developed a "Genuine Progress Indicator" that it believes can measure such a shift at the national level instead of the gross domestic product which, as is often pointed out, rises significantly during environmental disasters, such as the Exxon Valdez oil spill, since large sums are spent on clean-up and public relations (Talberth et al. 2006). At the local level, efforts to rethink growth should not take the form of exclusionary growth controls designed to keep out lower income residents by restricting the amount of multi-family housing, but should be a more comprehensive rethinking of how the community will coexist with local, regional, and global resource limits in the long run.

A final theme implicit within sustainable development is the need for active leadership by planners, politicians, and other community development professionals. In the past, these players have sometimes facilitated unintended consequences of development. More active and passionate engagement by professionals is needed to address current sustainability problems, often seeking new alternatives to the status quo. In this quest it is important for community development professionals to work actively with elected leaders, community organizations, businesses, and the general public to develop public understanding and political support for action.

Thus the concept of sustainability may be seen to have roots going back more than 35 years, a variety of different perspectives taken by different advocates, and some themes that can guide professionals in seeking real-world applications. With this background, some specific areas of sustainable community development planning are presented below. Since fully considering the topic would require a very large space, the intent here is simply to suggest some possible directions for action.

## Action areas

### Environment

Sustainability is often thought of as primarily an environmental concern, and certainly environmental initiatives are important within any sustainable community development agenda. These can be of many sorts, but one of the most timely and challenging types of initiatives aims to reduce greenhouse gas emissions. Global warming initiatives at the local level are increasingly common, thanks in part to the Cities for Climate Protection campaign coordinated by the International Council for Local Environmental Initiatives (ICLEI), and require a very broad and interdisciplinary rethinking of many local government policies. In the U.S., some 27 percent of greenhouse gas (GHG) emissions stems from transportation uses, another 27 percent is related to building heating, cooling, and electrical use, and about 20 percent results from industry (World Resources Institute 2007). Local governments can affect all of these areas.

Communities can best reduce private motor vehicle use – and resulting GHG emissions – through three types of initiatives: better land-use planning, better alternative travel mode choices for local residents, and revised economic incentives for travel. All of these types of initiatives are discussed later in this chapter. Local governments can also set an example by converting their own vehicle fleets, including buses, to cleaner technologies such as hybrid engines and use of compressed natural gas or biodiesel.

In terms of building heating, cooling, and electricity use, communities can modify building codes to require passive solar design of structures, higher degrees of energy efficiency, use of energy- and water-efficient appliances, and recycling of construction waste and debris. Subdivision ordinances can be modified to require solar orientation of lots in new subdivisions (with the long dimensions of lots and buildings facing south), and zoning codes can be amended to ensure solar access to each lot (by restricting the height of structures on adjoining lots near the southern property line). Other eco-friendly strategies such as handling rainwater runoff on-site, using graywater (lightly used wastewater) for irrigation or toilets, minimizing asphalt paving, promoting the use of alternative construction materials and green roofs, providing incentives for solar hot water or electricity, and encouraging shade trees to provide summer cooling may also be incorporated into these codes. In terms of electric power, communities may require that a certain percentage of electricity they purchase be generated from renewable sources. Some cities and towns have historically owned their own electric utilities which gives them an even greater ability to lower GHG emissions and promote green practices.

To reduce the 20 percent of GHG emissions stemming from industry, local governments can seek to identify such sources within their jurisdictions and work with them to reduce emissions, for example, by providing technical assistance, grants, or favorable tax treatment for eco-friendly practices. Giving priority to reducing emissions may also affect economic development policy choices, as discussed

later in this chapter. In short, a local greenhouse gas reduction program must address many different aspects of policy, integrating these initiatives together. Each of these steps will have other sustainability advantages, however, whether in terms of reducing traffic congestion and driving, lowering home heating costs, or developing more efficient industry and businesses.

Other types of materials use can also be extensively regulated at the local level. Of the three Rs – reduce, reuse, recycle – recycling has attracted the most attention in terms of municipal programs, but much greater energy and materials savings are likely in the long run from the first two. Reusing wooden shipping pallets or replacing them with more durable shipping materials offers many advantages over recycling them as chipped wood for mulch or throwing them away, as has been done in the U.S. until recently. A system of washing and reusing glass bottles, for example, as exists in many European countries and once existed in the U.S. until the widespread use of plastic containers in the 1970s, offers far greater energy savings than collecting, crushing, and recycling them. Communities may want to eliminate some materials altogether. Cities such as San Francisco have banned the use of non-biodegradable plastic bags. Portland, Oakland, and about 100 other cities have banned the use of styrofoam.

Ecosystem protection and restoration offers another main area for environmental initiatives. Whereas conservation was a main goal of previous generations of environmentalism, restoration has become a key objective in recent decades, especially in urban areas. Efforts to restore creeks, shorelines and wetlands, replant native vegetation, re-create wildlife corridors, and preserve existing habitat can form centerpieces of local environmental initiatives. Traditional forms of local government regulation, such as zoning codes and subdivision ordinances, can be amended to ensure that such features are protected within new development. For example, a community can require a substantial buffer (30–100 feet or more) along waterways, thus preserving both ecologically valuable riparian corridors and opening up the possibility for a recreational trail system. Cities and towns may also require developers to preserve heritage trees and important areas of wildlife habitat on project sites.

## Land use

Local governments in the U.S. have influence through regulation and investment over the development of land within their boundaries, and land use in turn can influence everything from how much people need to drive to how much farmland and open space remains near cities. Managing the outward expansion of communities is one main sustainability priority. "Smart growth" has been a rallying cry among U.S. local governments since the 1990s, especially since suburban sprawl often increases local costs for infrastructure and services (Burchell et al. 1997; Ewing et al. 2002).

Smart growth is generally defined as development that is compact, contiguous to existing urban areas, well connected by a grid-like network of through streets, characterized by a diverse mix of land uses, and relatively dense. Internationally there is some debate over just what degree of density or compactness is desirable in order to create more sustainable communities (Jenks et al. 1996). Certainly cities and towns need not approve high-rise buildings, although Vancouver, British Columbia provides a good example of how well-designed high rises can work well within residential neighborhoods. However, there is little question that most U.S. communities can use land far more efficiently than at present. In many cases this will require local governments to guide much more precisely where development will go and in what form rather than maintaining a reactive mode to proposals. Two major ways that cities and towns can do this is through area plans that contain precise design requirements for new development and through subdivision regulations that require connecting street patterns, neighborhood centers, greenways, and other community design elements.

Infill development, which includes reuse of existing built land as well as construction on vacant or leftover parcels, is one main smart growth strategy. The tens of thousands of old shopping malls, business parks, and industrial sites in American communities offer prime opportunities for infill and for creating new, walkable, mixed-use centers for existing neighborhoods. But infill is often more difficult for developers than greenfield projects and may require substantial municipal assistance. Community development staff can facilitate dialogue between developers and local constituencies, assemble land through redevelopment powers, develop design guidelines or a specific plan for the area in question, and provide infrastructure and amenities to complement new development. In the past, much urban redevelopment in U.S. communities was not done with sufficient respect for the historical context and existing residents, but more context-sensitive approaches in the future can help ensure that such intensification efforts work well. For example, ensuring that historic preservation guidelines are in place would be an appropriate approach.

A good mix of land uses is a further goal frequently cited within the sustainable communities literature as well as by advocates of the new urbanism and smart growth. Since the 1910s, Euclidean zoning has generally sought to separate land uses within American communities, leading to the creation of vast housing tracts in one location, large commercial strips and malls in another, and office or industrial development in yet others. One result is that Americans need to drive long distances to get to basic destinations in life. Separation of land uses also makes it very difficult for anyone to walk anywhere, or for motorists to "trip-chain" – carry out a number of different tasks with one relatively short trip.

Improving land-use mix requires fundamentally rethinking local zoning codes, community, and economic development. Many more neighborhood centers can be included within new development on the urban fringe, while downtowns and office parks can have new infill housing added. Zoning can be changed for existing neighborhoods to allow a greater variety of local uses, including home offices, second units within or behind existing homes, and apartments or mixed-use buildings along commercial streets. For example, the latter is especially important for allowing residential uses in the top floors of commercial buildings to use space more sustainably.

The scale of new development should be reconsidered as well. Size has been a defining feature of recent American land development whether residential, commercial, or industrial, but large scale is not necessarily desirable from a sustainability viewpoint. Such development often provides little diversity, interest, or sense of place, and can generate community impacts such as large amounts of traffic. Communities need to consider issues such as scale. Local sustainability planning is likely to emphasize smaller local businesses, more incremental growth of new neighborhoods, and more detailed specifications for development. Such modest-scale land development can potentially create more diverse, interesting, and vibrant communities in the long run, with fewer long-distance commuting needs.

Park and greenspaces planning is a final area of land use that is essential for more sustainable communities. Although such planning has gone through a number of eras, historically (Cranz 1982), many communities today emphasize networks of parks and greenways with a variety of environments for different user groups. Increasingly, native vegetation and restored wildlife habitats are part of the concept instead of the English-style trees-and-lawn planting scheme. The idea is to reconnect residents to the landscape on a daily basis, both through small-scale parks and landscape design near homes and through larger networks of greenways and wildlife preserves throughout urban areas.

## Transportation

A community's transportation systems determine much about its resource consumption, greenhouse gas emissions, civic environment, and quality of life. For the past 80 years, both infrastructure priorities and patterns of land development in the United States have emphasized mobility via the automobile and that the per capita amount driven annually has risen about 2 percent a year. Retrofitting communities to make other modes of transportation more possible — and to reduce the amount of travel needed in daily life — will be a long process. But everywhere some steps can be taken to encourage alternative modes of transportation.

Improving the pedestrian environment is one important step. This means not just adding or improving sidewalks in a given place, but coming up with a comprehensive package of street and urban design improvements to enhance the walking environment. Such a package may include street landscaping, pedestrian-scale lighting, narrower lanes and roadways, lower traffic speeds, improved medians, sharper curb radii at intersections, and better connected street patterns within new development. Pedestrian-friendly boulevard designs can be employed in place of unsightly and dangerous arterials in some communities (Jacobs et al. 2003). Traffic-calming strategies can be employed in residential neighborhoods to slow traffic. These employ a range of design strategies including speed humps, traffic circles, chicanes (staggered parking), and extensive landscaping.

In general, the street design philosophy in many communities is shifting from one of increasing the capacity and speed of streets, common several decades ago, to one of promoting slow-and-steady motor vehicle movement. Street design nationally is also moving toward "context-sensitive design" that respects existing historical, cultural, and ecological environments and promotes walking, bicycling, public transit, and neighborhood use of streetscapes (Federal Highway Administration 2007).

In the past couple of decades, an increasing number of communities have developed bicycle and pedestrian plans to coordinate investment and policies for these two modes of transportation. Ever since the passage of the Intermodal Surface Transportation Efficiency Act of 1991 (ISTEA), federal transportation funding has been more flexible, allowing resources to be used for these purposes. An increasingly creative mix of public transportation modes is also appearing in cities and towns (Cervero 1998). Large-scale metro and light-rail systems have been built in cities ranging from Dallas to Denver, Portland to Phoenix. But "bus rapid transit" systems, in which high-tech buses provide light-rail-style service, provide a less expensive alternative to rail systems in places such as Los Angeles and Albuquerque. Some communities are experimenting with "ride-on-demand" service using small vehicles such as vans, while others have built old-fashioned streetcars with very frequent stops in urban areas. The ideal is to provide residents with a web of interwoven transit options. "Transit-oriented development" (TOD) land-use strategies can then seek to cluster new development around transit routes, increasing ridership, and providing a range of destinations and residences close to transit.

Pricing is a final, and controversial, piece of the transportation planning puzzle. The aim is to make both transit and ride-sharing attractive and discourage long-distance drive-alone commuting. Car pools ("high-occupancy vehicles") are often given their own, toll-free lanes on urban freeways, while a few places have made transit use cheap or free. Portland, Oregon's "fareless square," including most of the city's downtown, is one example. Cities and towns can provide economic incentives for residents to drive less. For example, some communities raise the cost of parking (Shoup 2005) and develop employer-based trip reduction programs. Internationally, a number of large cities, including London, have established toll zones requiring motorists to pay a substantial sum to enter city centers. Most European cities also have at least some parts of their downtowns that are pedestrian-only zones.

## Housing

A community's housing stock affects its sustainability in several ways. For one thing, large amounts of energy and materials are required to construct and maintain housing. As previously mentioned, communities can revise local building codes to require more energy-efficient structures and appliances as well as water-efficient plumbing fixtures. But on a larger scale, imbalances of housing with

jobs and shopping generate high levels of motor vehicle use, traffic, pollution, and greenhouse gas emissions. "Jobs–housing balance" has become a mantra for many communities. Ideally, communities will provide slightly more than one job per household (since many households have more than one worker). The price and size of the available housing must also balance with the needs of workers employed in the community. One typical problem is lack of affordable housing for service workers, teachers, firefighters, nurses, and other essential professions. These personnel must either pay a large percentage of their salary for housing or commute from more affordable communities further away.

There is no easy solution to a community's housing affordability problems, but several strategies taken together can potentially make a difference. One basic step is to ensure that sufficient land is zoned for apartments, condominiums, townhouses, duplexes, and other forms of housing that tend to be less expensive. Another common strategy is "inclusionary zoning" in which developers are required to include a certain percentage (often 10 to 20 percent) of units affordable to households making a certain percentage (typically 80 percent) of the county median income. Other strategies include legalizing and encouraging the creation of secondary units on existing single-family home lots, encouraging creation of land trusts that will lease housing units at below-market rates, and subsidizing nonprofit affordable housing providers to build affordable housing.

## Economic development

Economic development strategies are among the most challenging to revise from a sustainable community perspective, in part because in the past they have been so often focused on what might now be seen as unsustainable development. Some cities and towns have traditionally sought any available form of economic growth regardless of impact — rapid land development, malls, big box development, and casinos. Although substantial municipal subsidies are often offered to such businesses, gaining them does not necessarily guarantee the community a stable and sustainable future. Multinational firms may move their jobs elsewhere. Big box retailers may negatively impact smaller locally owned businesses. Rapid suburban expansion can bring traffic, overburdened local services, and loss of local culture and identity.

Sustainable economic development is instead likely to emphasize the nurturing of green and socially responsible employers within a community. These businesses will use local resources, have clean production practices, pay decent wages, and contribute back to the community through civic involvement. They will be of a range of sizes, including many relatively small, locally owned enterprises with deeper community roots than current employers (Shuman 1998, 2006). Rather than seeking rapid overnight expansion, such firms will add employment at a slower and more sustainable rate.

If this sounds like an unachievable ideal given the nature of the economy, it may well be. However, local community development efforts can help bring this vision about in a number of ways. One is by supporting the existing local businesses and encouraging them to undertake both innovation and greener production practices. Another strategy is to grow new businesses of desirable types, frequently through the creation of business incubators that provide affordable office space and shared services for start-ups, and the preferential issuance of public contracts to green businesses. Investing in public education and training is a further municipal commitment to its economic future. Finally, in recent years many jurisdictions have passed "living wage" laws requiring that workers be paid significantly more than the federal minimum wage. This policy improves both social equity and potentially increases workers' spending power within the community.

## Social equity

As a symbol of the integrating approach common within sustainable development, advocates have often spoken of the "three e's" — environment, economy, and equity. Of these, equity is by far the least well developed and perhaps the most difficult to bring about in practice. Such rising inequality brings about many sustainability problems — from the degradation of ecosystems by impoverished people struggling to survive, to the loss of social capital and mutual understanding essential for healthy democracies.

Ensuring social equity is in substantial part the responsibility of federal and state levels of government which can promote it through tax policy, funding of social services, establishment of decent minimum wages, and guarantees of fair treatment and civil rights. Local communities can promote equity goals as well. Providing adequate amounts of affordable housing, livable minimum wages, and a supportive environment for local businesses are among the ways to do this. Ensuring that underprivileged neighborhoods receive excellent services, schools, parks, and other forms of municipal investment is also important. Environmental justice is another key equity theme; too often lower income neighborhoods and communities of color have borne the brunt of pollution, toxic contamination problems, and unwanted facilities. Cities and towns can address environmental justice through active steps to protect those most at risk, improve siting of hazardous land uses, bring about fairer and more transparent decision making, and include at-risk populations in local government processes.

Additional services important for social equity include adult education and literacy programs, preschool and after-school activities, drug and alcohol abuse treatment programs, and assistance for those with disabilities, mental health issues, or a history of homelessness. Good public education in general, of course, is also crucial. Such initiatives can help build the human capital important for healthy communities in the long run. The problem of funding always exists, but grant opportunities are available for certain types of programs, and creative, sustained attention by community development officials and political leaders can help build better support in the long run. Building a "healthy" community in all aspects supports long-term sustainability.

## Process and participation

A healthy democracy is an important element of sustainable communities in that it can enable informed decision making, meet the needs of diverse constituencies, and fulfill ideals of fairness and equity. For this reason, community sustainability groups have emphasized a variety of process indicators that reflect the health of our political system and society. Sustainable Seattle, for example, included "voter participation," "adult literacy," and "neighborliness" in its set of sustainability indicators (Sustainable Seattle 1998). The Jacksonville Community Council included not just voter registration, but "Percentage of people surveyed who are able to name two current City Council members" in its quality-of-life indicators, which have been updated for nearly 25 years now (Jacksonville Community Council, Inc. 2006).

For a healthy democracy, three things are needed: a clean, open system; real choices in elections; and an informed, active electorate. In particular, conflicts of interest, often around land development, are rife within U.S. local government. Historically, "growth coalitions" of developers, landowners, real estate interests, construction companies, and politicians have dominated local politics in many communities (Logan and Molotch 1987). These interests have often funded electoral candidates, and their members have frequently held elected or appointed office. Ending such conflicts of interest and improving the transparency and visibility of local government processes is important, as is making public office attractive to a wider variety of candidates including those without wealthy backers. Ensuring high participation rates in elections and citizen knowledge of development issues is a related challenge.

As anyone involved in community development knows, public participation in local government decisions is great in the abstract but difficult to ensure in practice. It can be hard to get people to turn out for meetings, ensure participation of diverse constituencies (especially lower income groups and communities of color), and facilitate involvement that is constructive rather than oppositional. From a local residents' point of view, public involvement exercises often do not seem to include real opportunity to shape decisions.

From a sustainability perspective, it is vitally important to establish a creative and collaborative local government decision-making environment in which participants can agree on positive, proactive strategies, "think outside the box," and learn to respect each others' points of view. Too often in recent years community events have been oppositional in tone, involving mutual suspicion and animosity as well as NIMBY ("not in my backyard") groups simply trying to stop projects that are not in their own self-interest. In order to enable a constructive, collaborative planning environment instead, a number of procedural reforms can help. Transparent and well-publicized government processes can ensure that residents understand what is going on and do not feel excluded by back-room deals. Strong conflict-of-interest regulation can alleviate citizens' sense that officials are just out for special interests. Workshops and charrettes (design workshops) can be conducted with a collaborative and collegial tone rather than the top-down and patronizing styles sometimes found in reality. And local residents' ideas can be very consciously incorporated into planning alternatives and reflected back to them so that it is clear what their input has been.

That being said, it is very important for local residents to understand that they are not the only stakeholders involved in public decisions. "Community-based planning" is frequently seen as focusing just on the local neighborhood or town. But in line with the sustainability themes discussed earlier, any given decision affects multiple overlapping communities at different scales including regional, national, and global levels. From a sustainability perspective, the practitioners' role is to take into account the needs and concerns of *all* of these different communities, including the needs of the planet itself, and to help local residents understand this complex picture.

Operating as a professional with a concern for sustainable community development may require a great many skills. It may require active efforts to frame debates, develop background information, and outline alternative courses of action. It may require careful organizing both within government and within the community to pull different constituencies together and develop institutional and political backing. It may require specific intervention in debates to call attention to the long-term implications of decisions. It may require constant efforts to weave together all aspects of community development, including physical planning, urban design, economic development, social welfare policy, and environmental planning, so that the public understands the interconnections. This is the challenge of working within local government and communities.

Good communication, networking, facilitation, presentation, and political skills can help in this regard as does passion and a sense of humor.

## Conclusion

Sustainable community development is clearly a major challenge in the early stages of a process that will take hundreds of years in order to figure out how to live indefinitely into the future on a small planet, in reasonable harmony with both natural ecosystems and each other. Although sustainable community development may at times seem like an overwhelming task, it is also one that can make the job of planners and community development professionals potentially very rewarding and meaningful.

---

### CASE STUDY: CITY OF SANTA MONICA, CALIFORNIA'S SUSTAINABLE CITY PROGRAM

One of the communities that understands the importance of integrating sustainability into overall community development is Santa Monica, California. Beginning in 1994, the City Council adopted a Sustainable City Program to address issues and concerns of sustainability. In 2003, the Sustainable City Plan was adopted, built on guiding principles of sustainability and focused on eight goal areas:

1  Resource conservation
2  Environment and public health
3  Transportation
4  Economic development
5  Open space and land use
6  Housing
7  Community education and participation
8  Human dignity

An innovative approach to guiding and monitoring success toward these goals is a comprehensive community indicator system that provides measurements of goal attainment. For example, the indicator set for economic development includes tracking specific issues such as quality job creation, resource efficiency of local businesses, and the balance between housing and jobs. The indicator system for this goal area can be seen at: www.smgov.net/epd/scpr/EconomicDevelopment/EconomicDevelopment.htm. Specific targets have been designed for many of the indicator areas. This allows for further progress tracking and provides information to consider adjusting policies or programs that may or may not be working as planned.

Having the Sustainable City Plan as a guiding document, the Sustainable City Program strives to integrate sustainability into every aspect of City government and all sectors of the community (City of

Santa Monica 2006). Collaboration among sectors as well as intensive evaluation and monitoring encourage desirable outcome achievement. This plan is also coordinated with overall comprehensive planning as well as capital budgeting and other local government tools to aid in decision making for investments and service provision.

The Editors

## Keywords

Sustainability, sustainable development, community development, planning.

## Review questions

1 Why is there "no single ideal" of a sustainable community?
2 How does a sense of place affect the context for local sustainability efforts?
3 Using the six action areas, describe one recommendation from each and explain how it influences community development outcomes.

## Bibliography

Barlow, M. and Clarke, T. (2001) *Global Showdown: How the New Activists are Fighting Corporate Rule*, Toronto: Stoddart.

Barnes, P. (2006) *Capitalism 3.0: A Guide to Reclaiming the Commons*, San Francisco, CA: Berrett-Koehler.

Blowers, A. (ed.) (1993) *Planning for a Sustainable Environment: A Report by the Town and County Planning Association*, London: Earthscan.

Brown, L.R. (1981) *Building a Sustainable Society*, New York: Norton.

Burchell, R.W. et al. (1997) *Costs of Sprawl Revisited: The Evidence of Sprawl's Negative and Positive Impacts*, Washington, DC: Transportation Research Board.

Capra, F. (1996) *The Web of Life: A New Scientific Understanding of Living Systems*, New York: Anchor Books.

Cervero, R. (1998) *The Transit Metropolis: A Global Inquiry*, Washington, DC: Island Press.

City of Santa Monica (2006) Santa Monica Sustainable City Plan. Santa Monica, CA: City of Santa Monica.

Congress for the New Urbanism (1999) *Charter of the New Urbanism*, New York: McGraw-Hill.

Costanza, R. (ed.) (1991) *Ecological Economics: The Science and Management of Sustainability*, New York: Columbia University Press.

Cranz, G. (1982) *The Politics of Park Design: A History of Urban Parks in America*, Cambridge: MIT Press.

Daly, H.E. (ed.) (1973) *Toward a Steady-state Economy*, San Francisco, CA: W.H. Freeman.

—— (ed.) (1980) *Economics, Ecology, Ethics: Essays Toward a Steady-state Society*, San Francisco, CA: W.H. Freeman.

—— (1996) *Beyond Growth: The Economics of Sustainable Development*, Boston, MA: Beacon Press.

Daly, H.E. and Cobb, Jr., J.B. (1989) *For the Common Good: Redirecting the Economy Toward Community, the Environment, and a Sustainable Future*, Boston, MA: Beacon Press.

Devall, B. and Sessions, G. (1985) *Deep Ecology: Living as if Nature Mattered*, Salt Lake City, UT: G.M. Smith.

Ewing, R., Pendall, R. and Chen, D. (2002) *Measuring Sprawl and Its Impact*, Washington, DC: Smart Growth America.

Federal Highway Administration. (FHWA) (2007) "FHWA and Context Sensitive Solutions" (CSS). Available online at http://www.fhwa.dot.gov/csd/index.cfm (accessed April 1, 2007).

Goldsmith, E. (1993) *The Way: An Ecological World View*, Boston, MA: Shambhala.

Goldsmith, E. et al. (1972) *Blueprint for Survival*, Boston, MA: Houghton Mifflin.

Goldsmith, E., Khor, M., Norberg-Hodge, H. and Shiva, V. (1992) *The Future of Progress: Reflections on Environment & Development*, Bristol, UK: The International Society for Ecology and Culture.

Hawken, P., Lovins, A. and Lovins, L.H. (1999) *Natural Capitalism: Creating the Next Industrial Revolution*, London: Earthscan.

Jacksonville Community Council, Inc. (JCCI) (2006) *2006 Quality of Life Progress Report*, Jacksonville, FL.

Jacobs, A., Macdonald, E. and Rofe, Y. (2003) *The Boule-*

*vard Book: History, Evolution, Design of Multiway*, Cambridge, MA: MIT Press.

Jenks, M., Burton, E. and Williams, K. (1996) *The Compact City: A Sustainable Urban Form?* London: E & FN Spon.

Leopold, A. (1949) *A Sand County Almanac*, New York: Oxford University Press.

Logan, J. and Molotch, H. (1987) *Urban Fortunes: The Political Economy of Place*, Berkeley: University of California Press.

Meadows, D., Randers, J. and Meadows, D. (2004) *Limits to Growth: The 30-Year Update*, White River Junction, VT: Chelsea Green.

Meadows, D., Meadows, D.L., Randers, J. and Behrens III, W.W. (1972) *The Limits to Growth*, New York: Universe Books.

Pearce, D., Barbier, E. and Markandya, A. (1989) *Blueprint for a Green Economy*, London: Earthscan.

Repetto, R. (ed.) (1985) *The Global Possible: Resources, Development, and the New Century*, New Haven, CT: Yale University Press.

Shiva, V. (2005) *Earth Democracy: Justice, Sustainability, and Peace*, Cambridge, MA: South End Press.

Shoup, D. (2005) *The High Cost of Free Parking*, Chicago, IL: Planners Press.

Shuman, M. (1998) *Going Local: Creating Self-reliant Communities in a Global Age*, New York: Free Press.

—— (2006) *The Small-mart Revolution: How Local Businesses are Beating the Global Competition*, San Francisco, CA: Berrett-Koehler.

Sustainable Seattle (1998) *Indicators of Sustainable Community*, Seattle, WA.

Talberth, J., Cobb, C. and Slattery, N. (2006) *The Genuine Progress Indicator 2006: A Tool for Sustainable Development*, Oakland: Redefining Progress.

Wheeler, S.M. (2004) *Planning for Sustainability: Creating Livable, Equitable, and Ecological Communities*, New York: Routledge.

World Commission on Environment and Development (1987) *Our Common Future*, New York: Oxford University Press.

World Resources Institute (WRI). (2007) "U.S. GHG Emissions Flow Chart." Available online at http://cait.wri.org/figures.php?page=/US-FlowChart (accessed March 28, 2007).

# 24 Conclusions and observations on the future of community development

## Rhonda Phillips and Robert H. Pittman

As indicated by the variety of material covered in this book, community development is a broad discipline, having grown considerably from its more narrowly focused genesis. Today's global challenges call for the application of community development principles in even more places and ways, and the discipline must continue to evolve and draw from many different fields of study.

## Introduction

As stated at the very beginning of this book, community development is a boundary-spanning field of study addressing not only the physical realm of community, but also the social, cultural, economic, political, and environmental aspects as well. Because of the interconnection and complexity of these dimensions of community, the field is constantly evolving to face new challenges of maintaining and improving quality of life.

We have taken quite a journey in this book. We started with the foundations of community development, studying frameworks and theoretical constructs, and then core principles such as asset-based community development, social capital, and capacity building. We moved on to preparation and planning, focusing on community visioning, community-based organizations, leadership skills, and community assessments. Next we considered some key programming techniques and strategies such as workforce training, community marketing, business development, tourism-based development, neighborhood renewal, and progress measurement. Finally, we focused on several topical issues including community development finance, securing grants, the impacts of the global economy on communities, and sustainability. Along the way we learned that community and economic development are inextricably linked – success in one leads to success in the other. We also learned that community and economic development are both processes and outcomes.

As the wide and varied territory covered by this book attests, community development is a broad field of study and practice. It has evolved from its roots in neighborhood renewal, housing, and related topics to a multidisciplinary field with applications in all sectors of society, all across the globe. Today, community development is being used to gain new insights into some of the most important contemporary issues such as global migration and international diplomacy. As awareness and knowledge of the discipline of community development grows, so will its applications.

## Future challenges

It has often been said that the only constant is change. While this may be trite, it remains true. At the dawn of the twenty-first century we are witnessing social, economic, technological, and political change that is arguably of historic proportions. As Chapter 22 discusses, global economic integration and interdependency has profound implications for communities of all sizes throughout the world. For those still skeptical that global economic trends directly affect them, they need travel only as far as the local gas station or grocery store to disabuse themselves of this quaint notion. Oil and its derivative products, critical to any community's economy, are supplied, demanded, and traded in a global marketplace. Growing demand for petroleum products in newly industrialized countries such as China and India is helping propel oil and gasoline prices to unprecedented high levels, with little likelihood that they will ever return to historical norms. Likewise, the prices of many agricultural commodities and food products are being driven higher by increasing global demand due to growth in population and wealth in countries around the world. Contributing to the increase in commodities demand is the growth in alternative fuels technologies. Corn and other agricultural products are being diverted from the food chain for use in bio-fuel production. Increased demand for fuel in Chicago, Tokyo, and London therefore affects the price of tortillas in Tijuana and Guatemala City.

Even the smallest, most remote communities must adapt and learn to compete in the global economy or see their quality of life deteriorate. In the late nineteenth and early twentieth centuries, the industrial revolution and growth of mass production precipitated demographic and economic changes that fundamentally altered communities. The old paradigm of self-sufficient communities producing their own agricultural products and manually produced goods gave way to centralized production and economies of scale. People moved *en masse* from rural communities to urban areas where production and distribution were concentrating. As illustrated by the case study of Tupelo, Mississippi in Chapter 1, many rural communities adapted to this change by shifting their economic base from agriculture to manufacturing. Now, as Chapter 22 also documents, manufacturing operations are moving to less developed countries once again. Not only are manufacturing jobs being "outsourced" and "offshored" to different locations around the globe, service jobs are as well (Pittman and Tanner 2008). One prominent economist, Alan Blinder, estimates that between 30 million and 40 million U.S. jobs (around a quarter of total U.S. employment) are potentially subject to offshoring, and no sector, even college teaching, is immune (*Wall Street Journal* 2007).

In this ever-changing environment, communities must literally reinvent themselves to survive and prosper. There is no better instrument to help meet this challenge than community development. Communities and regions must embrace the process of community development and the full set of tools described in this book to adapt to the changing environment. They must learn about community assessment and SWOT analysis to understand their current situation and external environment – the assets they can build on and the weaknesses they must address. They must learn how to create a vision for the future and develop a strategic plan to take them there. They must learn how to cultivate leaders to face the challenges of today and tomorrow. They must learn how to re-tool their workforce and adapt to changing skill requirements. They must learn new ways to create jobs from within, including retention and expansion of existing businesses and the facilitation of new business start-ups. And, they must learn how to measure their progress, learn from their successes and failures, and change their game plans as necessary. To add to the challenge, community and economic transformation must be undertaken with sustainability in mind in light of increasing demands for limited non-renewable resources. As mentioned previously, community development started out as a way to address a more narrowly defined set of issues, both socially and geographically. It is apparent now that the stage on which the future of community development will unfold is much broader in scope.

## The ongoing evolution of community development

To embrace its destiny and achieve its potential, community development must continue to evolve and improve both as a theoretical and applied discipline. As we have seen, community development is inherently multidisciplinary since it covers virtually all aspects of community. While community development has benefited greatly from its close association with sociology, there are many other disciplines that can continue to make important contributions to the field including political science, economics, business (marketing, organizational behavior, finance), and psychology. Community development offers a unique laboratory in which researchers from various backgrounds can collaborate to create new theories and tools based on multiple disciplines. Hopefully, even more scholars from these different fields will embrace community development not only as a promising area of research, but also as a way to help improve society.

As we have seen, the real benefit to community development is in its applications. The study and practice of community development is not just for students and professional community developers (as described in Chapter 1); it is also for elected officials, board members, community volunteers and citizens from all walks of life who want to make their communities better places to live, work, and play. This is the broad audience that needs to learn about community development. This book was undertaken with the intention of helping this entire constituency understand and embrace the theory and practice of community development. It is the profound hope of the editors and chapter authors that we have made some small progress toward this goal.

## Keywords

Change, process, outcome, paradigm, global, evolution.

## Review questions

1   What are some of the key areas covered by the field of community development?
2   How/why is it applicable to a wide variety of global issues today?
3   How do you think the field of community development will evolve in the future?

## Bibliography

Pittman, R. and Tanner, J. (2008), "Is Offshoring or Farmshoring Right for Your Company?", *Business Xpansion Journal*, March: 10–11.
*Wall Street Journal* (2007) "Job Prospects: Pain Free Trade Spurs Second Thoughts," March 28: A1.

# Index